Climate Resilient Animal Agriculture

Climate Resilient Animal Agriculture

Editors

G.S.L.H.V. Prasada Rao

G. Girish Varma

V. Beena

CRC Press

Taylor & Francis Group

Boca Raton London New York

CRC Press is an imprint of the
Taylor & Francis Group, an **informa** business

NEW INDIA PUBLISHING AGENCY

New Delhi – 110 034

CRC Press
Taylor & Francis Group
6000 Broken Sound Parkway NW, Suite 300
Boca Raton, FL 33487-2742

First issued in paperback 2023

© 2020 by New India Publishing Agency

CRC Press is an imprint of the Taylor & Francis Group, an informa business

No claim to original U.S. Government works

ISBN 13: 978-1-03-265427-0 (pbk)
ISBN 13: 978-0-367-42032-1 (hbk)
ISBN 13: 978-0-367-81747-3 (ebk)

DOI: 10.1201/9780367817473

Print edition not for sale in South Asia (India, Sri Lanka, Nepal, Bangladesh, Pakistan or Bhutan)

Publisher's Note
The publisher has gone to great lengths to ensure the quality of this reprint but points out that some imperfections in the original copies may be apparent.

Library of Congress Cataloging-in-Publication Data
A catalog record has been requested

Visit the Taylor & Francis Web site at
http://www.taylorandfrancis.com

and the CRC Press Web site at
http://www.crcpress.com

Foreword

 The Kerala Veterinary and Animal Sciences University (KVASU) recognize the relevance and importance of the Science of Climate Resilient Animal Agriculture under the projected climate change scenario since its inception. In 2013, KVASU under the leadership of former Vice Chancellor Dr B. Ashok established the "Centre for Animal Adaptation to Environment and Climate Change Studies" (CAADECCS) with financial support from Indian Council of Agricultural Research (ICAR) during XI Plan. The Centre has prepared a large project to strengthen education, research and extension in climate change adaptation and animal production. As a part of the capacity building in this important area, a training proposal was submitted entitled "Fundamentals of Livestock Meteorology" to the Science and Engineering Research Board (SERB) of the Department of Science and Technology (DST), Govt. of India, New Delhi. The resource personnel were invited from various prestigious institutes/universities across the country based on their expertise in the field of Livestock Meteorology to address the challenges and identity opportunities related to climate change and animal production. The resource personnel were requested to contribute chapters in the field of their specialization with an aim of understanding the current status in climate resilient animal agriculture. The responses from the participants were excellent. The University of Western Australia (UWA) has a MoU and ongoing collaboration with KVASU to develop course content for postgraduate programs and research in climate change adaptation and animal agriculture. I am pleased to see that Professor GSLHV Prasada Rao and his colleagues made outstanding efforts to bring out this publication. Climate change or seasonal variations (e.g. heat stress and thermoregulation) can lead to lower production and poor health in livestock. Studies on interaction between climate and livestock production is an emerging area of research and training in India. Therefore, the efforts made by Professor Rao and his team are timely.

This publication contains applied aspects in the field of Livestock Meteorology and Methods and Approaches to address the issues related to Climate Resilient Animal Agriculture. Under the focus of global warming and climate change, this publication is timely and informative. This book will be highly relevant to students, researchers, teachers, farmers and policy makers interested in livestock production and management against climate change risk. It is more so for students and researchers in Veterinary and Animal Sciences Universities and Agricultural Universities. I anticipate this volume will be a ready reference in climate resilient animal agriculture and will reinforce the understanding for its utilization to climate smart and profitable animal production systems in India and elsewhere.

Hackett Professor Dr Kadambot Siddique
Director
The UWA Institute of Agriculture
The University of Western Australia
Perth, Australia

Preface

Weather and climate play an important role in animal husbandry and livestock production. While climate determines the adaptability of a particular animal in a given region, weather determines animal health day-to-day. Polar bears and Penguins of polar and temperate zones and Kangaroo rats and camels of deserts are few examples of climate dependant. The temperate and tropical animals possess the optimum thermoregulatory mechanisms for adaptability in their respective environmental zones, though they are having more or less constant body temperature. When they are moved from their respective habitats the production performance is primarily compromised to cope up with change in weather conditions. Though the crossbred of cattle reared in the tropical zones have partially inherited the genetic back up of high producing temperate cattle, the production is not up to the expected level in the tropical climate. Rise in global temperature is likely to be between 2 and 2.5°C by the end of this century with regional uncertainties in rainfall. It is a threat to the society linked sectors viz., Agriculture, Animal Husbandry, Water Resources, Forestry, Biodiversity (both land and ocean), Infrastructure and Health. The adverse impact of climate change is already noticed across the World in the above society linked sectors due to weather related disasters in the form of cyclones, floods, droughts, cold and heat waves and sea level rise.

Livestock farmers are encountering new challenges in terms of shortage of labourers, rise in production costs, uncertain markets and more recently, increased weather / climate risks. Extreme weather events such as floods and droughts, heavy rainfall, avalanche, landslides, heat and cold waves, cyclonic storms, thunder storms, hail storms, sand storms and cloud bursts are not uncommon and likely to be frequent in ensuing decades under projected climate change scenario. Deficit monsoon during 1965, 1966, 1972, 1984, 1987, 1997, 2002, 2004, 2009, 2012, 2014 and 2015 led to drought and adversely affected the national economy. The monsoon 2016 is no way different since northern regions of India experienced floods while drought in southern states due to deficit monsoon rainfall. Occurrence of sun burns, heat waves and heat bursts are not uncommon in recent years and such weather extremes are likely to be frequent in ensuing decades. Prolonged summer drought, followed by heavy floods during the monsoon season are detrimental to dairy cattle, pig, goat and poultry farming directly or indirectly to a considerable extent. Climate change/variability is likely to influence animal reproduction cycles and diseases due to thermal stress and related issues. The human and wildlife conflicts due to deforestation and temperature rise across the High Ranges are likely to emerge under the projected climate change scenario. The dynamics and interaction among animal insect pest species may vary and the scenario of major and minor animal diseases is likely to change. Therefore, it is important to understand the impact of climate variability or climate change on animal agriculture as a proactive measure to sustain animal health and livestock production in the event of climate change and global warming. Realizing its importance, an attempt has been made to compile the work done from various parts of the Country to focus on impact of climate change in animal husbandry and its management. This publication is in an abridged format with the title as "Climate Resilient Animal Agriculture".

Altogether twenty three chapters are included in this publication, covering applied aspects of livestock meteorology in the field of animal agriculture. The very first three chapters cover over- all climate change impacts and adaptations in animal agriculture across the Country. Chapter 4 and 5 cover heat stress and management in dairy cattle and also covered various existing dairy housing systems and its suitability in different agro-climatic regions of Tamil Nadu. Management of lactating dairy cow during heat stress is focused in chapter 6. Chapter 7 deals with thermoregulation and its management in camel while physio-genomic responses of pigs to heat stress and strategies for mitigation of climatic stress in chapter 8. Thermal adaptability of yaks is focused in chapter 9. Adaptation and mitigation in poultry production against climate change/variability is discussed in chapters 10 and 16. Strategies for sustainable production under climate change scenario in livestock production and health are highlighted in chapter 11. Nutrition management in ruminants and small ruminants against climate change is focused in chapters 12, 13 and 14. Chapter 15 deals with conservation of native breeds under projected climate change scenario. Futuristic strategies for fodder and waste management in climate change scenario and conceptual design of future farms are dealt with in chapters 17 and 18. Chapter 19 covers various aspects of climate change and agriculture in North East India while chapter 20 with one health and climate change. Chapter 21 highlights advanced biotechnological tools with potential applications in climate change studies in farm animals. Chapter 22 is dealt with weather based livestock insurance. Adaptive mechanisms and mitigation strategies for bovine in the Tropics are dealt with in final chapter 23.

In nutshell, this publication revolves around applied aspects of livestock meteorology, heat stress management, modification of microclimate, nutrition management, fodder and waste management, native breeds, biotechnology, livestock insurance, future livestock farms and one health. These emerging areas as part of climate resilient animal agriculture are focused for sustained animal agriculture in ensuing decades against climate change risk.

This unique publication contains immense content for the benefit of students, faculty members, researchers, scientists and in particular, students of animal husbandry, livestock meteorology, climate science, climate change and animal agriculture. It will be reference material to all those who are interested to understand the impact of climate change on animal husbandry and livestock management. The Editors' earnest hope is that this publication will be widely referred and discussed among the scientific community in the field of climate resilient animal agriculture.

G.S.L.H.V. Prasada Rao
G. Girish Varma
V. Beena

List of Contributors

Anil Kumar
Climate Resilient Livestock Research Centre
(CRLRC), National Innovation of Climate
Resilient Agriculture (NICRA)
ICAR-National Dairy Research Institute
(ICAR-NDRI), Karnal
Haryana-132001 India

A.P. Usha
Director of Farms
Kerala Veterinary and Animal Sciences University
Pookode, Wayanad, Kerala

A. Prasad
Assistant Professor and Head
Cattle Breeding farm
Kerala Veterinary and Animal Sciences University
Thumburmuzhi, Kerala

A. Natarajan
Professor and Head and Nodal Officer
Gramin Krishi Mausam Seva
Animal Feed Analytical and Quality Control
Laboratory
Veterinary College and Research Institute
Namakkal, Tamil Nadu

Asit Chakrabarti
ICAR Research Complex for Eastern Region
Patna, Bihar

Anjumoni Mech
ICAR – National Institute of Animal
Nutrition and Physiology
Adugodi, Bangalore – 560030

Arindan Dhali
ICAR-National Institute of Animal Nutrition
and Physiology, Adugodi
Bangalore – 560030

C. Balusami
Assistant Professor
Kerala Veterinary and Animal Sciences University
Mannuthy, Thrissur – 680 651
Kerala

C. Latha
Department of Veterinary Public Health
College of Veterinary and Animal Sciences
Mannuthy
Kerala Veterinary and Animal Sciences University
Mannuthy
Thrissur – 680 651, Kerala

Deepa Ananth
Assistant Professor
Directorate of Entrepreneurship
Kerala Veterinary and Animal Sciences University
Pookode, Wayanad

Deepak Mathew
Kerala Veterinary and Animal Sciences University
Mannuthy, Thrissur – 680 651
Kerala

Dharmendra Kumar
Krishi Vigyan Kendra, Kishanganj
Bihar Agricultural University
Bihar

D. Rajakumar
Tamil Nadu Agricultural University, TNAU
Coimbatore, Tamil Nadu

E.M. Muhammed
Centre for Animal Adaptation to
Environment and Climate Change Studies
Kerala Veterinary and Animal Sciences University
Mannuthy, Thrissur – 680 651
Kerala

Francis Xavier
Professor (Rtd.)
Kerala Veterinar and Animal Sciences University
Mannuthy, Thrissur – 680 651, Kerala

G.S.L.H.V. Prasada Rao
Centre for Animal Adaptation to
Environment and Climate Change Studies
Kerala Veterinary and Animal Sciences University
Mannuthy, Thrissur – 680651
Kerala

Gyanendra Singh
ICAR – Indian Veterinary Research Institute
Izatnagar, Bareilly
Uttar Pradesh – 243122

G. Girish Varma
Dean, College of Dairy Science and Technology
Kerala Veterinary and Animal Sciences University
Mannuthy, Thrissur – 680651
Kerala

G. Krishnan
ICAR – National Institute of Animal
Nutrition and Physiology
Adugodi, Bangalore – 560030

K. Karthiayini
Associate Professor and Head,
Department of Veterinary Physiology
College of Veterinary and Animal Sciences
Kerala Veterinary and Animal Sciences University
Mannuthy
Kerala – 680 651, India

K.C. Raghavan, Director (Rtd.)
Centre for Advanced Studies in Animal
Genetics and Breeding
College of Veterinary and Animal Sciences
Kerala Veterinary and Animal Sciences University
Mannuthy
Thrissur – 680 651
Kerala

K.S. Ajith
Kerala Veterinary and Animal Sciences University
Thiruvavazhamkunnu

K. Vrinda Menon
Department of Veterinary Public Health
College of Veterinary and Animal Sciences
Mannuthy
Kerala veterinary and Animal Sciences
University
Mannuthy, Thrissur – 680 651, Kerala

Kolli N.Rao, Senior Advisor
International Reinsurance & Insurance
Consultancy & Broking Services
Pvt. (IRICBS) Nirmal Bhavan, Nariman
Point, Mumbai

Lipismita Samal
College of Veterinary Science & Animal
Husbandry
OUAT, Bhubaneswar – 751 003, Odisha

M. Manoj
Centre for Advanced Studies in Animal
Genetics and Breeding
College of Veterinary and Animal Sciences
Kerala Veterinary and Animal Sciences University
Mannuthy, Thrissur, Kerala – 680651

Mahesh Chander
ICAR – Indian Veterinary Research Institute
Izatnagar – 243122, Bareilly (UP)

Marykutty Thomas
Assistant Professor
Livestock Research Station
Kerala Veterinary and Animal Sciences University
Thiruvavazhamkunnu

M. Bagath
ICAR – National Institute of Animal
Nutrition and Physiology
Adugodi, Bangalore – 560030

Mihir Sarkar
ICAR – Indian Veterinary Research Institute
Izatnagar, Bareilly
Uttar Pradesh – 243122

N. Maragatham
Tamil Nadu Agricultural University, TNAU
Coimbatore, Tamil Nadu

N.V. Patil
ICAR – National Research Centre on Camel
Bikaner, Rajasthan

N.H. Mohan
ICAR – National Research Centre on Pig
Guwahati, Assam – 781131

Prakash Kahate
Assistant Professor (AH & Dairy Science)
College of Agriculture
Dr. Panjabrao Deshmukh Krishi Vidyapeeth
Akola, Maharashtra – 444104

Prakashkumar Rathod
ICAR – Indian Veterinary Research Institute
Izatnagar – 243122, Bareilly (UP) India

P.J. Das
ICAR – National Research Centre on Yak
Dirang – 790101, Kameng
Arunachal Pradesh

P.T. Suraj
Cattle Breeding Farm
Thumburmuzhy
Kerala Veterinary and Animal Sciences University
Kerala – 680 721, Kerala

P.K. Malik
ICAR – National Institute of Animal
Nutrition and Physiology
Adugodi, Bangalore – 560030

R.S. Abhilash
Livestock Research Station,
Kerala Veterinary and Animal Sciences University
Thiruvavazhamkunnu

R.U. Suganthi
ICAR – National Institute of Animal
Nutrition and Physiology
Adugodi, Bangalore – 560030

R. Bhatta
ICAR – National Institute of Animal
Nutrition and Physiology
Adugodi, Bangalore – 560030

R. Mathivanan
Tamil Nadu Agricultural University, TNAU
Coimbatore, Tamil Nadu

R. Karthikeyan
Tamil Nadu Agricultural University, TNAU
Coimbatore, Tamil Nadu

S. Sankaralingam
University Poultry and Duck Farm
Kerala Veterinary and Animal Sciences University
Mannuthy, Thrissur – 680651, Kerala

Shibu K Jacob
Centre for Animal Adaptation to
Environment and Climate Change Studies
Kerala Veterinary and Animal Sciences University
Mannuthy, Thrissur – 680 651, Kerala

Stephen Mathew
Livestock Research Station
Kerala Veterinary and Animal Sciences University
Thiruvavazhamkunnu

Sajjan Singh
ICAR – National Research Centre on Camel
Bikaner, Rajasthan

S.S. Hanah
ICAR – National Research Centre on Yak
Dirang – 790101, Kameng
Arunachal Pradesh

Sohan Vir Singh
Climate Resilient Livestock Research Centre
(CRLRC), National Innovation of Climate
Resilient Agriculture (NICRA)
ICAR-National Dairy Research Institute
(ICAR-NDRI), Karnal
Haryana-132001 India

Simson Soren
Climate Resilient Livestock Research Centre
(CRLRC), National Innovation of Climate
Resilient Agriculture (NICRA)
ICAR-National Dairy Research Institute
(ICAR-NDRI), Karnal
Haryana-132001

T.S. Rajeev
Department of Veterinary and Animal
Husbandry Extension
College of Veterinary and Animal Sciences
Mannuthy, Thrissur – 680651, Kerala

T.V. Aravindakshan
Centre for Advanced Studies in Animal
Genetics & Breeding
Kerala Veterinary and Animal Sciences University
Mannuthy, Thrissur – 680 651 Kerala

T. Sivakumar
Faculty of Food Science and Technology
Tamil Nadu Veterinary and Animal Sciences
University
Tamil Nadu – 600 007

T.K. Biswas
ICAR – National Research Centre on Yak
Dirang – 790101, Kameng
Arunachal Pradesh

T. Unnikrishnan
Department of Statistics
Farook College, Calicut University, Kerala

V.U.M. Rao
Rtd. Project Coordinator (AICRPAM)
CRIDA, Santoshnagar
Hyderabad

V. Sejian
ICAR – National Institute of Animal
Nutrition and Physiology
Adugodi, Bangalore – 560030

V.P. Maurya
ICAR – Indian Veterinary Research Institute
Izatnagar, Bareilly
Uttar Pradesh – 243122

V. Paul
ICAR – National Research Centre on Yak
Dirang – 790101, Kameng
Arunachal Pradesh

Zahoor Ahmad Pampori
Division of Veterinary Physiology
SKUAST – K, Shuhama Alusteng
Srinagar, Kashmir – 190 011

Contents

1

Climate Change Adaptation in Animal Agriculture

Lipismita Samal[1], G. Krishnan[2], M. Bagath[2], V. Sejian[2], P.K. Malik[2] and R. Bhatta[2]

[1]*College of Veterinary Science & Animal Husbandry, OUAT Bhubaneswar – 751003, Odisha*
[2]*ICAR-National Institute of Animal Nutrition and Physiology, Adugodi Bangalore – 560030, Karnataka*

Recent increase in extreme climate events threatens disruptive impacts on both animal and agriculture (Battisti and Naylor, 2009). Climate change is a global phenomenon, but its negative impacts are more severely felt by the poor people in developing countries. Moreover, rural poor communities rely greatly for their survival on animal and agriculture which belong to the most climate-sensitive economic sectors. Climate change is expected to intensify existing problems and create new combinations of risks. The incidences of droughts, snow-storms and blizzard like events have increased rapidly. The situation has further worsened due to factors such as widespread poverty, over dependence on rain-fed agriculture, inequitable land distribution, limited access to capital and technology, inadequate infrastructure, long term weather forecasts and inadequate research and extension. The impact of climate change is expected to heighten the vulnerability of animal systems and reinforce existing factors that are affecting animal production systems in many parts of the world. Nevertheless, global demand for animal products is expected to double during the first half of this century, as a result of the growing human population, and its growing affluence. The extreme climatic conditions impose various stresses on animals which adversely affect their growth, production and reproduction status. During the past few decades, crop yields have also been reduced because of climate change, and the results of modelling studies suggest that climate change will reduce feed crop yield potential, particularly in many tropical and mid-latitude countries (Rosenzweig *et al*., 2014). Rising atmospheric CO_2 concentrations will decrease feed and forage quality. Price and yield volatility likely will continue to rise as extreme weather continues, further harming livelihoods and putting feed security at risk (Wheeler and von Braun, 2013). This chapter is an attempt to collate and synthesis information pertaining to different adaptation strategies to sustain livestock production in the changing climate scenario.

1.1 Climate Change Adaptation

Adapting to climate change entails taking the right measures to reduce the negative effects of climate change or to exploit the positive impacts by making appropriate adjustments and changes. The Inter-Governmental Panel on Climate Change (IPCC) defines adaptation as adjustments in natural systems in response to actual or expected climatic stimuli or effects, which moderates harm or exploits beneficial opportunities (IPCC, 2007). Adaptation has three possible objectives: to reduce exposure to the risk of damage; to develop the capacity to cope with unavoidable damages; and to take advantage of new opportunities.

1.1.1 Strategies for adapting animals against climate change

Animal production not only contributes to climate change via green-house gases (GHG) emissions but also suffers due to extreme weather events and diseases related to climate change. Direct and indirect challenges in adaptation include fluctuating feed prices, habitat changes, expansion of vector borne diseases in warm climates, impaired reproduction, pasture quality and availability and physiological heat stress (Thornton *et al*., 2009). Fig. 1.1 describes the broader strategies for adapting both agriculture and livestock production system. Different adaptation strategies are as follows:

Income and livelihood diversification by mixing crop and animal production

Risk adaptation by farmers may also involve changing from cultivated crops to animal, as crops may be more environmentally and spatially constrained in the pastoralists' home regions (Jones and Thornton, 2009).

Sustainable intensification through pasture regeneration or destocking

Diversifying animal feeds

Manipulation of rumen microbial composition

Flexibility in livelihood options for pastoralist, agro-pastoralist and ranching communities can increase a household's capacity to manage risk and adapt external stress (Thornton *et al*., 2007). Adaptation options depend on household objectives and attitude, local access to natural resources, inputs and output markets and sustainable intensification.

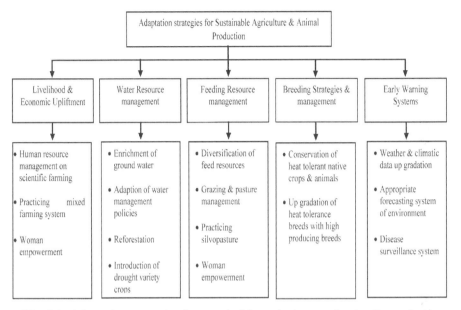

Fig. 1.1: Adaptation strategies for sustainable agriculture and animal's production system against climate change

1.1.2 Breeding strategies

Promoting breeds that have adapted to local climatic conditions and feed resources

Many breeds are already adapted to local climatic stress and feed sources. However, developing countries are usually characterized by a lack of technology in animal breeding programmes that might otherwise help to speed adaptation. Adaptation strategies address not only the tolerance of animals to climate change, but also their ability to survive, grow and reproduce in conditions of poor nutrition, parasites and diseases (Hoffmann, 2008). Several programmes are being undertaken such as genetic up-gradation of animals, encouraging breeding of animals for livelihood security etc.

Identifying and strengthening the breeds that have inherent resilient adaptive capacity to cope with climate change

Resilience is about maintaining diversity in genetic resources and in approaches. A diversity of farming practices allows animal keepers to cope with differences in local environments. The traits of inherent resilient and adptive capacity against heat stress are: long legs, short hair coat, higher sweating rate, large surface area, body conformation, higher capacity for maintenance of heat balance, lower metabolic rate, higher feed efficiency, higher tolerance to dehydration and adipose tissue depots and capacity to alter the hormone and biochemical profiles to adapt to a particular environment.

1.1.2.1 Improving local Genetics Through Cross-breeding with Heat and Disease-tolerant Breeds

If climate change is faster than natural selection, the risk to the survival and adaptation of the new breed is greater (Hoffmann, 2008). There are clear genetic differences in resistance to heat stress, with tropically-adapted breeds experiencing lower body temperatures during heat stress than non-adapted breeds. The superior fertility of tropically-adapted breeds during heat stress is a function in large part of the enhanced ability of animals from these breeds to regulate body temperature in response to heat stress. Even in non-adapted breeds, it is possible to perform genetic selection for resistance to heat stress. There are also specific genes that control cellular resistance to heat shock and confer increased thermoregulatory ability. These genes can be transferred into thermally-sensitive breeds through conventional or transgenic breeding techniques to produce an animal whose oocytes and embryos have increased resistance to elevated temperature. Genetic selection for major physiological defence mechanisms against rising body temperatures, such as reduced feed intake and metabolic rate and increased peripheral blood flow in alimentary tract tissues, offers little potential advantage. Selection for high producing animals which are also less sensitive to heat stress under existing natural conditions, would not seem plausible physiologically.

1.1.3 Women empowerment

The roles, relations, responsibilities, opportunities and constraints of pastoralists pose different challenges to women and men because of uneven access and control of resources. Using data on pastoralists' attitudes and perceptions related to climate change and variability on feed security, it was found that women's workloads and the pastoralists over-reliance on feed relief increased because of climate variability and change. Women play an important role in a community early warning system. They also aid in the recovery effort by securing animal and removing debris. With the increasing trend of feminisation of agriculture, women will play a crucial role in the adaptation strategies. Work undertaken by the International Centre for Integrated Mountain Development (ICIMOD) in the Himalayan region on gender perspectives in rangeland resources management and on climate change adaptation shows that women hold a rich knowledge and a variety of skills for maximizing the use of natural resources of the fragile mountain ecosystem and in farming practices (Leduc, 2010). However, the gender perspective is still rarely integrated in climate change policy and strategies both at international and national levels.

1.1.4 Early warning systems

It is the provision of timely and effective information, through monitoring various indicators affecting livelihood with a view to allow individuals exposed to hazard to take action to avoid or reduce their risk and prepare for effective response (Teshome, 2012). Information should be made available in good time to plan and implement appropriate intervention ahead of time to avoid or reduce the risk of abrupt climate change and prepare for effective response. The five dimensions of early warning systems (EWS) are: (a) the ability to make informed assessment and mapping of imminent threats and risks, (b) technical monitoring and warning service for the regional and emerging threats and risks, (c) dissemination and communication of understandable warnings to those at threats and risks, (d) knowledge and preparedness (skill, money, infrastucture) to act and respond and (e) implement the action (policy, governance, rights). Activities involved in this EWS include data collection, information development, dissemination methods and action triggering mechanisms. During drought, rainfall failure causes a sharp drop in available feed resources over a wide area. Animal condition deteriorates in long distance movements to find grazing. Production performance declines, disease incidence increases and the more vulnerable animals, such as the young and very old animals, begin to die. In order to buy feed, animal owners take their non-reproductive animals to market, but as supply quickly exceeds demand animal prices plummet. When the drought intensifies, more and more animals become emaciated and die. Effective drought cycle management calls for appropriate actions to prepare for drought, manage its impacts and assist affected animal owners to recover.

1.1.5 Strengthening meteorological services

The meteorological services should be strengthened for forecasting and warning of any type of climate change driven by a variety of socio-economic scenarios. The IPCC has shown that the earth temperature has been increased 0.74^0C between 1906 and 2005 due to increase in anthropogenic emissions of GHG. Temperature increase is likely to be 1.8-4.0^0C by 2100 (Aggarwal, 2008). The IPCC has projected 0.5-1.2^0C rise in temperature by 2020, 0.88-3.16^0C by 2050 and 1.56-5.44^0C by 2080 depending on the scenario of future development. There is evidence of increased global average air ($+0.74$°C) and ocean temperatures, widespread melting of snow and ice and rising global average sea level (by 3 mm annually) (IPCC, 2007). Moreover, the IPCC report estimates a confidence level of 90% that there will be more frequent warm spells, heat waves and heavy rainfall and a confidence level of 66% that there will be an increase in drought, tropical cyclones and extreme high tides. The magnitude of the events will vary depending on the geographical zones of India. Similar trend with varying magnitude has been shown by Indian Meteorology Department and the Indian Institute of Tropical Meteorology, Pune.

1.1.6 Disease surveillance measures

Increased temperatures due to climate change will affect the survival of pests like lice and ringworm in the winter and thus distribution of various diseases (e.g., zoonotic, endemic, emerging, feed borne and non-infectious diseases) in animals. Climatic conditions favourable for the growth of causative organisms during most part of the year due to temperature rise will facilitate spread of diseases in other seasons and also increase area of spread. Higher temperatures and changing rainfall patterns can enhance the spread of existing vector borne diseases (Bhattacharya *et al.*, 2006) and macro parasites, accompanied by the emergence and circulation of new diseases in animals. Climate change will modify the dispersal, reproduction, maturation and survival rate of vector species and consequently alter viral and bacterial disease transmission. In some areas, climate change is likely to generate new transmission models. Temperature and humidity variations could also have a significant increase in helminth infections, protozoan diseases such as Trypanosomiasis and Babesiasis. Some of the viral disease like Rinderpest may also reappear affecting animals. Change in pattern and range of vector-borne disease and helminth infections, increase in heat-related morbidity and mortality. During drought crisis, tick and skin diseases in animals are increasingly becoming common problems. As the animals have less access to pasture and become weak, they are more susceptible to different diseases when they are concentrated around a few water-points. Moreover, after a long dry season, when the rain finally comes, many animals die. So, primary health services such as deworming, vaccination and other preventive measures should be encouraged.

1.1.7 Water resource management

In addition to high temperature and deficient nutrition, water scarcity is another important limiting factor to animals during summer season in semi-arid tropical environment under changing climate scenario. Water is considered as an essential nutrient and is involved in every metabolic function of the body. In many regions, the main source of water is the nearby river. However, since most of the land on the river banks is occupied by a state farm, national park and private investors, it is difficult for pastoralists to access water for their animal. Measures should be taken for minimizing landslide, conserving forests and for making arrangements for safe landing of running water during rainy period. There should be improved management of water resources through introduction of simple techniques for localized irrigation (e.g., drip and sprinkler irrigation), accompanied by infrastructure to harvest and store rainwater, such as tanks connected to the roofs of houses and small surface and underground dams (IFAD, 2009).

1.1.8 Grazing management

There is a complex interaction between the grazing pressure and impact of climate change that will ultimately determine the limitations of the production system. Animal stocking rate should be inconsistent with the grazing capacity of the pastures. Strategic and rotational grazing system should be followed. The pressure on fodder supplementation can be reduced by supplementing feed.

1.1.8.1 Introducing Mixed Animals Farming Systems

Grazing is associated with daily activities considerably different than for confined animals, such as time spent in eating and distances travelled. These activities result in greater energy expenditure than in confinement, which can limit energy available for maintenance and production. So, mixed animals farming systems, such as stall-fed systems and pasture grazing should be encouraged. Maintenance of grazing productivity and high stocking rates through pasture reclamation and adoption of integrated crop animal systems, such as rotational grazing and introduction of legumes in pastures, buffers pressure on deforestation. Such pasture regeneration creates a potential for increasing soil C storage.

1.1.9 Reduction of animal numbers

A lower number of more productive animals lead to more efficient production and lower green house gas emissions from animal production (Batima, 2007).

1.1.10 Changes in animal/herd composition

Large animals should be selected rather than small animals.

1.1.11 Rotational lopping of vegetative biomass of fodder, trees, shrubs, herbs and grasses to enhance forage production

1.1.12 Introduction of silvi-pasture system

Plantation of fodder trees in grazing area can be a successful integrated-farming type appproach that provides feed as well as shade duing summer.

1.1.13 Formulation of proper range policy and establishment of competent authority for management of range lands

1.1.14 To complement farm based efforts, uniform and fair economic procurement and incentivized policies must be in place and enforced across the supply chain in order to establish supply and trade chains with low C footprints. Regional and national policies must contain mechanisms that balance market pressure to convert from low impact land uses (for example, forests) to relatively more intensive uses (for example, ranching).

1.2 Shade Management System

Physical protection with artificial or natural low-cost shade offers the most immediate and cost-effective approach for sustainable animals production. Evaporative cooling can also be effective. Bedded barn facilities appear to be useful for buffering animal against the adverse effects of the environment under hot conditions. It is recommended to start cooling strategies prior to animal showing signs of heat stress i.e. panting. Sprinkling animal in the morning is more effective than sprinkling in the afternoon. Sprinkling of pen surfaces may be more beneficial than sprinkling the animal. Minimal handling of animal is recommended for promoting animal comfort.

1.3 Migration

Animals which are more hardy and adopted to harsh climatic conditions may thrive well while others may either shift to more suitable region or suffer stressful environment. Pastoralist societies have habitually migrated, with their animals, from water source to grazing lands in response to drought as well as part of their normal mode of life. But it is becoming apparent that migration as a response to environmental change is not limited to nomadic societies. Temporary migration as an adaptive response to climate stress is already apparent in many areas. The ability to migrate is a function of mobility and resources. Pastoralists' nomadic mobility reduces the pressure on low carrying capacity grazing areas through the circular movement. This system of seasonal movement represents a local type of traditional ranching management system of range resources (Akinnagbe and Irohibe, 2015). There should be improvement in transport systems for animals to winter and summer grazing areas. Movements of animals and people may be restricted because of increase in conflicts among the pastoral community. On account of geographical location, a particular region has boundaries with different groups, such as pastoralists, farmers, state-owned farm and national park. Conflicts are not only because of competition over resources such as water and pasture but also because of animal raiding.

1.4 Role of Models

In support of agro-pastoral farming systems, models must integrate adaptation options, alternative intensification pathways, zoonotic disease and vector ecology (e.g. genetic shifts, patterns of emergence), mechanisms of effecting behavioural change and adaptation to future climate change scenarios. Some existing models include BEEFGEM of Ireland (Foley *et al.*, 2011), IFSM of USA (Rotz *et al.*, 2010) and SIMSDAIRY of UK~ (Del Prado *et al.*, 2011). Reisinger *et al.* (2012) recently evaluated different metrics on the integrated assessment model, MESSAGE, and the land use model, the Global Biosphere Management Model (GLOBIOM), to examine the global costs of abatement strategies used to reduce the magnitude of climate change and subsequent effects on regional feed production and supply prices for animal products and other agricultural commodities.

1.5 Marketing Facility

One of the transformative approaches to animal production include identifying the value of a blend of market-orientated small-holders vs. large scale farms, evaluating ecosystem services payments as a means of income diversification, forming institutional and market mechanisms for reaching smallholders to foster technological change, finding the best locations for both animal production and marginal land rehabilitation, and creating new capacity of the animal sector for adaptation in the face of climate change (Havlík *et al.*, 2014; Herrero *et al.*, 2014).

1.6 Capacity Building for Animal Keepers

There is a need to improve the adaptive capacity of animal keepers to understand and deal with climate changes. Training in agro-ecological technologies and practices for the production and conservation of fodder improves the supply of animal feed and reduces malnutrition and mortality in animal flocks. Experience with these strategies needs to be shared among communities. Techniques include:

- Building awareness of climate change and its consequences.
- Raising awareness of the value of different management strategies.
- Raising awareness of the value of different animal breeds.
- Creating awareness for proper use and appropriate management of pasture.
- Using forest products as a buffer against climate induced crop failure.
- Building knowledge on fodder production and conservation.
- Soil fertility improvement techniques.
- Soil moisture and water conservation practices.
- Decentralization of governance of resources and the manipulation of land use leading to land use conversion.
- Strengthening flock health and reducing mortality.
- Regular monitoring of the weather and accordingly increasing feeding in response to cold weather. Especially pregnant animals in the last trimester require additional feeding during periods when the effective temperature falls below the lower critical level.

- Animals should be protected from wind as wind reduces the effective temperature, increasing cold stress on animals.
- Animals should be kept clean and dry as wet coats greatly reduce insulating properties making them more susceptible to cold stress.
- Introduction of high yielding nutritious forages at selective sites.
- Animals should have access to ample amount of water as limiting water limits feed intake and make it more difficult to meet the energy requirements.

1.7 Adaptation Strategies to Improve Feed and Fodder Availability for Livestock

Adaptive actions address the challenge of meeting the growing demand for feed, fibre and fuel, despite the changing climate and fewer opportunities for agricultural expansion on additional lands. It focuses on contributing to economic development, poverty reduction and feed security~ maintaining and enhancing the productivity and resilience of natural and agricultural ecosystem functions. Effective management must involve stakeholders, address governance issues, examine uncertainties, incorporate social benefits with technological change and establish climate finance within a green development framework. Several adaptation strategies are as follows.

1.8 Planting of Drought Tolerant Varieties of Crops

Emphasis on more drought resistant crops in drought-prone areas could help in reducing vulnerability to climate change. For example, wheat requires significantly less irrigation water compared to dry season rice.

1.9 Changes in Cropping Pattern and Calendar of Planting

Climate change adversely affects crop production through long-term alterations in rainfall resulting in changes in cropping pattern and calendar of operations. In Tanzania, to avoid crop production risks due to rainfall variability and drought, staggered plating is very common to most farmers whereby crops are planted before rain onset on uncultivated land. Others were planted immediately after rain, while still other plots were planted a few days after the first rains (Akinnagbe and Irohibe, 2015).

1.10 Mixed Cropping

It involves growing two or m nzania where cereals (maize, sorghum), legumes (beans) and nuts (groundnuts) are grown together. The advantages of mixing crops with varying attributes are in terms of maturity period (e.g. maize and beans), drought tolerance (maize and sorghum), input requirements (cereals and legumes) and end users of the product (Akinnagbe and Irohibe, 2015).

Improving plant performance i.e. nutrition, yields, feed quality, etc. in response to elevated CO_2 and rising temperatures (Vermeulen *et al.*, 2012; Yin, 2013).

1.11 Avoiding Pest Damage and Feed Waste

Developing forecasting, management and insurance options to decrease the risk due to unexpected rainfall patterns, higher temperatures and shifting length in growth seasons (Vermeulen *et al.*, 2012).

Managing natural resources at the landscape and regional levels to assure the environmental quality and ecosystem services upon which agriculture depends (Minang *et al.*, 2014).

1.12 Feed and Fodder Banks

A specific and viable system has to be put in place for developing feed and fodder banks at strategic places. Biomass intensification specially targeting animals should receive highest priority. Much of the tree-biomass for animals can be enhanced easily with little effort and resources.

1.13. Molecular Approaches and Genetic Engineering

Molecular approaches will foster better understanding and manipulation of physiological mechanisms responsible for crop growth and development, as well as the breeding of stress adapted genotypes (Reguera *et al.*, 2012). High throughput phenotyping platforms and comprehensive crop models will lead to more rapid exploration of genetic resources, enabling both gene discovery and better physiological understanding of how crop improvement can increase tolerance to environmental stress (Araus and Cairns, 2014). Development of new crop genotypes to meet the need to thrive under future management and climate conditions, the expected increases in the frequency of climate shocks and the uncertainty of rates of climate change presents a challenge. Molecular approaches provide opportunities to establish linkages between biochemical pathways and physiological responses. In cereals such as rice, grain yield is highly dependent on the carbohydrate source (top leaves) and sink (florets) relationship, which is strongly influenced by the plant hormone cytokinin (Peleg and Blumwald, 2011). Cytokinin production also affects drought tolerance and senescence and isopentenyl transferase (IPT) expression controls up-regulation of pathways for cytokinin degradation. Therefore, it follows that tolerance of abiotic stress by delaying stress induced senescence through manipulation of IPT expression in transgenic lines could maintain optimal levels of cytokinin, resulting in greater fitness and more seed and grain production (Peleg *et al.*, 2011).

1.13.1 Combinations of multiple plant traits

Combinations of plant traits may produce more resilient crop production in the face of climate change. Survival strategies employed by plants include early flowering to escape drought periods, stomatal control to prevent water loss, enhanced root growth in deeper soil layers to access water and reduced leaf growth to minimize the transpiring surface. These adaptations come at a cost, where reductions in the growth cycle, light interception and carbon (C) assimilation by photosynthesis are often accompanied by a higher C requirement to build additional plant roots, especially under nutrient stress (Wang and Taub, 2010).

1.14 Introduction of New Plant Traits by Breeding

Selection and introduction of new plant traits must be considered for specific types of environmental stress. By examining the genetic basis of physiological mechanisms and environmentally induced stress responses, crops such as maize, wheat and other cereals can be bred to produce better yields and tolerances through targeted accumulation of alleles that confer robust responses to environmental stressors such as drought (Roy *et al.*, 2011). Trait based breeding programmes will be most effective when approaches are developed to simultaneously screen a broad array of genotypes for phenotypic responses to environmental stresses quickly.

1.15 Germplasm Screening Tool

Canopy temperature (CT) is an example of a widely used, high throughput germplasm screening tool. The CT is linked to stomatal conductance, an indirect indicator of water uptake by roots, especially under drought and heat stress (Rebetzke *et al.*, 2013).

1.16 Managing Forest Biodiversity to Increase Ecosystem Services and Resilience

Tree species and densities for each type are selected by desired ecological processes, farmers' criteria and land use policies. Trees and forests buffer microclimates, regulate water quality and flows, store C and provide habitat for plants and animals in protected areas and corridors (Mbow *et al.*, 2014). Forest loss and degradation cause GHG emissions and loss of C stocks, biodiversity and ecosystem services. When landscapes are managed to contain a mosaic of forestry and agro-forestry ecosystems, the diversification of feed, feed and timber production, income sources, and markets promotes greater resilience to environmental uncertainty. Examples of agro-forestry types in agricultural landscapes include remnant forest or savanna, agro-forests, tree crops, home gardens and boundary plantings.

1.17 Water Management

The effects of climate change on hydrology are far more uncertain than temperature change. Global irrigation water demand will likely increase by approximately 10% by mid-century (Wada *et al.*, 2013). IPCC models for irrigated areas indicate that the gap between potential evapo-transpiration and effective rainfall will be about 17% by 2050 under a high emission scenario, placing extra stress on demand for irrigation water (Sood *et al.*, 2013).

- Forests and woodland cover can support water quality and assist in reducing dry land salinization and water quality decline in semiarid environments.
- Massive abstraction of groundwater and redistribution to agricultural land (nearly 70% of global freshwater withdrawal and 90% of consumptive water use for irrigation) has led to groundwater depletion in regions with primarily ground water fed irrigation. Moreover, with projected increases in drought incidence and severity, changes in rainfall patterns and intensification and decreases in snowpack, agricultural areas that are currently irrigated with surface water will become heavily reliant on groundwater.
- Water policy and management practices should focus on efficient and equitable water rights and allocation~ increasing water productivity via more and better irrigation storage, conveyance and delivery systems that reduce evaporative losses~ in field water use efficiency improvements~ and technologies that reduce seawater intrusion in coastal environments. Responses to the spatial and temporal shifts in water quantity and quality due to climate change involve many scales and stakeholders, and the need for coordinated planning at regional and national scales will increase with growth in the urban and industrial sectors.
- Approaches to increasing the efficiency of water used for feed supply must employ drought tolerant crops and irrigation technology (for example, water conserving irrigation systems, crop coefficients and surface renewal. They also need to address both consumptive behaviour and waste incurred during postharvest and along the supply chain.
- Involvement of communities and government agencies in increasing storage capacity via small scale reservoir projects, rainwater harvesting, groundwater banking through artificial and/or natural aquifer recharge and flood harvesting and restoration of coastal vegetation to promote opportunities for aquaculture.
- Reduction in end user demand, de-engineering and reoperation of water systems to create adequate supply and distribution, improved wastewater treatment plants to facilitate wastewater reuse, desalination plants and targeted water conservation projects (Mukherji *et al.*, 2012).

1.18 Sequestration of Soil Organic Carbon

Soil resource degradation has led to loss of functions and ecosystem services, such as water availability, water holding capacity, soil organic C (SOC), mitigation of GHG emissions and sustained agricultural productivity. Soil degradation limits resilience to climate change and extreme events, such as drought, and therefore impacts feed security and augments susceptibility to poverty, especially in vulnerable regions. Better understanding of the biophysical capacity of agricultural landscapes to act as C sinks through capture and storage of atmospheric CO_2 in soils and perennial vegetation leads to strategic design and operational management for adaptation actions (Cochard, 2013).

1.19 Biophysical Models

Several models that can be used to examine the limits to crop adaptation as well as the impacts of climate change on biodiversity, land use and ecosystem services are available. Modelling can be used to identify climate change impacts and sensitivities as well as possible adaptation strategies. However, they contain much uncertainty due to (a) the ability of process models to accurately simulate the growth and development of crops when exposed to very high temperatures and elevated CO_2 levels, (b) the rate and degree to which agricultural productivity and development can progress in concert with reductions in GHG emissions and (c) the ramifications of successful agricultural adaptation to climate change for land use change and associated ecosystem services (Del Prado et al., 2013). Many climate modelling studies are focused on yield variations in response to changes in mean climate conditions~ yet, this approach overlooks several key factors, like the occurrence of extreme events in which variance is changing. Empirical approaches that capture the effects of extreme temperatures can be used to more efficiently assess climate impacts and adaptation.

1.20 Energy and Biofuels

Bioenergy is the native energy resource embedded within agriculture, but, more fundamentally, agriculture is itself an energy conversion process with the capacity to develop a rich portfolio of products for diverse markets, including markets for feed and energy. The role of biofuels in meeting future energy needs, as well as their impact on feed commodity prices, remains a global issue (LotzeCampen et al., 2014). Increased future demands for feed, fibre and fuels from biomass can only be met if the available land and water resources on a global scale are used and managed much more efficiently than they are now.

1.20.1 Effective outreach strategies

Effective strategies will manifest with greater understanding of farmers' beliefs about climate change and their readiness to respond to climate change through adaptation. Little is known about farmers' and their advisors' willingness to use outreach tools, their information needs with respect to climate change or their ability to incorporate this knowledge into existing decision making processes. The advice given to farmers has been based predominately on historical weather trends and focused on short term operational decisions rather than on long term strategies. For climatic data to be useful to such populations, designing outreach strategies that target extension agents and other professional advisors will increase the potential to influence beliefs and practices of farmers. Adoption of best management practices can be promoted by focusing on implementation among farmers most likely to adopt them, followed by leveraging social networks to inform other farmers about the benefits of adoption.

1.21 Insurance Instruments

Implementation of insurance instruments requires appropriate technical innovation, building awareness and trust, ensuring viable market demand and enhancing local capacity building among local financial institutions (Traerup, 2012). Index insurance is one such instrument that effectively reduces farmers' risk under a changing climate and generally has many advantages. With index insurance, indemnity payments are decoupled from actual crop losses, instead of linking payments to changes in attributes that impact or reflect crop growth or survival over a given spatial extent. This reduces transaction costs associated with verifying ownership and losses, removes the opportunity for individuals to change their risk behaviours to increase the likelihood of receiving a payout and allays the problem of adverse selection, in which high risk individuals are disproportionately represented in the insured pool. The rural poor are no longer widely excluded from insurance by the need to demonstrate assets as a prerequisite to purchasing a policy.

1.22 Constraints

The constraints that farmers face when making decisions, such as whether to use conservation agricultural techniques, may create barriers to practices that could improve resilience to climate change. Conservation agriculture includes practices such as minimum mechanical soil disturbance, permanent organic soil cover and crop rotation, all of which typically increase soil C storage, especially when applied in concert. Constraints to adoption include strong competition for mulched crop residues for animal feeding, increased labour demand for weeding and lack of access to use of herbicides and other inputs. Although adoption decisions are not strongly or explicitly based on labour constraints, farmer age or education level, farmers in districts that experience more rainfall variability are more likely to adopt conservation agriculture practices and to implement those practices with greater intensity. Because conservation agriculture allows planting to occur as soon as the rains begin, it offers an adaptive response to changing rainfall regimes. An existing lack of feed security and farmers' concerns about poor health may counteract incentives to their adoption of new farming technology.

1.23 Conclusions

Climate change is rapidly emerging as a global critical development issue affecting animal and agricultural sectors in the world. Adapting to climate change entails taking the right measures to reduce the negative effects of climate change or to exploit the positive impacts by making appropriate adjustments and changes. Efforts are needed to develop strategies associated with early warning system, promoting thermo-tolerant indigenous breeds, women empowerment, grassland management, disease surveillance and water resource management. The meteorological services should be strengthened for forecasting and warning of any type of climate change driven by a variety of socio-economic scenarios. Physical protection with artificial or natural low-cost shade offers the most immediate and cost-effective approach for sustainable animals production. Strategic and rotational grazing system should be followed. There should be improved management of water resources through introduction of simple techniques for localized irrigation, accompanied by infrastructure to harvest and store rainwater, such as tanks connected to the roofs of houses and small surface and underground dams. In support of agro-pastoral farming systems, models must integrate adaptation options, alternative intensification pathways, zoonotic disease and vector ecology (e.g. genetic shifts, patterns of emergence), mechanisms of effecting behavioural change and adaptation to future climate change scenarios. Capacity building in agro-ecological technologies and practices for the production and conservation of fodder improves the supply of animal feed and reduces malnutrition and mortality in animals flocks. The key point to be noted to counter climate change impact is by developing efficient and affordable adaptation technologies for the rural poor who are unable to afford expensive adaptation practices.

References

Aggarwal, P.K. 2008. Global climate change and Indian agriculture: impacts, adaptation and mitigation. Indian Journal of Agricultural Sciences, 78(10): 911-919.

Akinnagbe, O.M. and Irohibe, I.J. 2015. Agricultural adaptation strategies to climate change impacts in Africa: a review. Bangladesh Journal of Agricultural Research, 39(3): 407-418.

Araus, J.L. and Cairns, J.E. 2014. Field high throughput phenotyping: the new crop breeding frontier. Trends Plant Sci, 19: 52-61.

Batima, P., 2007. Climate change vulnerability and adaptation in the livestock sector of Mongolia. Assessments of impacts and adaptations to climate change. International START Secretariat, Washington DC, US.

Battisti, D.S. and Naylor, R.L. 2009. Historical warnings of future food insecurity with unprecedented seasonal heat. Science, 323: 240-244.

Bhattacharya, S., Sharma, C. et al. 2006. Climate change and malaria in India. NATCOM Project Management Cell, National Physical Laboratory, New Delhi, Current Science, 90(3).

Cochard, R. 2013. Natural hazards mitigation services of carbon rich ecosystems. Ecosystem Services and Carbon Sequestration in the Biosphere. Edited by: Lal, R., Lorenz, K., Hüttl, R.F., Schneider, B.U., von Braun, J., 221-293. New York: Springer Science

Del Prado, A., Crosson, P., Olesen, J.E. and Rotz, C.A. 2013. Whole farm models to quantify greenhouse gas emissions and their potential use for linking climate change mitigation and adaptation in temperate grassland ruminant based farming systems. Animal, 7: 373-385.

Del Prado, A., Misselbrook, T., Chadwick, D., Hopkins, A., Dewhurst, R.J., Davison, P., Butler, A., Schröder, J. and Scholefield, D. 2011. SIMSDAIRY: a modelling framework to identify sustainable dairy farms in the UK framework description and test for organic systems and N fertiliser optimisation. Sci Total Environ, 409: 3993-4009.

Foley, P., Crosson, P., Lovett, D.K., Boland, T.M., O'Mara, F.P. and Kenny, D. 2011. Whole farm systems modelling of greenhouse gas emissions from pastoral suckler beef cow production systems. Agric Ecosyst Environ, 142: 222-230.

Havlík, P., Valin, H., Herrero, M., Obersteiner, M., Schmid, E., Rufino, M.C., Mosnier, A., Thornton, P.K., Böttcher, H., Conant, R.T., Frank, S., Fritz, S., Fuss, S., Kraxner, F. And Notenbaert, A. 2014. Climate change mitigation through livestock system transitions. Proc Natl Acad Sci U S A, 111: 3709-3714.

Herrero, M., Thornton, P.K., Bernués, A., Baltenweck, I., Vervoort, J., van de Steeg, J., Makokha, S., van Wijk, M.T., Karanja, S., Rufino, M.C. and Staal, S.J. 2014. Exploring future changes in smallholder farming systems by linking socioeconomic scenarios with regional and household models. Glob Environ Change, 24: 165-182.

Hoffmann, I. 2008. Livestock genetic diversity and climate change adaptation. Livestock and Global Change Conference proceeding. May 2008, Tunisia.

IFAD (International Fund for Agricultural Development) 2009. Livestock and Climate change.https://www.ifad.org/documents/10180/48b0cd7b-f70d-4f55-b0c0-5a19fa3e5f38.

Intergovernmental Panel on Climate Change (IPCC). 2007. Climate Change 2007: Impacts, Adaptation and Vulnerability. Contribution of Working Group II to the Fourth Assessment Report of the IPCC. In: M.L. Parry, O.F. Canziani, J.P. Palutikof, P.J. van der Linden and C.E. Hanson, eds. Cambridge University Press, Cambridge, UK, 976 pp.

Jones, P.G. and Thornton, P.K. 2009. Croppers to livestock keepers: livelihood transitions to 2050 in Africa due to climate change. Environ Sci Policy, 12: 427-437.

Leduc, B. 2010. Counting on women: climate change adaptation in the Himalayas. In: Community champions: adapting to climate challenges. H. Reid, S. Huq and A. Laurel (eds.) pp. 76.

LotzeCampen, H., von Lampe, M., Kyle, P., Fujimori, S., Havlik, P., van Meijl, H., Hasegawa, T., Popp, A., Schmitz, C., Tabeau, A., Valin, H., Willenbockel, D. and Wise, M. 2014. Impacts of increased bioenergy demand on global food markets: an AgMIP economic model intercomparison. Agric Econ, 45: 103-116.

Mbow, C., van Noordwijk, M., Prabhu, R. and Simons, T. 2014. Knowledge gaps and research needs concerning agroforestry's contribution to sustainable development goals in Africa. Curr Opin Environ Sustain, 6: 162-170.

Minang, P.A., Duguma, L.A., Bernard, F., Mertz, O. and van Noordwijk, M. 2014.Prospects for agroforestry in REDD + landscapes in Africa. Curr Opin Environ Sustain, 6: 78-82.

Mukherji, A., Facon, T., de Fraiture, C., Molden, D. and Chartres, C. 2012. Growing more food with less water: how can revitalizing Asia's irrigation help?. Water Policy, 14: 430-446.

Peleg, Z. and Blumwald, E. 2011. Hormone balance and abiotic stress tolerance in crop plants. Curr Opin Plant Biol, 14: 290-295.

Peleg, Z., Reguera, M., Tumimbang, E., Walia, H. and Blumwald, E. 2011. Cytokinin-mediated source/sink modifications improve drought tolerance and increase grain yield in rice under water stress. Plant Biotechnol J, 9: 747-758.

Rebetzke, G.J., Chenu, K., Biddulph, B., Moeller, C., Deery, D.M., Rattey, A.R., Bennett, D., BarrettLeonard, G. and Mayer, J.E. 2013. A multisite managed environmental facility for targeted trait and germplasm phenotyping. Funct Plant Biol, 40: 113.

Reguera, M., Peleg, Z. and Blumwald, E. 2012. Targeting metabolic pathways for genetic engineering abiotic stress tolerance in crops. Biochim Biophys Acta, 1819: 186-194.

Reisinger, A., Havlik, P., Riahi, K., Vliet, O., Obersteiner, M. and Herrero, M. 2012. Implications of alternative metrics for global mitigation costs and greenhouse gas emissions from agriculture. Clim Change, 117: 677-690.

Rosenzweig, C., Elliott, J., Deryng, D., Ruane, A.C., Müller, C., Arneth, A., Boote, K.J., Folberth, C., Glotter, M., Khabarov, N., Neumann, K., Piontek, F., Pugh, T.A.M., Schmid, E., Stehfest, E., Yang, H. and Jones, J.W. 2014. Assessing agricultural risks of climate change in the 21st century in a global gridded crop model inter-comparison. Proc Natl Acad Sci U S A, 111: 3268-3273.

Rotz, C., Montes, F. and Chianese, D.S. 2010. The carbon footprint of dairy production systems through partial life cycle assessment. J Dairy Sci, 93: 1266-1282.

Roy, S.J., Tucker, E.J. and Tester, M. 2011. Genetic analysis of abiotic stress tolerance in crops. Curr Opin Plant Biol, 14: 232-239.

Sood, A., Muthuwatta, L. and McCartney, M. 2013. A SWAT evaluation of the effect of climate change on the hydrology of the Volta River basin. Water Int, 38: 297-311.

Teshome, S. 2012. Analysis of traditional drought early warning indicators in use in five villages of Elidaar Woreda. Afar Pastoralist Development Association (APDA).

Thornton, P.K., Boone, R.B., Galvin, K.A., Burnsilver, S.B., Waithaka, M.M., Kuyiah, J., Karanja, S., GonzálezEstrada, E. and Boone, B. 2007. Coping strategies in livestock dependent households in East and Southern Africa: a synthesis of four case studies. Hum Ecol, 35: 461-476.

Thornton, P.K., van de Steeg, J., Notenbaert, A. and Herrero, M. 2009. The impacts of climate change on livestock and livestock systems in developing countries: a review of what we know and what we need to know. Agric Syst. 101: 113-127.

Traerup, S. 2012. Informal networks and resilience to climate change impacts: a collective approach to index insurance. Glob Environ Change, 22: 255-267.

Vermeulen, S.J., Aggarwal, P.K., Ainslie, A., Angelone, C., Campbell, B.M., Challinor, A.J., Hansen, J.W., Ingram, J.S.I., Jarvis, A., Kristjanson, P., Lau, C., Nelson, G.C., Thornton, P.K. and Wollenberg, E. 2012. Options for support to agriculture and food security under climate change. Environ Sci Policy, 15: 136-144.

Wada, Y., Wisser, D., Eisner, S., Flörke, M., Gerten, D., Haddeland, I., Hanasaki, N., Masaki, Y., Portmann, F.T., Stacke, T., Tessler, Z. and Schewe, J. 2013. Multimodel projections and uncertainties of irrigation water demand under climate change. Geophys Res Lett, 40: 4626-4632.

Wang, X. and Taub, D.R. 2010. Interactive effects of elevated carbon dioxide and environmental stresses on root mass fraction in plants: a meta analytical synthesis using pair-wise techniques. Oecologia, 163: 111.

Wheeler, T. and von Braun, J. 2013. Climate change impacts on global food security. Science, 341: 508-513.

Yin, X. 2013. Improving ecophysiological simulation models to predict the impact of elevated atmospheric CO_2 concentration on crop productivity. Ann Bot, 112: 465-475.

2

Impacts of Climate Change in Animal Agriculture

V.U.M. Rao

Project Coordinator(Agromet) CRIDA, Santoshnagar, Hyderabad

Increasing evidence over the past few decades indicate that significant changes in climate are taking place worldwide as a result of enhanced human activities. The inventions that were discovered during last few centuries, more so in the last century has altered the concentration of atmospheric constituents that lead to global warming. The major cause to climate change has been ascribed to the increased levels of greenhouse gases like carbon dioxide (CO_2), methane (CH_4), nitrous oxides (NO_2), chlorofluorocarbons (CFCs) beyond their natural levels due to the uncontrolled activities such as burning of fossil fuels, increased use of refrigerants, and enhanced agricultural related practices. The temperature increase is widespread over the globe and is greater at higher northern latitudes. Land regions have warmed faster than the oceans. January 2000 to December 2009 was the warmest decade on recordreported by NASA. These activities accelerated the processes of climate change and increased the average global temperatures by about 0.8°C (1.5°F) since 1880 (NASA,2010). Global average sea level has risen since 1961 at an average rate of 1.8 [1.3 to 2.3] mm/yr and since 1993 at 3.1 [2.4 to 3.8] mm/yr, with contributions from thermal expansion, melting glaciers and ice caps, and the polar ice sheets. Satellite data since 1978 show that annual average Arctic sea ice extent has shrunk by 2.7 % [2.1 to 3.3%] per decade, with larger decreases in summer of 7.4% [5.0 to 9.8%] per decade. Mountain glaciers and snow cover on average have declined in both hemispheres. It has also induced increased climatic variability and occurrence of extreme weather events in many parts of the world.

In India, agricultural production is mainly dependent on the monsoonal rains. Evidences also indicate that large-scale climatic variations are prevalent at micro-regional level influencing the rainfall distribution in different parts of Asia. Since climate is closely related to human activities and economic development including agricultural system, there is a serious concern about its stability (Sinha *et al*., 2000). Rice yields have shown declining trend during last three decades in the northern parts of India and is attributed to increasing temperatures (Aggarwal, 2007). In view of the above there is an immediate need to address the impacts of climate change in different sectors in India, particularly agriculture. Further, Global warming could increase water, shelter and energy requirement of live stock for meeting projected milk demands. For agriculture, climate change is expected to affect productivity through increased temperatures, changed rainfall patterns, increased levels of carbon dioxide and increased climate variability. There may also be indirect effects through changes in diseases and pests, and increased rates of soil erosion/degradation.The effects of climate on animal agriculture are largely direct. The yearly productivities of cropland and rangelands vary widely from year to year, and drought and floods are common recurrences. For a developing country, however, livestock can be a better hedge than crops to extreme weather events such as drought. Animals are better able to survive than cultivated crops; they have a lower sensitivity to climate change. Thus, animal agriculture can serve as a buffer to the effects of variable crop production, reducing risks.This paper deals with the impact of climatic variability and change on agriculture and animal agriculture in India and some adaptation strategies to cope with.

2.1 Temperature and Its Variability

The temperature increase is widespread over the globe and is greater at higher northern latitudes. Land regions have warmed faster than the oceans, which has already started affecting the climatic phenomenon in different parts of the world and India is no exception. The mean annual temperature for India as a whole has risen by 0.56 °C (Fig. 2.1) over the period 1901-2009, generally above normal (normal based on period, 1961-1990) since 1990(IMD, 2010). This warming is primarily due to rise in maximum temperature across the country, over larger parts of the data set (Fig. 2.2). However, since 1990, minimum temperature is steadily rising (Fig. 2.3) and rate of its rise is slightly more than that of maximum temperature (IMD, 2009).

Spatial pattern of trends in the mean annual temperature (Fig. 2.4) showssignificant positive (increasing) trend over most parts of the country except overparts of Rajasthan, Gujarat and Bihar, where significant negative (decreasing) trendswere observed (IMD, 2009).

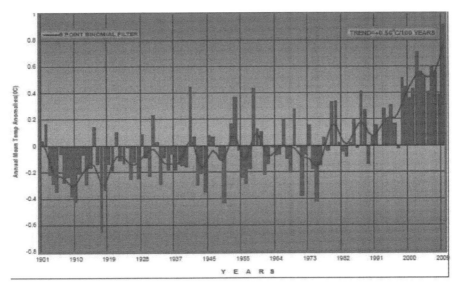

Fig. 2.1: All India mean annual temperature anomalies for the period 1901-2009 (based on1961-1990 average) shown as vertical bars. (*The solid blue curve show sub-decadal time scale variations smoothed with a binomial filter*)
Source: IMD, 2010

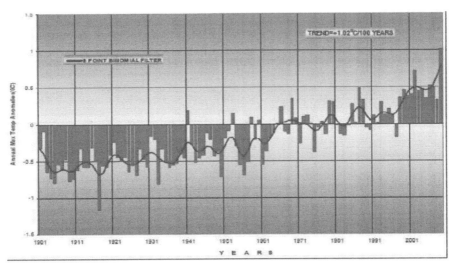

Fig. 2.2: All India annual maximum temperature anomalies for the period 1901-2009 (based on 1961-1990 average) shown as vertical bars (*The solid blue curve show sub-decadal time scale variations smoothed with a binomial filter*)
Source: IMD, 2010

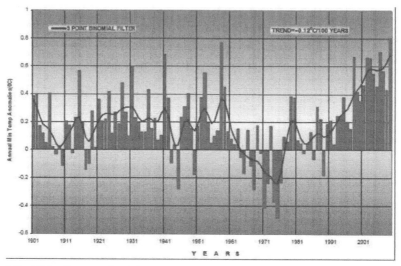

Fig. 2.3: All India annual minimum temperature anomalies for the period 1901-2009 (based on 1961-1990 average) shown as vertical bars (*The solid blue curve show sub-decadal time scale variations smoothed with a binomial filter*)
Source: IMD, 2010

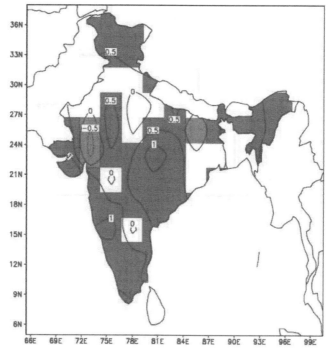

Fig. 2.4: Spatial pattern of trend (°C/ 100 years) in mean annual temperature anomalies (1901-2009). Areas where trends are significant are shaded (red: warming, blue cooling)
Source: IMD, 2009

Season wise, maximum rise in mean temperature (Fig. 2.5) was observed during the post-monsoon season (0.77 °C) followed by winter season (0.70 °C), premonsoonseason (0.64 °C) and monsoon season (0.33 °C).

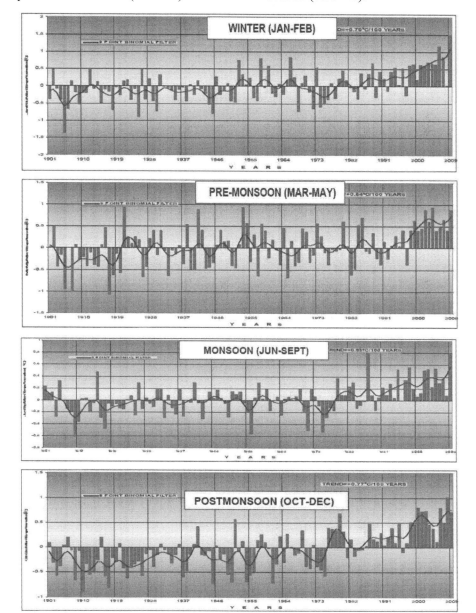

Fig. 2.5: All India Mean Temperature Anomalies for the four seasons for the period 1901-2009 (based on 1961-1990 average)
Source: IMD, 2010

A recent study conducted by All India Coordinated Research Project on Agrometeorology(AICRPAM) (Anonymous, 2013) showed that minimum temperatures on annual and seasonal (*Kharif* & *Rabi*) basis showed strong increasing tendency in majority of the locations in the country (Fig. 2.6 a to c). Warming tendency is found to be strong during *Rabi* season compared to *Kharif*. This rise in minimum temperature is expected to have an influence on the agricultural production in the country.

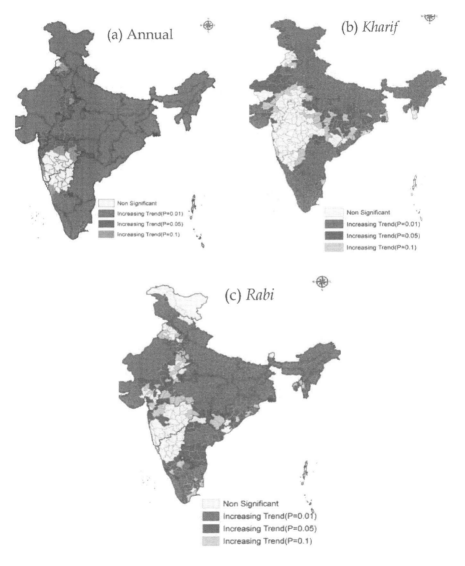

Fig. 2.6: Trends in minimum temperature over India (1971-2009)
(See colour version on page 439)

2.2 Rainfalland Its Variability

Indian agriculture is highly dependent on the spatial and temporal distributionof monsoon rainfall.The average rainfall (1871- 2010) of the country is 848 mm with a standard deviation of 84 mm (Fig. 2.7). Considering rainfall as a deficit or in excess if all-India monsoon rainfall for that year is less than or greater than the mean standard deviation.Over a 140-year period there are total 24 deficient, 20 excess and remaining are normal monsoon years. During the period 1871-1920, the occurrence of deficient monsoon rainfall years (9) are more than the excesses (8), whereas during the period 1921-1960 excess years (5) are more than deficient years (2). After 1961 to 2010, deficient monsoon rainfall years are 13 and excess are only 7.

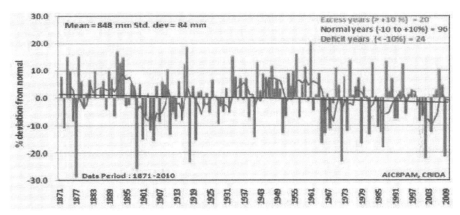

Fig. 2.7 : Inter-annual variability of Indian monsoon rainfall 1871-2010. Bars denote per cent age departure from normal (blue) with excess (green) and deficient (red) years. The long term trend is denoted by the black line. The violet curve denotes decadal variability of Indian monsoon rainfall. (See colour version on page 440)

Though there is no significant trend in the monsoonal rain on a country basis, the spatial variability accounted for slightly higher trend values in northwest, west coast and peninsular India monsoon rainfall. Pockets of increasing / decreasing trends in 36 meteorological subdivisions over India are seen (Fig. 2.8a and b).

Northwest India, west coast and peninsular India shows increasing trends though not statistically significant. Coastal Andhra, West Bengaland Punjab show significant increasing trends (INCCA, 2010). Central India shows adecreasing trend, which is significant over Chhattisgarh and East Madhya Pradesh. About 14sub-divisions show decreasing and 22 sub-divisions show increasing trends. In the recent decades (Fig.2.8), 16 sub-divisions show decreasing and 20 sub-divisions show increasing trends. East central India shows positive trends, which were decreasing based on the entire period 1871-2008. Only West Bengal showed a significant increasing trend in the recent period.

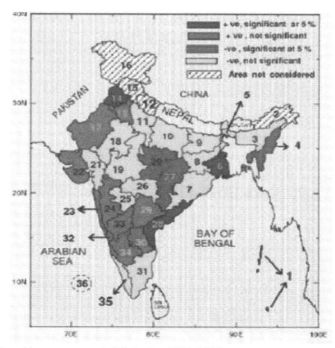

(a) Trend in monsoon rainfall during 1871-2008 (See colour version on page 440)

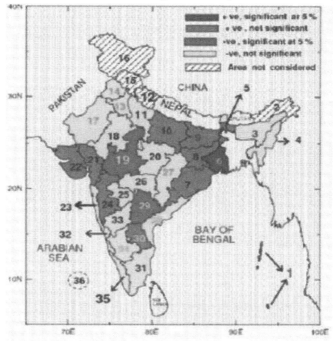

(b) Trend in monsoon rainfall during 1951-2008 (See colour version on page 441)

Fig. 2.8: Trends in summer monsoon rainfall for 1871-2008 (a) and 1951-2008, (b) for 36 meteorological subdivisions (INCCA, 2010)

2.3 Retreat of Himalayan Glaciers

The glaciers and the snowfields in the Himalayas are on the decline as a result of climate variability. The rate of retreat of the snow of Gangotri glacier demonstrated a sharp rise in the first half of the 20th century. This trend continued up to around the 1970s, and subsequently there has been a gradual decline in its rate of retreat. The diminishing rate of retreat of the snout of the Gangotri glacier could be a consequence of the diminishing rate of rise in the temperatures. The retreat of some of the Himalayan glaciers of India is presented in the following Table 2.1.

Table 2.1: Retreats of Important Glaciers in the Himalayas

Glacier	Location	Period	Avg. retreat (Meters/yr)	Reference
Milam	Uttaranchal	1849-1957	12.5	Vohra (1981)
Pindari	Uttaranchal	1845-1966	23.0	Vohra (1981)
Gangotri	Uttaranchal	1935-1976	15.0	Vohra (1981)
Gangotri	Uttaranchal	1985-2001	23.0	Hasnain, S. I., et. al. 2004
Bada Shigri	Himachal Predesh	1890-1906	20.0	Mayekwski, & Jeschke (1979)
Kolhani	Jammu & Kashmir	1857-1909	15.0	Mayekwski, and Jeschke, (1979)
Kolhani	Jammu & Kashmir	1912-1961	16.0	Mayekwski, and Jeschke, (1979)
Machoi	Jammu & Kashmir	1906-1957	8.1	Tiwari, (1972)
Chota- Shigri	Himachal Pradesh	1970-1989	7.5	Surender et al. (1994)

Ref: WWF Nepal Program (2006)

Composite ASTER image showing retreat of the Gangotri Glacier terminus in the Garhwal Himalaya since 1780.Glacier retreat boundaries courtesy of the Land Processes Distributed Active Archive Center.

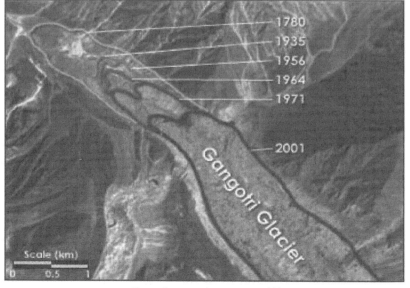

See colour version on page 441

Although the warming processes continue unabated, the rate of rise in temperatures in the Gangotri glacier area has nevertheless demonstrated a marked gradual decline since the last quarter of the past century. However, SamudraTapu, one of largest glaciers in Chandra Basin in Lahul and Spiti receded by 862 m between 1963 and 2006, at a rate of 18.5 m in a year, with the rapid rate retreat being observed during past six years compared to earlier decades (India Today, 2006).Glaciers in the Himalayan mountain ranges will retreat further, as temperatures increase: they have already retreated by 67% in the last decade. Glacial melt would lead to increased summer river flow and floods over the next few decades, followed by a serious reduction in flows thereafter.

2.4 Droughts and Desertification

Drought is a regular part of the natural cycles affecting productivity and leading to desertification. Impacts of shifts in climatic pattern becomes more prominent when one considers the climatic spectrum of the Dryland and arid regions (Ramakrishna et al, 2000) as these marginal areas provide early signals of the impacts of climate variability and change (Sinha et al, 2000). The rainfed regions encompassing the arid, semi-arid and dry sub-humid regions (covering regions less than 1150 to 1200 mm) are more prone to climatic variability (Ramakrishna et al., 2007) as in these eco-systems. Long term data analysis for India concludes that major parts of rainfed areas of India is having the probability of three to four year drought in every ten year period in which, there is again a probability of getting one to two years moderate and half year to one year severe droughts (Fig. 2.9) is visible from the following figures (Rao et al., 2009). Due to the variability of climate in the recent past more intense and longer droughts have been observed over wider areas since 1970's, particularly in tropics and subtropics (IPCC, 2007).

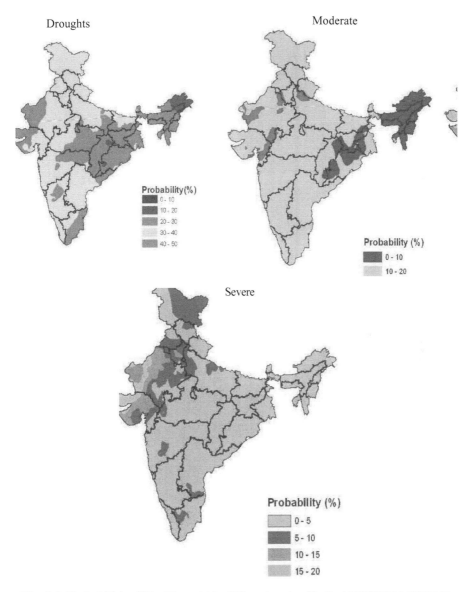

Fig. 2.9: Probabilities (%) of drought in different parts of India (AICRPAM-CRIDA)

2.5 Observed Evidences of Climate Variability / Change

There is preliminary evidence to indicate that decrease in rice yields in recent past in Indo-Gangetic plains was associated with a slight rise in minimum temperatures. Wheat yields and hence production is showing losses of 4-6 million tons in recent years due to increased heat in February-March of 2004. Increasing temperatures in Himachal Pradesh has resulted in a decrease in apple productivity and the apple belt is gradually shifting upwards (higher elevation)(NPCC,2004-07). Intensity of cyclones increased. The cyclone that hit Orissa State (India) in October 1999 affected the livelihoods of 12.9 million people and resulted in the loss of 1.6 million houses, nearly 2 million hectares of crops and 40,000 livestock. An intense heat wave in May 2002 followed by drought and heat wave in 2003,caused loss to life of human, animal and heavy Crop loss. Continuous floods caused heavy crop loss in different parts of Andhra Pradesh during 2005, 06 and 07 after continuous deficit rainfall for a period of four years led to significant economic losses to these regions(Fig. 2. 10).

Fig. 2.10: Floods in Andhra Pradesh

2.6 Projected Climate Change Scenarios

The climate change scenarios for the Indian subcontinent as inferred by Lal *et al.* (2001) from simulation experiments using atmosphere-ocean GCMs under the four SRES marker scenarios are presented below. These results suggest an annual mean area-averaged surface warming over the Indian subcontinent to range between 3.5 and 5.5°C over the region by 2080. These projections showed more warming in winter season over summer monsoon. The spatial distribution of surface warming suggests a mean annual rise in surface temperatures in north India by 3°C or more by 2050. The study also suggests that during winter the surface mean air temperature could rise by 3°C in northern and central parts while it would rise by 2°C in southern parts by 2050. In case of rainfall, a marginal increase of 7 to 10 percent in annual rainfall is projected over the sub-continent by the year 2080. However, the study suggests a fall in rainfall by 5 to 25% in winter while it would be 10 to 15% increase in summer monsoon rainfall over the country. It was also reported that the date of onset of summer monsoon over India could become more variable in future.

Table 2.2: Seasonal Projections of Temperature and Rainfall in India

Year	Season	Temperature Change (°C)		Rainfall Change (%)	
		Lowest	Highest	Lowest	Highest
2020s	Annual	1.00	1.41	2.16	5.97
	Rabi	1.08	1.54	-1.95	4.36
	Kharif	0.87	1.17	1.81	5.10
2050s	Annual	2.23	2.87	5.36	9.34
	Rabi	2.54	3.18	-9.22	3.82
	Kharif	1.81	2.37	7.18	10.52
2080s	Annual	3.53	5.55	7.48	9.90
	Rabi	4.14	6.31	-24.83	-4.50
	Kharif	2.91	4.62	10.10	15.18

(*Source:* Lal *et al.*, 2001)

2.7 Impacts of Climate on Animal Agriculture

The effects of climate on animal agriculture are largely direct. The yearly productivities of cropland and rangelands vary widely from year to year, and drought and floods are common recurrences. Livestock can be a better hedge than crops to extreme weather events such as drought. Animals are better able to survive than cultivated crops; they have a lower sensitivity to climate change. Thus, animal agriculture can serve as a buffer to the effects of variable crop production, reducing risks and utilizing grain in excess of human food needs.

- Heat waves, which are projected to increase under climate change, could directly threaten livestock. Heat stress affects animals both directly and indirectly. Over time, heat stress can increase vulnerability to disease, reduce fertility, and reduce milk production.
- Drought may threaten pasture and feed supplies. Drought reduces the amount of quality forage available to grazing livestock. Some areas could experience longer, more intense droughts, resulting from higher summer temperatures and reduced precipitation. For animals that rely on grain, changes in crop production due to drought could also become a problem.
- Climate change may increase the prevalence of parasites and diseases that affect livestock.

- The earlier onset of spring and warmer winters could allow some parasites and pathogens to survive more easily. In areas with increased rainfall, moisture-reliant pathogens could thrive.
- Animal agricultures in developed countries, however, are not immune from climatic phenomena. The productivities animal agriculture, can be affected by long-term changes in climate as well as sporadic, widespread droughts.

2.7.1 Impacts on livestock

India owns 57% of the world's buffalo population and 16% of the cattle population. It ranks first in the world in respect of cattle and buffalo population, third in sheep and second in goat population. The sector utilizes crop residues and agricultural by-products for animal feeding that are unfit for human consumption. Livestock sector has registered a compounded growth rate of more than 4.0% during the last decade, in spite of the fact that a majority of the animals is reared under sub-optimal conditions by marginal and small holders and milk productivity per animal is low. Increased heat stress associated with rising temperature may, however, cause distress to dairy animals and possibly impact milk production. A rise of 2 to 6 °C in temperature is expected to negatively impact growth, puberty and maturation of crossbred cattle and buffaloes. The low producing indigenous cattle are found to have a high level of tolerance of these adverse impacts than high yielding crossbred cattle. Therefore, high producing crossbred cows and buffaloes will be affected more by climate change.

Upadhyay and his co-workers at National Dairy Research Institute, Karnal studied the temperature and humidity induced stress level on Indian livestock. Livestock begins to suffer from mild heat stress when Temperature Humidity Index (THI) reaches higher than 72, moderate heat stress occurs at THI 80 and severe stress is observed after THI reaches 90. In India, huge variations in THI are observed throughout the year (Table 2.3). In most of the agroclimatic zones of India and the average THI are more than 75. More than 85% places in India experiences moderate to high heat stress during April, May and June. THI ranges between 75 and 85 at 2.00 PM in most part of India. The THI increases and exceed 85 i.e. severe stress levels at about 25% places in India during May and June. Even during morning THI level remains high during these months. On an average THI exceed 75 at 75-80% places in India throughout the year. As can be seen from the data, the congenial THI for production i.e. 70 is during Jan and Feb at most places in India and only about 10-15% places have optimum THI for livestock productivity i.e. during summer and hot humid season. Climate change scenario constructed for India revealed that temperature rise of about or more than 4 °C is likely to increase uncomfortable days (THI>80) from the existing 40 days (10.9%) to 104 days (28.5%) for Had CM 3 - A2 scenario and 89 days for B2 scenario for time slices 2080-2100. The results further indicate that number of stress days with THI >80 will increase by 160%.

Table 2.3: Distribution of THI (%) in India at 7:20 and 14:20 hrs

Months	<70		70-75		75-80		80-85		85-90	
	7:20 hrs	14:20 hrs	7:20 hrs	14:20 hrs	7:20 hrs	14:20 hrs	7:20 hrs	14:20 hrs	7:20 hrs	14:20 hrs
January	85	58	12	25	3	17	-	-	-	-
February	82	40	12	21	6	39	-	-	-	-
March	59	12	26	16	15	57	-	15	-	-
April	13	11	41	1	34	22	12	65	-	1
May	9	9	8	2	49	11	34	53	-	25
June	9	8	10	5	33	15	48	47	-	25
July	6	8	20	10	45	30	29	49	-	3
August	6	8	20	10	57	37	17	45	-	-
September	8	8	20	8	57	36	15	48	-	-
October	16	10	50	10	34	74	-	6	-	-
November	73	23	14	49	13	48	-	-	-	-
December	83	56	13	30	4	14	-	-	-	-

It is estimated that global warming is likely to result in a loss of 1.6 million tones in milk production by 2020 and 15 million tones by 2050. Based on THI, the estimated annual loss in milk production at the all-India level by 2020 is valued at Rs. 2661.62 crores at current prices. The economic losses were highest in Uttar Pradesh followed by Tamil Nadu, Rajasthan and West Bengal. Stressful THI with 20h or more daily THI-hrs (THI >84) for several weeks, affect animal responses. Under the climate change scenario, increased number of stressful days with a change in maximum and minimum temperature and decline in availability of water will further impact animal productivity and health in Punjab, Rajasthan and Tamil Nadu (Upadhyay et al., 2009).

A rise of 2-6 °C due to global warming (time slices 2040-2069 and 2070-2099) is likely to negatively impact growth, puberty and maturity of crossbreds and buffaloes and time to attain puberty of crossbreds and buffaloes will increase by one to two weeks due to their higher sensitivity to temperature than indigenous cattle. Lactating cows and buffaloes have higher body temperatures and are unable to maintain thermal balance. Body temperature of buffaloes and cows producing milk is 1.5 - 2 °C higher than their normal temperature, therefore more efficient cooling devices are required to reduce thermal loads of lactating animals as current measures are becoming ineffective (Upadhyay et al., 2009).

2.7.2 Impacts on poultry

The analysis of mortality data from 2004 to 2009 at the Project Directorate of Poultry, Hyderabad revealed that the overall mortality was increased as the ambient temperature rises in broiler, layer and native chickens (Fig.2.11). The mortality started increasing when the temperature reaches 32 °C and the peak was observed at 38 to 39 °C (13.5%).

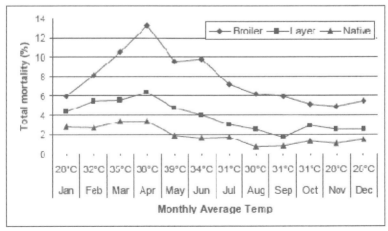

Fig.2. 11: Effect of ambient temperature on the survivability of meat type (broiler), egg type (layer) and native (desi) chicken

The mortality was highest in broiler type chickens followed by layers and native chicken. The mortality due to heat stress in broiler type birds was started appearing at the ambient temperature 30 °C, while in the layer and native chicken the heat stress related mortality was observed at the ambient temperature of 31 °C. The deaths due to heat stress were 10 times more in broiler type chickens as compared to layer and native type chickens (Fig. 2.12). The mortality due to heat stress was negligible in native (Desi type chickens) which may be due to low metabolic rate and natural heat tolerance.

Another study was conducted to find the influence of high ambient temperature on the feed intake body temperature and respiratory rate in commercial layers for 13 weeks. The consumption which was 108 g / bird /day at 28 °C was reduced to 68 g / bird / day at the shed temperature 37.8 °C (Fig. 2.13).

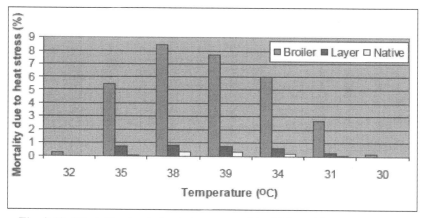

Fig. 2.12: Mortality due to heat stress caused by high ambient temperature

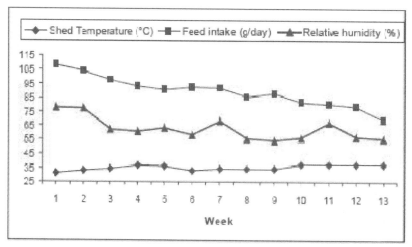

Fig. 2.13: Influence of elevated ambient temperature on feed consumption

2.7.3 Impacts of climate change on marine fisheries

A rise in temperature as small as 1 °C could have important and rapid effect on the mortality of fish and their geographical distributions. Oil sardine fishery did not exist before 1976 in the northern latitudes and along the east coast of India as the resource was not available/and sea surface temperature (SST) were not congenial. With the warming of sea surface, the oil sardine is able to find the temperature to its preference especially in the northern latitudes and eastern longitudes, thereby extending the distributional boundaries and establishing fisheries in larger coastal areas as shown in Fig.12.14. (Vivekanandan *et al.*, 2009a).

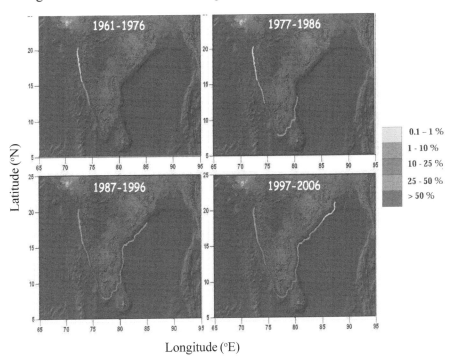

Fig. 2.14: Extension of the northern boundary of oil sardine (% in colour code indicates the % contribution of oil sardine catch from each 2° grid to the total oil sardine catch along the entire sea coast) (See colour version on page 442)

The dominant demersal fish, the threadfin breams have responded to increase in SST by shifting the spawning season off Chennai. During the past 30 year period, the spawning activity of *Nemipterusjaponicus* reduced in summer months and shifted towards cooler months (Fig. 2. 15a). A similar trend was observed in *Nemipterus*mesoprion too (Fig. 2.15b). Analysis of historical data showed that the Indian mackerel is able to adapt to rising in sea surface temperature by extending distribution towards northern latitudes, and by descending to depths (Vivekanandan *et al.,* 2009b).

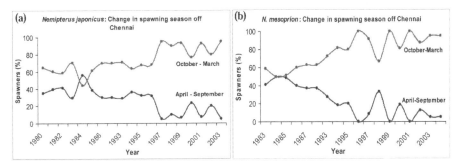

Fig. 2.15: Change in spawning season of the threadfin breams (N*emipterus japonicus* and *N. Mesoprion)* off Chennai

Central Marine Fisheries Research Institute, Cochin has studied the vulnerability of 75 coastal fishing villages of Maharashtra which are located within 100 m from the high tide line to sea level rise. Among the 75 coastal villages five coastal districts (Thane, Mumbai, Raigad, Ratnagiri and Sindhudurg) of Maharashtra, it was found that 35 villages in Raigad and Ratnagiri districts would be affected due to rise in sea level by 0.3 m (Table 2.4).

Table 2.4: Area (km²) of coastal fishing villages in Maharashtra likely to be submerged due to sea level rise by 0.3 to 1.0 m

District	No. of villages	Sea level increase by		
		0.3 m	0.6 m	1 m
Thane	8	0.95	1.04	1.45
Mumbai	5	0.12	0.19	0.28
Raigad	19	0.29	0.38	2.03
Ratnagiri	16	0.005	0.05	0.332
Sindhudurg	27	0	0	0.905
Total	75	1.365	1.66	4.997

2.7.4 Impacts of climate change on inland fisheries

In recent years, the phenomenon of Indian Major Carps maturing and spawning as early as March is observed in West Bengal with its breeding season extending from 110-120 days (Pre 1980-85) to 160-170 days (2000-2005). Consequently, it has become possible to breed them twice in a year at an interval ranging from 30-60 days. A prime factor influencing this trend is elevated temperature, which stimulates the endocrine glands and help in the maturation of the gonads of Indian major carp. The average minimum and maximum temperature throughout the state has increased in the range of 0.1 to 0.9 °C (Das, 2009; NPCC 2009).

Recent climatic patterns have brought about hydrological changes in the flow pattern of river Ganga. This has been one major factor resulting in erratic breeding and a decline in fish spawn availability. As a result of this, the total average fish landing in the Ganga river system declined from 85.21 tonnes during 1959 to 62.48 tonnes during 2004. In the middle and lower Ganga, 60 genera of phytoplankton were recorded during 1959 which declined to 44 by 1996. During the same period the Zooplankton number diminished from 38 to 26. A number of fish species, which were predominantly only available in the lower and middle Ganga in 1950s, are now recorded from the upper cold-water stretch up to Tehri (Das, 2009; NPCC 2010).

2.7.5 Adaptation and mitigation strategies

Successful adaptation to climate change requires long-term investments in strategic research and new policy initiatives that mainstream climate change adaptation into development planning. As a first step, we need to document all the indigenous practices rainfed farmers have been followed over time for coping with climate change. Secondly, we need to quantify the adaptation and mitigation potential of the existing best bet practices for different crop and livestock production systems in different agro-ecological regions of the country. Thirdly, a long-term strategic research planning is required to evolve new tools and techniques including crop varieties and management practices that help in adaptation.

More recently during 2010, ICAR has launched the National Initiative on Climate Resilient Agriculture (NICRA) as a comprehensive project covering strategic research, technology demonstration and capacity building. Targeted research on adaptation and mitigation is at nascent stage in India but based on knowledge already generated, some options for adaptation to climate variability induced effects like droughts, high temperatures, floods and sea water inundation can be suggested. These strategies fall into two broad categories viz., (i) crop based and (ii) resource management based.

2.8 Conclusion

Climate change seems to be imminent and agriculture is going to be most affected by it. To mitigate the climate change effects on agricultural production and productivity a range of adaptive strategies need to be considered. Changing cropping calendars and pattern will be the immediate best available option with available crop varieties to mitigate the climate change impact (Rathore and Stigter, 2007). The options like introducing new cropping sequences, late or early maturing crop varieties depending on the available growing season, conserving soil moisture through appropriate tillage practices and efficient water harvesting techniques are also important. Developing heat and drought tolerant crop varieties by utilizing genetic resources that may be better adapted to new climatic and atmospheric conditions should be the long-term strategy. Genetic manipulation may also help to exploit the beneficial effects of increased CO_2 on crop growth and water use (Rosenzweig and Hillel, 1995). One of the promising approaches would be gene pyramiding to enhance the adaptation capacity of plants of plants to climatic change inputs (Mangala Rai, 2007).

There is thus an urgent need to address the climate change and variability issues holistically through improving the natural resource base, diversifying cropping systems, adapting farming systems approach, strengthening of extension system and institutional support. Latest improvements in biotechnology and information technologies need to be used for better agricultural planning and weather based management to enhance the agricultural productivity of the country and meet the future challenges of climate change in the dryland regions of the world.

Several experts have identified research areas that would reduce uncertainty and improve knowledge to face the consequences of climate change and provide improved planning. The following are some of the points for consideration.

- Quantitative assessment of specific crop responses at different crop stages to enhanced levels of GHG, precipitation and UV-B radiation.
- Breeding agricultural crops for tolerance to high temperatures.
- New area that is made available for agriculture is to be properly categorized and mapped to avoid chances of in appropriate land-use choices.
- Probabilities of occurrence of extreme weather events (droughts & floods) and their impacts on plant growth.
- The impacts of elevated CO_2 on plant soil-water balances and the corresponding crop growth should be linked.
- Water balance for drought or flood prone regions in different parts of the world for changing climatic conditions.
- The quality of global modeling projections is further improved with suitable modifications in the global circulation models.
- The databases for all the parameters need to be strengthened.

References

Aggarwal P.K. 2007. Climate Change: Implications for Indian Agriculture, Jalvigyan Sameeksha, Vol. 22, pp40.

Das, M. K. 2009. Impact of recent changes in weather on inland fisheries in India. *In: Global Climate Change and Indian Agriculture: Case Studies from the ICAR Network Project*(Ed. Aggarwal, P. K.). ICAR Publication. pp.101-103.

Hingane, L. S., Rupa Kumar, K. and RamanaMurty, B.V. 1985. Long-term trends of surface air temperature in India. Journal of Climatology 5: 521-528.

INCCA. 2010. Climate Change and India-A 4 X 4 Assessment - A sectoral and regional analysis for 2030s. INCCA Report No 2. Ministry of Environment and Forests, New Delhi. 164p

IPCC (Intergovernmental Panel on Climate Change Working Group I) Climate Change. 2007. The Physical Science Basis IPPCC Working Group I.

IPCC. 2001. Climate Change 2001. The Scientific Basis. Contribution of Working Group-I to the Third Assessment Report of the Intergovernmental Panel on Climate Change [Houghton, J.T., Y. Ding, D.J. Griggs, M. Noguer, P.J. van der Linden, X., Dai, K. Maskell and C. A. Johnson (eds.)], Cambridge University Press, Cambridge, UK and New York, USA.

IUCC. 1992. Information unit on climate change, Fact Sheet No. 101. Impacts of Climate Change, UNEP, Polaisdes Nations. CH 1211, Geneva.

Lal, M. 1999. Growth and yield responses of soybean in Madhya Pradesh, India to climate variability and change. Agric. and For. Meteorol. 93: 65-66.

Lal, M., Nozawa, T., Emori, S., Harasawa, H., Takahashi, K., Kimoto, M., Abe-Ouchi, A., Nakjima, T., Takemura, T. and Numaguti, A. 2001. Future climate change: Implications for Indian summer monsoon and its variability. Current Science. 81: 1196-1207.

MangalaRai, 2007. Presidential Address at the National Conference on Climate Change and Indian Agriculture, held at NASC Complex, New Delhi during 11-12 October 2007.

Morey, D.K. and Sadaphal, M.N. 1981. Effect of weather elements on yield of wheat at Delhi. PunjabraoKrishiVidyapeeth. Res. Journal. 1: 81-83.

NASA 2010 http://www.giss.nasa.gov/research/news/20100121/

Ramakrishna, Y. S., Kesava Rao, A. V. R., Nageswara Rao, G. and Aggarwal, P. K. 2002. Impacts of climate change scenarios on Indian agriculture: Evidences, *South Asia Expert Workshop on Adaptation to Climate Change for Agricultural Productivity*, 1-3 May 2002, New Delhi India Today, 6th November 2006.

Rao G.G.S.N., Rao A.V.M.S., and Rao V.U.M. 2009. Trends in rainfall and temperature in rainfed India in previous century. In Global Climate change and Indian agriculture: Case studies from the ICAR Network Project.Ed.by Dr. P.K.Aggarwal and published by Indian Council of Agricultural Research(ICAR), New Delhi. pp.71-73.

Rao, G.S.L.H.V.P. 2011. Climate Change Adaptation Strategies in Agriculture and Allied Sectors. Scientific Publishers, Jodhpur (India). 330p.

Rao, G.S.L.H.V.P.,Rao, G.G.S.N. and Rao, V.U.M. 2010. Climate Change and Agriculture over India. PHI Learning Private Limited, New Delhi. 352p.

Rathore, L.S. and Stigter, C.J. 2007. Challenges to coping strategies with agrometeorological risks and uncertainities in Asian regions. In: Sivakumar MVK and Motha R P. (eds.). Managing Weather and Climate Risks in Agriculture, Springer Berlin Heidelberg New York. pp. 53-70.

Rosennzweig, C. and Parry, K. L. 1994. Potential impact of climate change on world food supply. *Nature*. 367: 133-138.

Sinha, S. K., Kulshrestha, S. M. and Ramakrishna, Y. S. 2000. Climatic variability and climate change impact on agriculture. Invited Plenary lecture at the *Intl. Conf. on Managing Natural Resources for Sustainable Agricultural Production in the 21st Century*, New Delhi, February 14-18. Extended Summaries Vol. 1: Invited papers, 13-15.

TERI (Tata Energy Research Institute). 2002. India specific impacts of climate change, *http://www.teriin.org/climate/impacts.htm*, as viewed on July 2, 2002.

Upadhyay, R. C., Ashutosh, V. S., Raina. and Singh, S. V. 2009. Impact of climate change on reproductive functions of cattle and buffaloes. *In: Global Climate Change and Indian Agriculture: Case Studies from the ICAR Network Project* (Ed. P.K. Aggarwal). ICAR Publication. pp.107-110.

Vivekanandan, E., Hussain Ali, M. and Rajagopalan, M. 2009b. Impact of rise in seawater temperature on the pawning of threadfin beams. *In: Global Climate Change and Indian Agriculture: Case Studies from the ICAR Network Project* (Ed. P.K. Aggarwal). ICAR Publication. pp.93-96.

Vivekanandan, E., Rajagopalan, M. and Pillai, N. G. K. 2009a. Recent trends in sea surface temperature and its impact on oil sardine. *In: Global Climate Change and Indian Agriculture: Case Studies from the ICAR Network Project* (Ed. P.K. Aggarwal). ICAR Publication. pp.89-92.

WWF Nepal Program. 2006. An overview of glaciers, glacier retreat and subsequent impacts in the Nepal, India and China. Published by WWF Nepal program, pp.32.

3

Climate Change and Animal Agriculture

G.S.L.H.V. Prasada Rao, G. Girish Varma, A. Prasad S. Sankaralingam A.P. Usha, T.S. Rajeev, Deepa Ananth and T. Unnikrishnan

Centre for Animal Adaptation to Environment and Climate Change Studies, Kerala Veterinary and Animal Sciences University, Mannuthy, Kerala – 680 651 India

Global warming is the biggest long term threat to life on earth. Rise in temperature may drive thousands of species to extinction, trigger more frequent floods and droughts and sink low lying islands and coastal areas by rising sea levels. It is the result of rising atmospheric content of CO_2 mainly owing to burning of hydrocarbons or fossil fuels like as petrol and diesel. Destruction of forests and their degradation too contribute to rise in carbon dioxide levels. The IPCC (2006) projected the rate of warming for the 21st century to be between 0.8 and 4.4°C at various stabilized CO_2 levels in atmosphere and it is most likely to be 3°C by the end of this century. It could cost global economy almost $7 trillion by 2050, is equivalent to a 20% fall in growth if no action is taken on greenhouse gas emissions. If action is taken, it will cost only $350 billion due to climate change already taken place, just 1% of the global GDP. The winter 2007 was the warmest and recorded 0.85°C above average of 12°C and the previous highest was 0.71°C, which occurred in 2002 in Northern Hemisphere. The entire Europe Union recorded the warm winter, having more than 2°C above average. New York experienced the highest temperature of 21.7°C on a day in January, 2007 and the second highest was recorded as 17.2°C in 1950. The year 2007 was the warmest winter in the NHS. However, floods and excess rains were also noticed due to hurricanes and tropical storms worldwide in 2007. The year 2010 was the warmest year in India, followed by 2009. It was the second warmest year globally after 1998. It was also one of the wettest years globally in recent years and it was a landmark in annals of climatology.

Out of 10 years in the first decade of this century, 8 warm years took place in India except in 2005 and 2008 while 9 globally. 1998 was the warmest, followed by 2010 across the World. Several continents experienced floods too while heat wave and drought during summer in Russia. Cloud bursts in Pakistan and India led to devastating floods in August. Snow storms in the United States and the European Union were noticed. Heavy rains poured during the Northeast monsoon in southern States of India.

Increase in all-India mean temperatures is almost solely contributed by increase in maximum temperature (0.6°C/100years) with minimum temperature remaining practically trendless. Consequently, there is a general increase in diurnal range of temperature. The rate of increase was more during the post monsoon season (0.87°C), followed by 0.72°C in winter. Across different zones of the Country, the rate of increase was more in West Coast of India, followed by the Western Himalayas of India. In rainfall, there was a decrease since last 50 years. It appears that rainfall cycle is advanced by two weeks since increase in rainfall was noticed during May and June while declined in July and August in Northwest, West and Eastern parts of the Country. A marked increase in rainfall and temperature is projected in India during the current century. The maximum expected increase in rainfall is likely to be 10-30% over central India. Temperature is likely to increase by 3°C towards end of the Century. It is more pronounced over Northern parts of India. In recent years, increase in night temperature during winter months was noticed in southern states of the Country. Consecutive droughts during monsoon in 2014 and 2015 across the Country directly or indirectly affected the Indian economy. Untimely and erratic distribution of rainfall is the concern as seen in 2016.

3.1 Climate Change and Animal Agriculture

The State of Kerala always appears to be greenery due to high monsoon rainfall and thick vegetation. At the same time, the central and northern districts of Kerala experience prolonged dry spell from November to May if pre-monsoon showers fail. For the first time, the unprecedented summer drought was noticed across the State of Kerala in 1983. Similar drought was noticed in summer 2004 and 2013.The water levels were very low in major reservoirs wherever summer drought was prevalent. Surface water resources were dried up. Dairy farming is adversely affected due to lack of fodder and water in several districts across Kerala during the summer drought in 2013. The deficit rainfall during monsoon season followed by scanty rains during northeast monsoon as seen in recent years is a threat to Agriculture including Animal Agriculture. The deficit monsoon rainfall across the State of Kerala was the order of more than 30 per cent. Such deficit rainfall was seen for the first time in Kerala.

Decline in wetlands, forest area and increase in forest fires, indiscriminate land filling and sand mining, groundwater depletion, drying of streams, rivers and surface wells, floods and droughts, landslides, cloud bursts, rainfall decline,

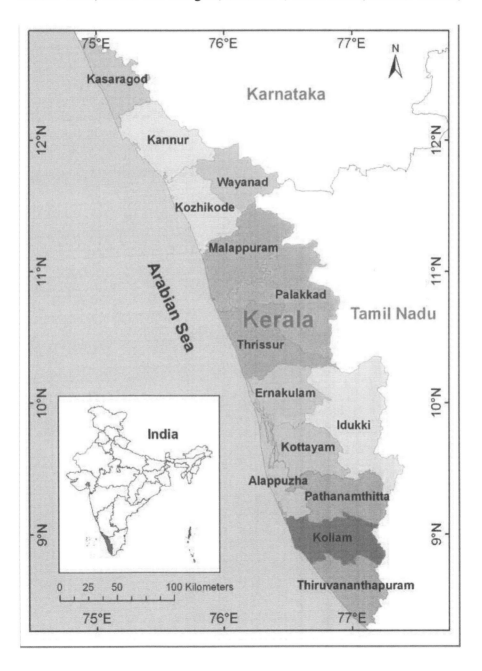

temperature rise (Fig.3.1) are the issues related to climate change. The State of Kerala was moving from wetness to dryness within the humid climates (Fig.3.2). It is likely to be a threat to the society linked sectors like animal agriculture, agriculture, forestry, biodiversity, water resources and health. Sea level rise is another important climate change related issue along the Kerala Coast. Saline water intrusion and water quality is deteriorated in many parts of Kerala along the Coast and their impact in Animal Agriculture need to be understood. The occurrence of summer droughts and floods during monsoon season is not uncommon across the State of Kerala. Reports indicate that such weather abnormalities are likely to occur and reoccur under the projected climate change scenario. Thermal stress is noticed on cattle, poultry and elephants during summer due to high maximum temperature that prevails between 35 and

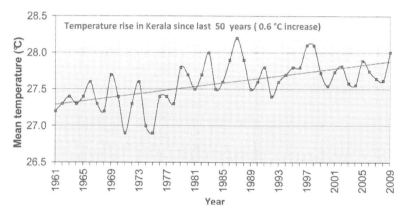

Fig. 3.1: Temperature trends over Kerala from 1961 to 2009

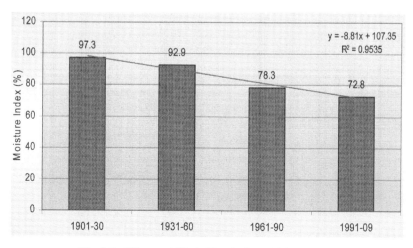

Fig. 3.2: Climate shifts in Kerala from 1901 to 2009

40 degree Celsius. The seasonal fluctuations are predominant in fodder availability and milk production (Fig.3.3). One of the reasons for low milk production during summer in cattle is due to thermal stress and poor intake. The mortality rate in poultry could be explained again due to heat load though the rate of mortality is relatively less in Kerala when compared to other States, where heat wave during summer is a threat. However, the egg size, egg number, fertility, hatchability and feed conversion ratio are adversely affected. In detail studies in this direction is already on.

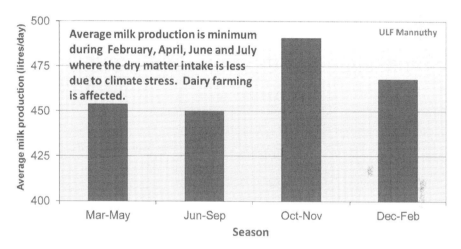

Fig. 3.3: Average milk production at ULF Mannuthy

Initial studies indicated that installation of sprinklers on roof-top reduces inside temperature, by which the birds feels comfortable during summer and the mortality is minimised in poultry. The disease outbreak (like duck plague, duck cholera, fowl cholera) was noticed during summer 2013 and heavy mortality was noticed up to 50 per cent.

In contrast, high rainfall with high relative humidity for more number of days during monsoon season leads to diseases. Of course, animal diseases that are prevalent mostly season bound and detail studies are

Egg size and quality are likely to be adversely affected due to global warming and climate change

Heat stress symptoms in dairy (See colour version on page 442)

to be taken up on mission mode for sustenance of animal agriculture under the Humid Tropics.

Thermal stress is noticed on cattle, poultry and elephants during summer due to high maximum temperature that prevails between 35 and 40 degree Celsius. The seasonal fluctuations are predominant in fodder availability and milk production. One of the reasons for low milk production during summer in cattle is due to thermal stress and poor intake. The mortality rate in poultry could be explained again due to heat load though the rate of mortality is relatively less in Kerala when compared to other States, where heat wave during summer is a threat. However, the egg size, egg number, fertility, hatchability and feed conversion ratio are adversely affected. In detail studies in this direction is already on. Initial studies indicated that installation of sprinklers on roof-top reduces inside temperature, by which the birds feels comfortable during summer and the mortality is minimised in poultry. The disease outbreak (like duck plague, duck cholera, fowl cholera) was noticed during summer 2013 and heavy mortality was noticed up to 50 per cent. In contrast, high rainfall with high relative humidity for more number of days during monsoon season leads to diseases. Of course, animal diseases that are prevalent mostly season bound and detail studies are to be taken up on mission mode for sustenance of animal agriculture under the Humid Tropics.

3.2 Direct and Indirect Effects

The effect of animal production in the hot humid climate has global significance in the light of climate change threats. Effects of increasing effective temperature, droughts, increasing salinity of inland water bodies, deforestation and pollution are concerns of subtropical and even temperate regions due to possible fallout of global climate change.

3.2.1 Direct effects

Direct effects include multiple stresses due to temperature, humidity, radiation, low plane of nutrition, heavy rains, pests and diseases. The cyclical phenomenon of drought and floods has additional effects. The indirect effect of climate is the reduced availability of quality feed ingredients. Direct and indirect effects cause stress which cause depletion of body reserves and there by reduced production, growth and reproduction. Among all stresses that the climate offers, the thermal stress due to effective environmental temperature in the most important. Challenge before the scientific community of the tropical world is to find ways to enhance milk production in the prevailing climatic conditions. Historically the traditional livestock production largely depended on heat tolerant native breeds that produced less milk compared to temperate exotic breeds. The dairy sector now largely comprises of extensive and expanding crossbred population in Kerala. For crossbreds, increased air temperature, and humidity measured as Temperature Humidity Index (THI) above critical thresholds are related to low dry matter intake (DMI) and to reduced efficiency of milk production cause significant heat stress. The daily mean Temperature Humidity Index (THI) at selected locations over Kerala is depicted in Fig. 3.4. Results indicated that THI may not be conducive from February to May.

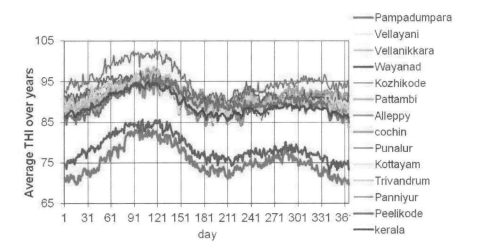

Fig. 3.4: Daily mean Temperature and Humidity Index (THI) at selected locations across Kerala

The Zone of Thermo-neutrality with in which no additional energy above maintenance is expended to heat or cool the body- for livestock is between -0.5 to 20℃ and the upper critical temperature (B on the right side) may reach to 25-26℃ (West, 2003).The ambient temperature of hot humid region is above this critical temperature during several months of each year. Effective environmental temperature is a combined effect of ambient temperature and humidity (Fig. 3.5). The combined effect is quantified as Temperature Humidity Index (THI) $db°C - \{0.31 - 0.31RH)(db°C - 14.4)\}$. The normal THI to maintain production in dairy cattle is 72. In our state, during most days in a year it is hot and humid and hence the THI is high enough to cause significant heat stress.

C = lower critical temperature

D= point of reduction of metabolic heat

B on the left side = lower point of zone of thermal neutrality below which chemical regulation is needed to maintain homeothermy.

B on the right side = upper critical temperature.

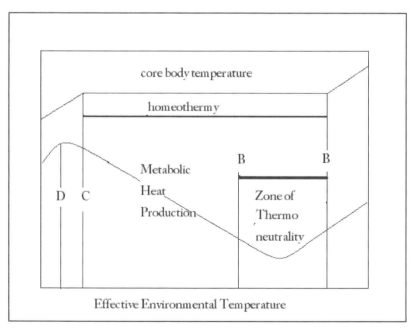

Fig. 3.5: Kleiber's law of metabolic heat production and core body temperature as influenced by environment temperature

Heat distress on animals will reduce the rate of animal feed intake and causes poor performance growth (Rowlinson, 2008). Among the direct effects of climate change most significant is the reduced milk production at higher temperatures. The following chart adapted from Esmay 1961 based on studies conducted in Psychrometric chamber shows reduction in milk production in breeds which incidentally forms the parent sires used for crossbreeding activity of Kerala (Fig.3.6).The following chart adapted from Esmay 1961 based on studies conducted in Psychrometric chamber shows reduction in milk production in breeds which incidentally forms the parent sires used for crossbreeding activity of Kerala.

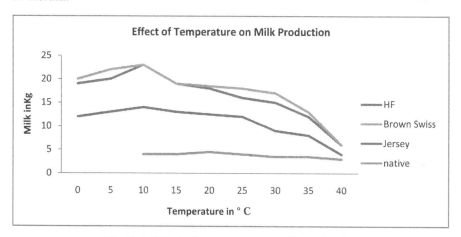

Fig. 3.6: Effect of temperature on milk production

3.2.2 Indirect effects

Earth's Climate is undergoing noteworthy changes. The frequency and severity of risks in agriculture particularly in last few decades have increased on account of climate variability. The dependence of Indian agriculture on climate, weather, rainfall and the timeliness of rainfall is substantial. The principal evidence of climatic change has been rising temperatures, erratic rainfall pattern, and increase in the severity of droughts, floods, and cyclones which have caused huge losses in agricultural production and the livestock population. The climate change, especially global warming is now well documented Projected climate change over India based on various models suggest steady increase in temperature and at a later stage slight increase of rainfall. Climate change will have far-reaching consequences for dairy, meat and wool production mainly via impacts on grass and range productivity. Changes in rainfall patterns may translate to an increased spread of existing vector-borne diseases and macro parasites of animals as well as humans with the emergence and spread of new diseases. In some areas, climate change may also cause new transmission models; these effects has already experienced in Kerala with the increasing density of mosquito population and a variety of fever outbreaks. Inspite of the major impact of climate change on livestock sector and the direct contribution of this sector to climate change not many studies has been undertaken in this area in Kerala. A few studies have been reported on the effect of extreme climates on production performance of livestock. Studies on the physiological sturdiness with respect to adaptability of cross bred animals (sahiwalbrownswiss crossbreds) in comparison to native zebu cattle indicated that the cross bred could face the thermally adverse climate in tropics successfully and can perform in hot humid areas. Based on this research he recommended some simple effective management procedures for the cross bred animals under the tropical and sub tropical conditions like providing shelter during day time, night feeding, feeding roughages in the evening. From his study he could evolve a modified tolerance index for tropics. He has made elaborate studies on the problem of dairying in hot humid climate and developed a formula for Temperature Humidity Sunshine Index (THSI) which is a modification of THI (Temperature humidity Index (Thomas, 1969). In another study by Rajagopalan (1975) to evaluate the performance of Jersey cattle in Indian conditions, it was found that Jersey animals are quite suitable for cross breeding programme in view of their high average performance level under local climatic conditions in India.Thiagarajan (1989) in his work has attempted to investigate the effects of housing and feeding on growth and production of B. Taurus * B. indicus cross bred cattle. He has studied the beneficial effects of open air conditions in hot humid tropical environment and advised loose housing system in hot humid tropical environment in which cattle have continuous access to open paddock shaded by trees. Under the hot humid conditions high wind velocity in the open seems to favour the cows considerably. Study by Suraj and Sivakumar, (2012) suggests that the environment in which the dairy cattle are reared affects its ability to maintain thermal balance. It was found that the climatic condition of the different zones and the choice of materials played a major role in deciding the comfort level of the animal in the individual housing system.The analysis based on Temperature Humidity Index of Kerala for the last 10 years has revealed that climatic stress is very high in most parts of the state throughout the year (unpublished data, Prasad, 2013). Indo Swiss Project in 1963 contemplated to involve a new multipurpose breed suitable for the climatic conditions prevailing in Kerala. Integrated Cattle development project (ICDP) plan was mooted for the fourth five year plan. Increase in animal population has also been attributed as a cause for climatic change. The increase in global average temperature is due to the observed increase in anthropogenic greenhouse gas concentrations. The global livestock sector generates directly or indirectly 18% of global greenhouse gas

(GHG) emissions as measured in CO_2 equivalent (FAO, 2007). Considering the large population of livestock, most of the developed countries were complaining about the GHG emission from Indian livestock. Commercial livestock farming has not emerged as a successful model in Indian condition due to various reasons. The varying environmental factors and the inability to devise and to popularize cost effective adaptation methodologies contribute to this factor. However, the small holder production system prevailing in the country has the advantage of reduced GHG because of dung distributed in the pasture during grazing.

3.3 Climate Change and Poultry Industry

The poultry farming is very sensitive to heat stress. Egg production and size is likely to reduce in high summer temperature. It may not be that much severe in the Humid Tropics when compared to that of arid and semi-arid regions. However, increase in the egg production is hindered and it is reduced by 5% due to thermal stress even in the Humid Tropics. If there is no sprinkler, reduction in egg production could have been around 10%. The sprinkler reduces the poultry house temperature by 2-3°F. High summer temperature led to reduction in egg weight from 52 grams to 45 grams. This problem is reduced by providing more protein in the feed and also supplementation of aminoacids like methionine. The fertility of the eggs received at Revolving fund hatchery was reduced from 95 % to 75 % due to the hot summer. The hatchability also reduced from 85 % to 50 % on total egg set in the Revolving fund hatchery. It led to the closure of the Revolving fund hatchery in the remaining period of summer.

3.4 Disease Outbreak and Mortality

A poultry farmer 'Basheer of Thrissur district' although he vaccinated all 8000 ducklings for duck plague, due to summer stress the immunity did not develop which led to the outbreak of duck plague and loss of 50% (4,000) of his ducklings (45 days old) due to the outbreak of duck viral hepatitis (duck plague) in April and May 2013. Another farmer 'Varghese of Thrissur' lost around 1000 ducklings (3 months old) out of 7000 due to pasteurellosis outbreak in April, 2013. There was a mortality of 8 pullets out of 700 pullets reared in cages due to heat stroke accompanied by AI (Artificial insemination) stress at Poultry Farm, Mannuthy. The same thing had happened in a commercial broiler breeder farm with the mortality of 250 birds due to heat stroke and AI stress. The heat stress can be alleviated by providing cool water by installing sprinklers at roof top.

3.5 Ozone Depletion and UV Radiation

The greatest concentration of ozone is at an average in height of 25 km and above in the stratosphere. The ozone molecule is made up of three atoms of oxygen. It is the most efficient absorber of ultraviolet radiation from sun and thus protects all life forms in planet earth. Ozone depletion due to industrialization in recent decades is the concern of humankind and biological activities. Release of compounds like chlorofluorocarbons (CFCs), carbon tetrachloride and methyl chloroform could significantly deplete ozone layer that shielded the planet from ultraviolet radiation. The CFCs are used in a variety of industrial, commercial, and household applications. These substances are non-toxic, non-flammable and non-reactive. They are used as coolants in commercial and home refrigeration units, aerosol propellants and electronic cleaning solvents. The global average thickness of ozone is 300 Dibson Units, equivalent to 3mm. In contrast, it is about 100 Dibson units, which alarm us that we are in great danger if precautionary steps are not taken up. The ozone level fell to 90 Dibson Units, which was the lowest value recorded on 30-09-98 nearly equalled to the lowest value ever recorded 88 Dobson Units on 28-09-94 over Antarctica. The ozone losses are caused by chlorine and bromine compounds released by chlorofluorocarbons and halons. Year-to-year variation of size and depth of ozone hole depend on variations in meteorological conditions. The unusual cold temperature by 5-9°F over Antarctic zone enables greater activation of reactive chlorine that ultimately causes more ozone loss and lower ozone levels. Increase in chlorine levels should peak in the Antarctic stratosphere within a few years (Shashi et al., 2002). Among UV radiations, UV-B radiation in the range of 280-320 nm is more sensitive to ozone fluctuations and reaches earth surface. A global network is created to monitor UV-B filtered radiation in terms of Minimum Erythema Dose (MED). The human-made interventions in industrial development lead to ozone depletion, thereby filtered UV radiation reaches the ground, resulting in various human, animal and crop diseases. The ozone loss has the potential to increase incidence of skin cancer, cataracts and damage to people's immune system, harm some crops and interfere with marine life. However, little is known on impact of ozone depletion and increasing UV-B radiation on ontogeny of tropical plants and human and animal diseases since studies in this direction are lacking. Because CFCs remain in the atmosphere for100 years, continued accumulation of these chemicals pose ongoing threats, even after their use is discontinued. Sun burns and heat strokes are noticed in recent years across Kerala.

3.6 Deforestation and Biodiversity

The natural forest utilizes about 686 Giga tonnes of carbon- about 50 per cent more than atmosphere – is being cleared at an average rate of about 13 million hectares per annum over the World. It makes deforestation responsible between 20 and 25 per cent of global greenhouse gas emissions. As many as 572 fires broke out in the forest of Kerala during 2004 summer and the average loss of forest per year during the last decade (1991-2003) came to about 2,093 ha. Severe droughts, high temperature and low relative humidity are the atmospheric factors responsible for forest fires in addition to human interventions. The Economic Review, prepared by the State Planning Board, 2003 warns that a third's of State's biodiversity would vanish or would be close to extinction by 2030 unless steps are taken to check extinction of species. Of the 300 rare, endangered or threatened species in the Western Ghats, 159 are in Kerala. Of these, 70 are herbs, 23 climbers, eight epiphytes, 15 shrubs and 43 trees. Besides, 10 species of fresh water fish are identified as most threatened. Kerala has a flora of 10,035 species, which represents 22 per cent of Indian flora. Of these, 3,872 are flowering plants of which 1,272 are endemic. As many as 102 species of mammals, 479 birds, 169 reptiles, 89 amphibians and 262 species of fresh water fish are reported from Kerala. Many of these are endemic. The review recalls that during the 20th century, at least 50 plant species have become extinct in the Country. Three species of birds-Himalayan Mountain Quail, forest spotted owlet and pink-headed duct-have become extinct. Besides, as many as 69 bird species have been categorised as extinct. The mammals, Indian Cheetah and lesser one horned rhinoceros have also perished. The Malabar civet is on the threshold of extinction and 173 species have been listed as threatened. It may be noted that nearly 23 per cent of the total endemic flora species of the Country are in Kerala. Describing a conservation strategy, the review says that ecologically sensitive areas have to be identified with reference to topography, hydrological regimes and this has to be networked with species diversity.

3.7 Impacts of Droughts and Floods

Unprecedented low water level in Peechi Dam due to Hydrological Drought during summer 2013

The occurrence of summer droughts and floods during monsoon season is not uncommon across the State of Kerala. Reports indicate that such weather abnormalities are likely to occur and reoccur under the projected climate change scenario. For the first time, the unprecedented summer drought was noticed across the State of Kerala in 1983. Similar drought was noticed in summer 2004 and 2013. The water levels are very low in major reservoirs wherever drought is prevalent. Surface water resources are dried up. Dairy farming is adversely affected due to lack of fodder and water in several districts during the current summer 2013. The deficit monsoon rainfall across the State of Kerala was the order of more than 30 per cent. Such deficit rainfall was seen for the first time in Kerala.

Though Kerala has achieved much in crossbreeding of cattle compared to other states of India, the average milk production of the cows is much less than its genetic potential. Improved production potential and sustainability in production in the prevailing climate should be supported by scientific research and this field will be of more importance in the background of climate change. Soil degradation through use of agro-chemicals is a serious issue that needs to be addressed on a priority basis in India during the next five-year plan. Imbalanced use of chemical fertilizer has led to declining fertilizer response in the fertile irrigated regions. Excess use of some nutrients, driven in part by imbalanced subsidies, has led to depletion of other nutrients from the soil leading to deterioration of the soil health. Alarmingly, the drastic decrease in the organic carbon content of the soil and the change in C: N ratio is one of the major reasons for soil degradation and decrease in agricultural productivity. The nature and extent of problem differ in different of parts of the country.

The carbon sequestration potential by cultivatable fodder plots could be used to partly mitigate the greenhouse gas emissions of the livestock sector. This will require avoiding land use changes that reduce ecosystem soil carbon stocks and a cautious management of fodder plots aiming at preserving and restoring soils and their soil organic matter content. Trade-offs between greenhouse gas emissions and animal production need to be better understood at the farm and regional scales, through a continued development of observational, experimental and modeling approaches (Soussana, 2010). Development of appropriate cost effective technologies for climate change mitigation, conservation of natural resources and ensuring the livestock productivity is very important from social and economic point of view.

Models on global warming indicate that the rise in temperature is likely to be around 3 ° C by end of this century. It is likely that the extreme weather events like droughts and floods, cold and heat waves increase in coming decades. The global economy will be adversely affected as mentioned in the latest report of IPCC. If the sea level increase as projected, the coastal areas which are thickly populated will be in peril and for the existing population, the safe drinking water will be a great problem. The whole climate change is associated with increasing greenhouse gases and human- induced aerosols and the imbalance between them may lead to uncertainty even in year-to-year monsoon behaviour over India. Therefore, there should be a determined effort from developed and developing countries to make industrialisation environment-friendly by reducing greenhouse gases pumping into atmosphere. In the same fashion, awareness programmes on climate change and its effects in various sectors viz., animal agriculture, agriculture, health, infrastructure, water, forestry, biodiversity and sea level and the role played by human beings in climate change need to be taken up on priority. In the process, life style of people should be changed so as not to harm Earth-Atmosphere continuum by pumping CFCs into atmosphere.

From the animal agriculture point of view, effects of extreme weather events on important animal species are to be documented so that it will be handy to planners and weather insurance agencies in such reoccurrence events for mitigating the ill effects. Also, there is need to guide planners on projected future animal products' scenarios including dairy farming based on climate change events, which will be more realistic at field level as models always overestimate the impacts. Finally, we have to foresee these extreme events and prepare ahead to combat them so that the losses can be minimised. In this direction, livestock advisory based on weather forewarning will go a long way for sustenance of animal agriculture. It is high time to institutionalize awareness programmes on climate change/variability effects in animal agriculture in varied climates on research mode as a part of World Universities Network

3.8 Future Line of Work

- To analyze and update the trends in climatic changes and to integrate the information to develop a database.
- To forecast extreme weather events which can be detrimental to livestock farming by using computer simulation tools.
- To study the impact of climatic change on production and reproduction performance of animals.
- To study the decline in disease resistance as well as emerging new diseases on basis of changing climate of Kerala.
- To sort out suitable mitigation measures to the changing climate so that the livestock farming sector of Kerala is not affected by the climatic variation.
- Assisting the farmers in developing and adopting appropriate measures and management schemes in changing climate scenario.
- Immediate reasons for climate change should be traced out and effort should be taken to develop measures to mitigate the production of greenhouse gas emissions from livestock.

References

Esmay ML 1978. Principles of Animal Environment, AVI Publishing Company, Connecticut, 351p.

IPCC (Intergovernmental Panel on Climate Change) 2007. Climate change 2007: the physical science basis. Contribution of Working Group I to the Fourth Assessment Report of the Intergovernmental Panel on Climate Change [S. Solomon, D. Qin, M. Manning, Z. Chen, M. Marquis, K.B. Averyt, M. Tignor& H.L. Miller, eds.]. Cambridge, UK, Cambridge University Press in The Hindu, pp 147-155.

IPCC (Intergovernmental Panel on Climate Change) 2007. The Economics of Climate Change: Stern Review. The summary of conclusions. Survey of the Environment 2007, The Hindu, pp 141-145.

IPCC (Intergovernmental Panel on Climate Change) 2007. Climate Change: The Physical Science Basis. Extracts from the IV Assessment Report. Survey of the Environment 2007, The Hindu, pp 147-155.

Rajagopalan, T. G. 1974. Study of some aspects of the performance of Jersey cattle. MSc Thesis submitted to Gujarat College of Veterinary Science and Animal Husbandry, Anand.

Rowlinson, P. 2008. Adapting livestock production systems to climate change – temperate zones. In: P. Rowlinson, M. Steele & A. Nefzaoui, eds. Livestock and global change.Proceedings of an international conference, Hammamet, Tunisia, 17–20 May 2008. Cambridge, UK, Cambridge University Press.61-63.

Shashi, K., Pathak and Mason, N.J. 2002. Our shrinking ozone layer: General perspectives and remedial measure. Resonance, December pp.71-80.

Shukla, P.R., Sharma, S.K and Ramana, V.P. 2002.*Climate Change and India - Issues, Concerns and Opportunities*.Tata – McGraw-Hill Publishing Company Ltd, New Delhi, 314p.

Soussana, J. F., Tallec, T. and Blanfort, V. 2010 Mitigating the greenhouse gas balance ofruminant production systems through carbon sequestration in grasslands. Animal 4:3, 334-350.

Suraj, P. T. 2011. Effect of dairy cattle under different housing pattern in north and western regions ofTamilnadu. PhD thesis submitted to Tamilnadu Veterinary and Animal Sciences University.

Thiagarajan, M. 1989. Effect of environmental heat stress on performance of crossbred dairy cattle. PhD thesis submitted to Kerala Agricultural University.

Thomas C.K. and Sastry, N.S.R. 1991. Dairy Bovine Production (first ed.). Kalyani Publishers, New Delhi, pp 106-125 .

West, J. W. 2003. Effects of heat-stress on production in dairy cattle. J. Dairy Sci., 86, 2131-2144.

4

Heat Stress in Dairy Cattle - Consequences and Management

K. Karthiayini, Shibu K. Jacob, Muhammed.E.M. and Zahoor Ahmad Pampori*

Kerala Veterinary and Animal Sciences University, Kerala
**Division of Veterinary Physiology, SKUAST-K, Shuhama Alusteng, Srinagar Kashmir-190 011*

Heat stress induced by climate change is a serious issue of livestock industries worldwide (Bajagai, 2011). When environmental temperatures move out of the thermo-neutral zone (or comfort zone) animals begins to experience either heat stress or cold stress. Both these stresses require the animals to increase the amount of energy used to maintain the body temperature so that less energy is available for productive processes. Atmospheric temperature, relative humidity, air movement and solar radiation are the major factors that determine the heat load on animal. Thermoregulation is the ability of the animals to maintain their body temperature in cold or hot environments. It consists of behavioural, physiological and anatomical responses that affect energy metabolism. At lower temperature as a physiological adjustment to maintain the body temperature the animal increases the metabolic heat production by increasing the basal metabolic rate and uncoupling of oxidative phosphorylation. Hence the maintenance energy requirement of animals increases in a cold environment, which reduces the amount of energy available for production. From a practical point of view higher temperatures are much more dangerous for producing animals than a cold environment. Temperatures exceeding the higher critical level compromise animal performance not only by changing the energy and nutrient metabolism, but also by upsetting the body homeostasis, with detrimental consequences both for immune competence and for product quality. In general, livestock with high production potential are at greatest risk of heat stress, and require most attention.

Because of their high metabolic rate and poor water retention mechanism in kidney and gastrointestinal tract dairy cattle are particularly more susceptible to increased ambient temperature than other ruminants (West, 2003). Effects of thermal stress vary among individuals according to breeds, production level, prior experience etc. Research show that *Bosindicus* (Zebu) cattle posses thermo-tolerant gene and are therefore more thermo-tolerant than *Bostaurus* cattle (Gaughan *et al.*, 1999).

4.1 Heat Stress Responses and Its Consequences

Thermally stressed animals show different physiological metabolic and behavioural changes inorder to maintain the core body temperature constant. These responses include increased respiratory rate and heart rate (Wheeloc *et al.* 2010), reduced thyroxine concentration and reduced basal metabolic rate, reduced feed intake, increased water intake etc. When these mechanisms are not sufficient to maintain the normal temperature, the animal's rectal temperature shoot up leading to a condition called hyperthermia. The increased respiration and heart rate are the body mechanism to improve the heat dissipation. Respiration rate of the animal can be used as an indicator of severity of thermal load but several other factors like animal condition, prior exposures to high temperature etc. should also be considered to interpret the respiration rate.

Process of adaptation and acclimation to thermal stress in animal is generally mediated by alteration in hormonal profile in the body. Thermal stress in animal has found to alter the activity of thyroid gland resulting reduced concentration of thyroxine (T_4) and increased concentration of triodothyronine (T_3) in plasma (Nardone *et al.*, 1997). It has been speculated that reduced thyroid activity reduces metabolic rate and therefore the metabolic heat production. Thermal stress reduce hepatic glucose synthesis and therefore blood glucose concentration (Wheelock *et al.*, 2010). Increased level of catecholamines and glucocorticoides released during stress cause lipolysis and mobilize adipose tissue and increase non esterified fatty acid (NEFA) level. Increased sodium and potassium loss through sweat during thermal stress result in electrolyte imbalance in the body. Decreased net mineral intake due to reduced appetite and reduced absorption of minerals during hot ambient temperature further exacerbate the situation. Hyper ventilation resulting from increased respiratory rate reduces the level of bicarbonate (HCO_3^-) in blood leading to respiratory alkalosis. Hyper ventilation results in increased secretion of bicarbonate (HCO_3^-) from kidney and decreased secretion of HCO_3^- in saliva (Nardone *et al.*, 2010). Bicarbonate is required to have buffering action in rumen to buffer the VFA produced by carbohydrate fermentation in the rumen. This is done by the HCO_3^- secreted through saliva. This HCO_3^- secretion is impaired in heat stress which results in disturbances in rumen pH. The imbalance in rumen PH may leads to rumen acidosis and resultant laminitis and reduction in milk fat production. Heat stress decreases rumination leading to digestive disturbances. Sweating and panting during thermal stress can lead to excessive water loss leading to cardiovascular disturbances (Silanikove,1994). Similarly, modification in glucose and fatty acid metabolism together with reduced liver function and oxidative stress during thermal stress causes more incidences of metabolic diseases. Thermal stress has both direct and indirect adverse effects on animal health (Bernabucci, *et al.*, 2010). Direct effects of thermal stress range from simple physiological disturbances to organ dysfunction and death. Indirect effect reduced feed quality, reduced feed intake, digestability etc. Feed consumption by exotic dairy cattle starts to decline when average daily temperature reaches 25 to 27⁰C. Reduction in feed intake together with diversion of more energy to maintain normal body function creates negative energy balance which compromise health and deteriorate animal body condition. Reduced Immune status of the heat stressed animals makes them prone to infectious diseases. Reduced disease resistance of the animals along with enhanced multiplication of microorganisms and altered vector population results in increased incidence of certain diseases like mastitis during summer when ambient temperature is high. High ambient temperature with high humidity also creates an environment suitable for fungal growth in feed and feedstuffs leading to mycotoxicosis. Reports show that mortality in dairy cows increases sharply when maximum and minimum temperature humidity index (THI) increases from 80 and 70 respectively. However, this again varies with the animal's prior exposure and productive status.

Development of resistance against disease in calves is largely influenced by amount of immunoglobulin present in colostrum. Exposure of cows in late pregnancy and early postpartum period to high ambient temperature reduces the concentration of immunoglobulins (IgG and IgA) in colostrums. Passive transfer of immunity from dam to neonatal calves through colostrum also decreases with increased ambient temperature (Donovan et al.1986). Altered metabolic status of the animals may also cause reduction in immunity making animals more susceptible to diseases. Different breeds show varying degrees of immunological response to high ambient temperature (Lacetera *et al.*, 2006). Stress hormones released also play a major role in reducing the immune status of thermally stressed animals. Thermal stress causes imbalance in secretion of reproductive hormones leading to reproductive problems like cystic ovary (Ronchi *et al.*, 2001). Plasma progesterone level in animals under high ambient temperature is low as compared to animals under thermo-neutral zone (Ronchi*et al.*, 2001). *Badinga et al.* (1993) have reported that high ambient temperature causes poor quality of ovarian follicles resulting poor reproductive performance in cattle. Reduced luteinizing hormone (LH) and estradiol secretion during thermal stress reduces the intensity of heat and the fertility rate of cattle. Reduced libido, decreased length and intensity of heat and increased embryonic mortality in heat stressed animals further reduce reproductive efficiency. Thermal stress also delays the returning of the animals to heat after calving which thereby increases the inter-calving interval. Impaired spermatogenesis during thermal stress lead to lower sperm concentration and increased Sperm abnormalities.

Heat stress reduces milk yield in all animals. owever the quantum of reduction depends on breed. The reasons of reduced milk yield in heat stressed cattle are many. Studies show that decrease in milk yield due to thermal stress is more prominent in Holstein than in Jersey cattle. Decreased synthesis of hepatic glucose in heat stressed animals reduces glucose supply to the mammary glands leading to low lactose synthesis which in turn reduces the milk yield. Reduction in milk yield is further intensified by decrease in feed consumption by the animals to compensate high environmental temperature. Apart minor share of reduced milk production is due to decreased feed intake while a major share is isdue to decreased nutrient absorption, altered rumen function and hormonal status and increased maintenance requirement resulting reduced net energy supply for production. Quality of milk is also decreased by heat. Quantity of milk protein and solid not fat (SNF) is also reduced during thermal stress in dairy cattle (Rhoads *et al.*, 2009).

4.2 Breed Variations in Thermo Tolerance

The environmental influence on animal largely depends on the internal heat load of the animal and the ability for efficient heat exchange between animal and environment. The heat exchange in turn dependent on the temperature and vapour pressure gradients between the animal and the environment. In natural environment breed evolved will have the inherent capacity to protect itself from the stressors to which it is constantly or very frequently exposed during its period of development. For the same reason the indigenous types of cattle found in India will therefore develops appropriate mechanisms that ensure their continued survival in the different agro-ecological areas found in the region. These genotypes possess an inherent capacity to sufficiently tolerate the stressors imposed by the physical environment like considerable resistance to high temperature, humidity as well as persistentdraught that occur in the region. Because of these qualities, they will have a comparative advantage over introduced high potential breeds and have a very crucial role in all the efforts to increase livestock productivity through breeding. Careful planning and rationalisation of the breeding strategies will be required to ensure optimal use of their adaptive qualities with due attention given in conserving them for sustained use. Fortunately we have around 37 indigenous breeds in India which can be improved for their productive and reproductive efficiency through advanced research techniques and properly planned breeding policies.

4.3 Methods to Alleviate Heat Stress in Dairy Cattle

Modern dairy breeds are of northern European origin, and are most comfortable at a temperature range of 41 to 77°F. These animals are tolerant of very low environmental temperatures (e.g., < 0°F) but intolerant of temperatures above 77°F, especially when the relative humidity is greater than 80%. Older, heavier, high-producing cows are more susceptible to heat stress than smaller, younger animals, and Holstein animals are more sensitive than Brown Swiss, Guernseys, Jerseys and Brahmans. Heat stress becomes a particular problem in Southern India where elevations in heat and humidity during the summer months have a negative effect on animals general well being and especially on udder health and production. To reduce heat stress on dairy cattle a multi-disciplinary approach should be developed involving nutritional modification, Physical modification of environment, and Genetic modification (to develope heat tolerant breeds using the local germplasm)· In the long term, dairy cattle can be made more tolerant to hot and humid weather conditions by selective breeding. Colored breeds such as Jerseys and Brown Swiss seem to show greater tolerance to heat stress than Holsteinanimals. In the short term, cows must be helped to withstand high temperatures by modifying their environment. One of the best practices to reduce heat stress is to provide adequate fresh, cool, clean drinking water *adlibitum*. In the case of animals let loose for grazing, enough shade should be given to take rest during hot hours. Shade is probably the easiest and least expensive option to help cows cool themselves.Use of fans is important, especially in confined structures, because fans help to move warm air from cows' bodies.

4.4 Nutritional Manipulations to Reduce Heat Stress

During heat stress farm animals reduce their feed intake so as to reduce the metabolic heat production, which produce significant consequences on their nutrient intake. The practice of feeding the daily ration in several smaller portions or during the cooler parts of the day will be useful to improve their feed intake

Nutritional manipulations to reduce total heat production of farm animals include

(i)   *Fat supplementation*: This is because, in comparison to other nutrients, fat generates the least heat, either when deposited as body fat or when used for energy. Thus high fat diets reduce the total heat production of the animals. Fat supplementation also boosts the energy density of the diet, as the energy content of fat is the highest compared to the other nutrients. By adding fat to the diet the energy requirement of the animals can be met accurately evenwith decreasedfeed intake.

(ii)   *Feeding low protein ration with synthetic amino acids according to the ideal protein concept*: The ideal protein ration refers to a well-defined amino acid pattern, which expresses the requirement of essential amino acids in percentage of lysine. Amino acid conversion and nitrogen excretion are the lowest when diets are formulated according to ideal protein concept. Excess amino acids that cannot be used in protein synthesis due to a limiting factor (such as a limiting amino acid, energy supply or genetic potential) are metabolized in the body. Compared to other nutrients, the oxidation of amino acids yields the most heat contributing to the total heat production. Consequently, the heat increment is higher when excess amino acids are present in the diet. The heat increment from protein metabolism is at the minimum if the dietary protein level meets the requirements of the animal, and if the amino acid content corresponds to the ideal protein concept.

(iii)   *Feeding high quality dietary fibre forages:* Dietary fibre increases the heat production of the animals. However, in ruminants high dietary fibre content supplied in the daily ration is indispensable for adequate rumen fermentation. Hence during heat stress the cattle should be fed with high quality forages and other highly digestible fibre sources. This will result in a somewhat lower heat production by the animals, in contrast to cows fed with poor quality forages. The major problem faced by the dairy farmersduring hot summer months is the scarcity of good quality forage at reasonable prices.Attempts should be made to develop draught resistant high quality forage using modern cultivation technologies so as to improve the quantity and quality of the forage.

As the heat stress impairs feed intake and the digestibility of nutrients, it is recommended to feed more concentrated diets with high levels of easily digestible nutrients. Hydrothermic treatments, micronization etc are some of the methodsemployed by feed manufacturers to improve the digestibility of feed stuffs. Increasing the level of dietary vitamins and minerals, as well as improving their bioavailability (by use of organicsources)also help to improve the feed digestibility. Enzyme supplementations can be used to improve the ileal digestibility of nutrients, such as amino acids, carbohydrates and Ca and P. Anti oxidants can also be used to alleviate the oxidative damages associated with the heat stress.

4.5 Management Alterations

Simple management alterations can produce drastic change in the thermal load on cattle. However, care should be given to choose environment friendly as well as animal friendly methods for a better and sustained effect. Also the management system adopted should be cost effective. When the levels of ambient humidity are high through the day and night, cows are unable to dissipate heat effectively unless they are exposed to at least three hours of cooling in a 24-hour period (Avendano-Reyes *et al.*, 2010). In such situations cooling systems will be effective to reduce the air temperature inside the cow shed, so that dissipation of heat between the animal's body and environment is made possible.

4.5.1 Use of sprinklers

Sprinklers can be used to wet the cow'sbody, allowing the loss of body heat via conduction. Use of fans and sprinklers together allow simultaneous conductive and evaporative cooling, as the fans help to vaporize the water that has been warmed by the release of body heat thereby increasing the efficiency of cooling.Cooling ponds are very effectiveprovided they are scientifically constructed and properly maintained. Disadvantage of most of the cooling systems is that they require huge amount of water during the period of water scarcity. Moreover, the excess water in the cow's environment, which, along with warm temperatures, provides ideal conditions for the growth of mastitis-causing bacteria.

4.5.2 Use of fans in cow sheds

Cows generate approximately 20% of their gross energy as body heat, which is released to the surrounding air, making their surrounding hot. Fans remove this body heat (via convection), thereby cooling down the surface of the animal. Use of thatched roof, false ceiling, wet gunny bags and palm leaves are few methods which are cheap and can be easily adopted by the marginal farmers to reduce the heat stress. Heat stress can be managed in a more efficient and economic way if dairy farming is changed from low and marginal farming system to a level of dairy farming industry.

References

Avendano-Reyes, L., Alvarez-Valenzuela, F.D, Correa-Calderon, A.,Algandar-Sandoval, A., Rodriguez-Gonzalez, E., Perez-Velazquez, R., Marcias-Cruz, U., Diaz-Molina, R., Robinson, P.H., Fadel, J.G. 2010. Comparison of three cooling management systems to reduce heat stress in lactating Holstein cows during hot and dry ambient conditions. Livestock Sci. 132: 48-52.

Badinga, L., Thatcher, W. W., Diaz, T., Drost, M. and Wolfenson, D. 1993. Effect of environmental heat stress on follicular development and steroidogenesis in lactating Holstein cows.Theriogenology. 39: 797-810.

Bajagai, Y.S. 2011. Global climate change and its impacts on dairy cattle. Nepalese Vet.J.30: 2-16.

Bernabucci, U., Lacetera, N., Baumgard, L., Rhoads, R., Ronchi, B. and Nardone, A. 2010. Metabolic and hormonal acclimation to heat stress in domesticated ruminants. Animal.4:1167-1183.

Donovan, G.A., Badinga, L., Collier, R. J., Wilcox, C. J. and Braun, R. K. 1986. Factors Influencing Passive Transfer in Dairy Calves. J Dairy Sci.69: 754-759.

Gaughan, J., Mader, T., Holt, S., Josey, M. and Rowan, K. 1999. Heat tolerance of Boran and Tuli crossbred steers. J. Anim. Sci.77: 2398.

Lacetera, N., Bernabucci, U., Scalia, D., Basiricò, L., Morera, P. and Nardone, A. 2006. Heat Stress Elicits Different Responses in Peripheral Blood Mononuclear Cells from Brown Swiss and Holstein Cows. J. Dairy Sci, 89: 4606-4612.

Nardone, A., Lacetera, N.,Bernabucci, U. and Ronchi, B. 1997. Composition of Colostrum from Dairy Heifers Exposed to High Air Temperatures During Late Pregnancy and the Early Postpartum Period. J. Dairy Sci.80: 838-844.

Nardone, A., Ronchi, B., Lacetera, N., ranieri, M. S. and Bernabucci, U. 2010. Effects of climate changes on animal production and sustainability of livestock systems. Livestock Sci.130: 57-69.

Rhoads, M., Rhoads, R., Vanbaale, M., Collier, R., Sanders, S., Weber, W., Crooker, B. and Baumgard, L 2009. Effects of heat stress and plane of nutrition on lactating Holstein cows: I. Production, metabolism, and aspects of circulating somatotropin. J Dairy Sci, 92: 1986-1997.

Ronchi, B., Stradaioli, G., Verini-Supplizi, A., Bernabucci, U., Lacetera, N., AccorsI, P. A., Nardone, A. and Seren, E. 2001. Influence of heat stress or feed restriction on plasma progesterone, oestradiol-17 beta, LH, FSH, prolactin and cortisol in Holstein heifers. Livestock Prodn Sci, 68: 231-241.

Silanikove, N. 1994. The struggle to maintain hydration and osmo regulation in animals experiencing severe dehydration and rapid rehydration: the story of ruminants. Experimental Physiol. 79: 281-300.

Vitali, A; Segnalini, M; Bertocchi, L; Bernabucci, U; Nardone, A and Lacetera, N.2009. Seasonal pattern of mortality and relationships between mortality and temperature-humidity index in dairy cows. J. Dairy Sci.92:3781-3790.

Wheelock , J. B., Rhoads R. P., VanBaale M. J., Sanders S. R and Baumgard, L. H. 2010. Effects of heat stress on energetic metabolism in lactating Holstein cows. J. Dairy Sci.93: 644– 655.

West, J.W . 2003.Effects of Heat-Stress on Production in Dairy Cattle. J. Dairy Sci.86:2131-2144.

5

Existing Dairy Housing Systems and Its Suitability in Different Agro-Climatic Regions of Tamil Nadu

Suraj P.T[1]and T. Sivakumar[2]

[1]*Cattle Breeding Farm Thumburmuzhy, Kerala Veterinary and Animal Sciences University, Kerala – 680 721, India*
[2]*Dean Faculty of Food Science and Technology, Tamil Nadu Veterinary and Animal Sciences University, Tamil Nadu – 600 007, India*

The annual milk production in Tamil Nadu during 2009-10 from crossbred cows, non-descript cows and buffaloes were 42.28,7.89 and 7.61 lakh tones, respectively. Tamil Nadu had 59.83 lakhs crossbred female cows in the year 2007 as against 41.47 lakhs in 2003 with an annual growth rate of 9.59 per cent (Anon, 2010). With the increasing crossbred cattle population and average annual production, better management of the high producers becomes a necessity. Production and health of animals depend mostly on environment in which they live. A satisfactory environment for any farm livestock is the one that ensures not only optimal productivity but also meets the health and behavioral needs of the animals. A satisfactory environment is the one that satisfies the following four criteria: thermal comfort, physical comfort, disease control and behavioral satisfaction. The most important environmental interventions done in recent days are those that have been done in housing and other attempts to ameliorate the thermal environment. The effort of the body to maintain a stable internal environment to challenges from widely variable environments was first described by Claude Bernard (1878) and later referred to as homeostasis (Cannon, 1932). An environment in which stressors are minimized would likely be favorable for efficient production of products derived from domestic farm animals and for helping ensure the well-being of those species. In order to ensure optimum productivity of the dairy cattle, understanding of the existing housing patterns and the environmental stress in various agro-climatic regions are very essential. There has been very limited research carried out on defining the causes of difference in performance among different housing systems. Hence, the present study was undertaken in the back ground of the challenges caused by impending climate change.

5.1 Materials and Methods

The research was carried out in two phases. The first one involved the base line survey to identify the existing dairy cattle housing system and the second one involved the seasonal stress assessment of the cattle in selected farms. The first phase of the research involved the data collection on the existing dairy housing patterns in the study area. The schedule for preliminary enquiry was developed in consultation with the extension personnel and subject matter specialists. The schedule was pre-tested and appropriate modification in the construction and sequence of questions were made. Pre-tested interview schedule were filled on the spot with observation and interview with the dairy farmers.The study was conducted in the four agro-climatic regions of Tamil Nadu viz. North Eastern zone, North Western zone, Western zone and Hilly zone. Based on the survey, the major housing patterns existing in the region were identified and categorized for further detailed investigation. Farmers with at least five cows were selected from each agro-climatic zone with the major types of housing pattern identified for conducting the field investigation. For each season a total of thirty samples were collected (five housing types with six replicates) for measuring bio-chemical profile, haematological studies, serum enzyme analysis, serum hormone estimation and electrolyte status from each agro-climatic region. A total of 120 samples were collected for each season with a grand total of 480 samples during the whole experimental period of one year. In each sampling season the number of animals sampled in each group was always greater than the five recommended by Whitaker (2000).

The collected data were statistically analyzed by one way analysis of variances (ANOVA) for finding the difference between the groups by using statistical package SPSS 17. The significance was tested using Duncan's multiple range test (Duncan, 1955). Factor analysis was used to rank the housing systems in the study area based on the significant climatic, physiologic and production parameters. The dairy housing comfort index was constructed based on the factor scores obtained by different housing systems and the housing systems were ranked accordingly by using statistical package SPSS 17.

5.2 Results and Discussion

5.2.1 Existing animal shelter system

The average number of animals kept in shelters in North Eastern zone, North Western zone, Western zone and Hilly zone of Tamil Nadu is presented in Table 5.1a. It can be seen from the table that the average number of animals per farm is 6.34, 7.62, 8.60 and 2.48 in North Eastern, North Western, Western and Hilly zones, respectively. The average milk production per farm is 45.35, 55.00, 71.07 and 27.35 litres per day in North Eastern zone, North Western zone, Western zone and Hilly zone respectively. It is seen from the table that the average milk production per animal is 6.83, 8.55, 7.52 and 11.15 litres per day in North Eastern, North Western, Western and Hilly zones, respectively.In all the four agro-climatic regions, the farmers were keeping their animals in separate shelters away from their own dwellings. Only a small percentage of the farmers were keeping their animals attached to their houses and the percentage was 10.1, 1.4, 2.6 and 20.0 in North Eastern, North Western, Western and Hilly zones, respectively. In all the four agro-climatic zones of the study area the farmers have mostly constructed the animal shelter as single row; 79.9, 70.0, 81.6 and 93.3 per cent, respectively in North Eastern, North Western, Western and Hilly zones. The same trend was reported by Ahirwar et al. (2009).

5.2.2 Type of roof

Roofing pattern followed in the study area is presented in Table5.1a. From the table it is seen that mostly the farmers were using thatch, tile, cement sheet and metal sheets for roofing the animal shelters. Thatched shelters were found more in Eastern zone (35.3 per cent) and least in Hilly zone (6.7 per cent). Tile roofing was seen 26 to 27 per cent in North Western, Western and Hilly zones. Cement sheet was used nearly 30 per cent in North Eastern, North Western and Western zones. In Hilly zone majority of the roofs (46.7 per cent) were of metal sheets and a few sheds were constructed with available plastic sheets.Regional differences in selection of roofing materials were evident from the study. Similar findings were reported by Singh et al. (2004), Sabapara et al. (2010) and Divekar and Saiyed (2010).

5.2.3 Type of floor

There were three flooring types in the cattle shelters of the study area. Its numbers and per cent are presented in Table 5.1a. Mud floors were most prevalent in Western zone (65.8 per cent). However, in North Eastern zone 43.2 per cent of the floors were of mud and more hygienic cement floorings were present in 44.6 per cent. In North Western zone majority of the cattle sheds were having cement floorings (52.9 per cent). In Hilly zone majority of

Table 5.1a: Details of animal shelters in the four agro-climatic zones

Agro-climatic Zones	North Eastern Zone	North Western Zone	Western Zone	Hilly Zone	Over all
No of observations (per cent)	139(100)	70(100)	38(100)	30(100)	277(100)
Milk production					
Average number of animals per farm	6.34	7.62	8.6	2.48	6.26
Average total milk production per farm (litres)	45.35	55.00	71.07	27.35	49.69
Average milk production per animal (litres)	6.83	8.55	7.52	11.15	8.51
Location of shed					
Separate housing	125 (89.9)	69 (98.6)	37 (97.4)	24 (80.0)	255 (92.1)
Attached to farmers house	14 (10.1)	1 (1.4)	1(2.6)	6 (20.0)	22 (7.9)
Number of rows					
Single row	111 (79.9)	49 (70.0)	31 (81.6)	28 (93.3)	219 (79.1)
Double row	28 (20.1)	21 (30.0)	7 (18.4)	2 (6.7)	58 (20.9)
Type of roof					
Thatch	49 (35.3)	17 (24.3)	11 (28.9)	2 (6.7)	79 (28.5)
Tile	18 (12.9)	19 (27.1)	10 (26.3)	8 (26.7)	55 (19.9)
Cement sheet	41 (29.5)	21 (30.0)	11 (28.9)	4 (13.3)	77 (27.8)
Metal sheet	31 (22.3)	13 (18.6)	6 (15.8)	14 (46.7)	64 (23.1)
Plastic	0 (0)	0 (0)	0 (0)	2 (6.7)	2 (0.7)
Type of floor					
Mud	60 (43.2)	20 (28.6)	25 (65.8)	1 (3.3)	106 (38.3)
Cement	62 (44.6)	37 (52.9)	10 (26.3)	7 (23.3)	116 (41.9)
Stones	17 (12.2)	13 (18.6)	3 (7.9)	22 (73.3)	55 (19.9)

Figures in parenthesis indicate per cent

the cattle sheds (73.3 per cent) were having stone floorings.The higher percentage of cement flooring found in North Eastern and North Western zones indicate the improvement in animal housing in these regions. This is contrary to the findings reported from other parts of the country by Ahirwar *et al.* (2009), Singh *et al.* (2004), Sabapara *et al.* (2010) and Divekar and Saiyed (2010). In Hilly zone 73.3 per cent of the cattle sheds were having stone floorings and the farmers were having the belief that the stone floor was comfortable to the animals and it would provide heat during cool night hours.

The details of floor space and dimensions of shed in the four agro-climatic regions are presented in Table 5.1b. The average length of the sheds were varying from 11.0 to 12.5 m in North Eastern, North Western and Western zones and in Hilly zone it was 5.73 m only. The average length of the sheds were positively correlated to the average number of animals (r value = + 0.9889) and was highly significant (p< 0.01). The width of the sheds also showed positive correlation with the average number of animals (r value = + 0.7594) but it was not significant (pe"0.05). The over all averageof height at ridge and at eves was 2.80 and 1.84m, respectively and the values were showing only slight difference among the agro-climatic regions. However, the height at ridge of the sheds was lower in Hilly zone (2.43 m). The average floor space per animal was found to be 7.16, 6.53, 5.58 and 7.66 m^2 per animal in North Eastern, North Western, Western and Hilly zones, respectively. The floor space was observed as sufficient in most of the sheds in North Eastern, North Western, Western and Hilly zones (86.3, 97.1, 84.2 and 96.7 per cent, respectively). This finding is similar to that of Sinha *et al.* (2009). The higher availability of floor space was attributed to the reduction in herd size by disposal of animals by the farmers.

5.2.4 Presence of walls and ventilation

Walls were present in all the cattle sheds of Hilly zone. The presence of walls on front, side and back portions of the cattle sheds under the study area and ventilation is presented in Table 5.1b. The cattle sheds without any walls were 46.0, 28.6, 21.1 and zero per cent in North Eastern, North Western, Western and Hilly zones respectively. The ventilation was found to be sufficient in 89.9, 92.9, 86.8 and 13.3 per cent of the cattle sheds in North Eastern, North Western, Western and Hilly zones respectively.

5.2.5 Mangers and manger materials

The details of mangers and manger materials of the cattle sheds under the study area are also presented in Table 5.1b. In North Eastern, North Western and Western zones majority of the cattle sheds were having mangers (57.6, 81.4 and 71.1 per cent). Though brick was the material used for construction of

Table 5.1b: Heat and cold amelioration measures the four agro-climatic zones

Agro-climatic Zones	North Eastern Zone	North Western Zone	Western Zone	Hilly Zone	Over all
No of observations (per cent)	139(100)	70(100)	38(100)	30(100)	277(100)
Dimensions of shed					
Average length (m)	11.00	11.80	12.53	5.73	10.26
Average breadth (m)	4.13	4.22	3.83	3.32	3.87
Average height at ridges (m)	2.89	2.94	2.97	2.43	2.80
Average height at eves (m)	1.83	1.85	1.86	1.82	1.84
Average floor space/cow (m²)	7.16	6.53	5.58	7.66	6.34
Adequacy of space					
Yes	120 (86.3)	68 (97.1)	32 (84.2)	29 (96.7)	249 (89.9)
No	19 (13.70)	2 (2.9)	6 (15.8)	1 (3.3)	28 (10.1)
Presence of walls (Nos)					
Front wall	69 (49.6)	41(58.6)	30 (78.9)	30 (100)	170 (61.4)
Side wall	56 (40.3)	32 (45.7)	17 (44.7)	30 (100)	135 (48.7)
Back wall	43 (30.9)	25 (35.7)	5 (13.2)	30 (100)	103 37.2)
No wall	64 (46.0)	20 (28.6)	8 (21.1)	0 (0)	92 (33.2)
Ventilation					
Sufficient	125 (89.9)	65 (92.9)	33 (86.8)	4 (13.3)	227 (81.9)
Not sufficient	14 (10.1)	5 (7.1)	5 (13.2)	26 (86.7)	50 (18.1)
Presence of manger (Nos)					
Present	80 (57.6)	57 (81.4)	27 (71.1)	8 (26.7)	172(62.1)
Absent	59 (42.4)	13 (18.6)	11 (28.9)	22 (73.3)	105 (37.9)
Manger material (Nos)					
Brick	63 (78.8)	49 (86.0)	15 (59.3)	3 (37.5)	131 (76.2)
Stone	4 (5.0)	2 (3.5)	8 (29.6)	0 (0)	14 (8.1)
RCC	7 (8.8)	3 (5.3)	0 (0)	2 (25.0)	12 (7.0)
Wood	6 (7.5)	3 (5.3)	4 (14.8)	3 (37.5)	16 (9.3)

Figures in parenthesis indicate per cent

mangers in majority of the sheds, stones, cement concrete and woods were other materials of choice among the farmers.

5.2.6 Drainage and cleanliness

In all the areas under the study the condition of drainage, disposal of manure and cleanliness in the sheds were found to be poor and its details are presented in Table 5.1c. Proper drainage was found in 54.0, 58.6, 50.0 and 36.7 per cent and manure pits in 20.9, 7.1, 10.5 and 6.7 per cent of the cattle sheds in North Eastern, North Western, Western and Hilly zones, respectively. In Hilly zone, 50 per cent of the sheds were not clean and in other zones it was more than one-fourth.The drainage facility reported in the study is far better than that reported by Sinha *et al.* (2009), Sabapara*et al.* (2010) and Ahirwar*et al.* (2009) in rural areas. However, the drainage facilities were very poor when compared to the observation of Modi *et al.* (2010). Majority of the farmers stored manure in open place. Balusamy (2004) observed the same trend among buffalo farmers in the North Eastern zone of Tamil Nadu. Cattle farmers need to be trained in proper storage and value addition of manures for reducing the environmental impacts and to generate additional income.

5.2.7 Heat and cold amelioration measures followed

In North Eastern, North Western and Westernzones some of the farmers were following heat amelioration measures and in the Hilly zone the farmers were using cold amelioration measures during winter. The details of heat and cold amelioration measures followed by the farmers are presented in Table 5.1c. A very few number of farmers were using heat amelioration measures in North Eastern, North Western and Westernzones (10.1, 14.3 and 13.2 per cent). This indicates the vulnerability of animals to heat stress during hot hours of the day, especially during summer season. The heat amelioration measures followed by farmers were provision of fan, sprinkler, shower, false ceiling and shade. Since the use was limited to a few farms, extensive measures are to be undertaken to familiarize the advantages of these additional fittings to farmers. During winter season 66.7 per cent of farmers were providing bedding as cold amelioration measures in Hilly zone. The large proportion indicates the familiarity of the farmers with the advantages experienced.

5.2.8 Housing types in the study area

Based on the survey, the major housing types in the selected four agro-climatic zones were classified into thatch roofed, tile roofed, metal roofed, cement sheet roofed and open housing. In Hilly zone, the poor farmers were having dairy housing made out of plastic sheet; this type of housing was taken as equivalent to the open type one in other agro-climatic zones.

Table 5.1c: Heat and Cold amelioration measures in the four agro-climatic zones

Agro-climatic Zones	North Eastern Zone	North Western Zone	Western Zone	Hilly Zone	Over all
No of observations (per cent)	139(100)	70 (100)	38 (100)	30 (100)	277 (100)
Drainage					
Yes	75 (54.0)	41(58.6)	19 (50.0)	11(36.7)	146 (52.7)
No	64 (46.0)	39 (55.7)	19 (50.0)	19 (63.3)	141 (50.9)
Cleanliness in shed					
Clean	86 (61.9)	50 (71.4)	25 (65.8)	15 (50.0)	176 (63.5)
Not clean	53 (38.1)	20 (28.6)	13 (34.2)	15 (50.0)	101 (36.5)
Manure pit					
Present	29 (20.9)	5 (7.1)	4 (10.5)	2 (6.7)	40 (14.4)
Absent	110 (79.1)	65 (92.9)	34 (89.5)	28 (93.3)	237 (85.6)
Heat/cold amelioration measures					
Yes	14 (10.1)	10 (14.3)	5 (13.2)	20 (66.7)	49 (17.7)
No	125 (89.9)	60 (85.7)	33 (86.8)	10 (33.3)	228 (82.3)
Type of heat/cold amelioration measures followed					
Fan	6 (4.3)	5 (7.1)	4 (10.5)	0 (0)	15 (5.4)
Sprinkler	3 (2.2)	0 (0)	1 (2.6)	0 (0)	4 (1.4)
Shower	1 (0.7)	1 (1.4)	0 (0)	0 (0)	2 (0.7)
False ceiling	2 (1.4)	4 (5.7)	0 (0)	0 (0)	6 (2.2)
Shade	2 (1.4)	0 (0)	0 (0)	0 (0)	2 (0.7)
Bedding	0 (0)	0 (0)	0 (0)	20 (66.7)	20 (7.2)

Figures in parenthesis indicate per cent

5.2.9 Suitability of existing different housing patterns

The multiple correlation between physiologic, climatic and production parameters were worked out of the values obtained from 480 samples collected during the whole experimental period of one year and given in Table 5.2. Based on the correlation coefficient 25 parameters were selected out of the total 39 parameters and used for working out the dairy housing comfort index and is given in Table 5.3. The climatic, physiologic and production parameters under the study were statistically analysed using factor analysis to find out the suitable housing system. The Ranking of housing systems and dairy housing comfort index is given in Table 5.3. It is evident that the animals under cement sheet and metal sheet roofed shelters in Hilly zone rank first and second. This is followed by tile roofed shelters in North Eastern zone (3rd rank) and thatched housing in Hilly zone (4th rank). Based on the ranking of dairy housing comfort index, in North Eastern zone tile roofed shelters were found to be the best one (dairy housing comfort index 1.17537). The next suitable types were thatched housing (dairy housing comfort index 0.6393) followed by cement sheet roofed (dairy housing comfort index 0.62732), metal sheet roofed (dairy housing comfort index -0.11986) and lastly open housing (dairy housing comfort index -0.01114).

In North Western zone tile roofed shelters (dairy housing comfort index 0.38899) were found to be the best one. The next suitable types were metal sheet roofed housing (dairy housing comfort index 0.37114) followed by cement sheet roofed (dairy housing comfort index -0.17183), open (dairy housing comfort index-0.47193) and lastly thatched housing (dairy housing comfort index -0.74321). In Western zone cement sheet roofed shelters (dairy housing comfort index -0.7844) were found to be the best one. The next suitable types were open housing (dairy housing comfort index -0.93999) followed by thatched (dairy housing comfort index -1.45285), metal sheet roofed (dairy housing comfort index -1.57605) and lastly tile housing (dairy housing comfort index -1.63564).

In Hilly zone cement sheet roofed shelters (dairy housing comfort index 1.83584) were found to be the best one. The next suitable types were metal sheet roofed housing (dairy housing comfort index 1.35513) followed by thatched (dairy housing comfort index 0.9113), plastic wrapped (dairy housing comfort index 0.90149) and lastly tile roofed housing (dairy housing comfort index -0.29897). In Hilly zone ventilation of the cattle housings were found to be very poor, 86.7 per cent of the houses were having insufficient ventilation. Hence all existing shelters should be improved with modifications to provide sufficient ventilation for improving the comfort of the animals.

The result shows that the climatic condition of the zone and the choice of materials played a major role in deciding the comfort level of the animal in the

Table 5.2: Multiple correlations between the parameters under study

	WBC	RBC	HGB	HCT	MCV	MCH	MCHC	RDW	PLT	MPV	PDW	PCT	AST	ALT	CHE	LDH	CORT	T3	T4	Na	K	RT	FR	RR	ST	BCS	MP	CMP	PPH	SP	SPC	T.Max	T.Min	TB	RH8	TH8	T2	RH2	THZ
WBC	1.00																																						
RBC	0.04	1.00																																					
HGB	-0.01	0.87	1.00																																				
HCT	0.08	0.86	0.93	1.00																																			
MCV	0.21	0.20	0.49	0.57	1.00																																		
MCH	0.18	0.25	0.52	0.40	0.65	1.00																																	
MCHC	0.03	0.04	0.24	-0.08	-0.01	0.59	1.00																																
RDW	0.34	-0.04	-0.31	-0.16	-0.40	-0.57	-0.32	1.00																															
PLT	-0.05	0.20	0.28	0.24	0.22	0.17	0.24	-0.15	1.00																														
MPV	0.18	0.16	0.35	0.44	0.64	0.47	-0.07	-0.32	0.20	1.00																													
PDW	0.08	0.16	0.38	0.32	0.61	0.68	0.33	-0.74	0.18	0.37	1.00																												
PCT	-0.30	0.20	0.36	0.40	0.40	0.23	0.04	-0.29	0.32	0.32	0.35	1.00																											
AST	0.26	0.16	0.41	0.26	0.54	0.50	0.40	-0.36	0.40	-0.09	0.32	0.41	1.00																										
ALT	0.26	0.33	0.40	0.49	0.56	0.25	-0.05	0.10	0.40	0.32	0.25	0.41	0.13	1.00																									
CHE	0.14	-0.03	-0.20	-0.14	-0.34	-0.44	-0.35	0.46	-0.35	-0.12	-0.45	-0.29	-0.30	0.18	1.00																								
LDH	0.35	0.57	0.73	0.61	0.52	0.49	0.00	-0.28	0.55	0.15	0.53	0.42	0.82	0.45	-0.49	1.00																							
CORT	0.26	-0.05	-0.29	-0.18	-0.23	-0.08	0.00	0.33	-0.49	-0.11	0.18	-0.50	-0.43	0.04	0.41	-0.41	1.00																						
T3	-0.31	-0.06	0.20	-0.02	0.00	0.28	0.49	-0.29	-0.12	0.07	-0.28	0.07	0.32	0.04	0.31	-0.58	0.31	1.00																					
T4	-0.22	-0.03	0.26	0.12	0.28	0.26	0.47	-0.45	0.45	-0.06	0.35	0.46	0.53	0.06	-0.66	0.47	-0.61	0.72	1.00																				
Na	-0.55	0.20	0.19	0.04	-0.06	0.03	0.21	-0.27	-0.23	0.03	-0.08	0.21	-0.03	-0.14	0.24	-0.13	0.05	0.23	-0.01	1.00																			
K	0.12	-0.35	-0.43	-0.44	-0.29	-0.16	0.11	0.44	-0.23	-0.12	-0.45	-0.36	-0.27	-0.14	0.24	-0.39	0.56	0.23	-0.13	-0.05	1.00																		
RT	-0.03	-0.37	-0.54	-0.48	-0.53	-0.39	-0.27	0.26	-0.44	-0.10	-0.55	-0.50	-0.64	-0.29	0.37	-0.74	0.72	-0.24	-0.41	-0.26	0.44	1.00																	
FR	0.11	-0.28	-0.49	-0.33	-0.32	-0.40	-0.45	0.44	-0.44	0.13	-0.58	-0.53	-0.82	-0.12	0.60	-0.76	0.66	-0.36	-0.16	-0.47	0.41	0.83	1.00																
RR	0.09	-0.33	-0.55	-0.41	-0.47	-0.44	-0.37	0.47	-0.56	-0.01	-0.60	-0.59	-0.80	-0.03	0.60	-0.80	-0.26	-0.29	-0.70	-0.29	0.52	0.89	0.94	1.00															
ST	0.03	-0.60	-0.72	-0.59	-0.46	-0.53	-0.33	0.48	-0.48	-0.03	-0.61	-0.48	-0.72	-0.30	0.54	-0.84	0.49	-0.42	-0.55	-0.23	0.56	0.80	0.85	0.92	1.00														
BCS	-0.71	0.35	0.43	0.26	-0.05	0.07	0.24	-0.45	0.04	-0.29	0.24	-0.25	0.21	0.14	0.13	0.33	0.45	0.33	0.44	0.53	-0.49	-0.39	-0.49	-0.47	-0.49	1.00													
MP	-0.25	0.44	0.54	0.41	0.37	0.41	0.42	-0.37	0.42	-0.05	0.60	0.47	0.57	0.19	-0.52	0.71	-0.26	-0.16	0.53	0.38	-0.56	-0.47	-0.82	-0.83	-0.84	-0.92	1.00												
CMP	0.11	0.26	0.19	0.28	0.14	-0.08	-0.17	0.24	0.03	0.33	0.12	-0.25	0.41	0.07	-0.05	0.11	-0.05	-0.27	-0.23	-0.13	-0.03	-0.21	0.09	0.00	0.00	-0.16	0.08	1.00											
PPH	0.21	0.03	-0.03	0.09	0.11	-0.35	-0.37	0.43	0.28	-0.20	-0.25	-0.20	0.14	0.13	0.32	0.17	-0.26	-0.48	-0.28	-0.42	-0.09	-0.42	0.24	0.21	0.55	-0.28	-0.60	0.27	1.00										
SP	0.38	-0.07	-0.17	-0.03	-0.15	-0.37	-0.44	0.40	0.22	-0.26	-0.35	-0.26	0.57	0.19	0.30	0.30	-0.18	-0.37	-0.37	-0.47	0.15	-0.47	0.70	0.80	0.57	-0.29	-0.65	0.43	0.81	1.00									
SPC	0.41	-0.14	-0.24	-0.13	-0.14	-0.27	-0.34	0.52	0.09	-0.18	-0.37	-0.29	0.18	0.41	0.17	0.45	-0.34	-0.15	-0.31	-0.34	0.33	0.34	0.72	0.85	0.38	0.35	-0.47	0.43	0.34	0.82	1.00								
T.Max	0.00	-0.52	-0.68	-0.51	-0.40	-0.41	-0.27	0.43	-0.07	-0.48	-0.52	-0.19	-0.78	-0.28	0.64	-0.87	0.55	-0.69	-0.28	-0.87	0.53	0.71	0.94	0.91	0.94	-0.38	-0.82	0.04	0.42	0.43	0.26	1.00							
T.Min	-0.12	-0.54	-0.69	-0.54	-0.45	-0.54	-0.39	0.38	-0.08	-0.48	-0.47	-0.81	-0.47	-0.33	0.62	-0.90	0.45	-0.53	-0.64	-0.18	0.47	0.70	0.88	0.94	0.93	-0.29	-0.78	0.04	0.40	0.35	0.18	0.99	1.00						
TB	-0.07	-0.54	-0.68	-0.54	-0.43	-0.52	-0.39	0.38	-0.08	-0.49	-0.49	-0.51	-0.77	-0.32	0.62	-0.88	0.51	-0.56	-0.65	-0.19	0.48	0.74	0.90	0.95	0.94	-0.31	-0.82	0.00	0.43	0.38	0.20	0.98	0.99	1.00					
RH8	-0.06	0.62	0.75	0.59	0.43	0.58	0.41	-0.50	0.42	0.04	0.51	0.36	0.70	0.48	-0.55	0.79	-0.36	0.48	0.79	0.28	-0.44	-0.76	-0.82	-0.82	-0.95	0.47	0.82	-0.14	-0.92	-0.92	-0.34	-0.93	-0.93	-0.93	1.00				
TH8	-0.08	0.62	0.68	0.58	0.41	0.51	0.37	-0.61	0.42	-0.08	-0.49	-0.51	0.70	0.61	-0.77	0.50	-0.88	-0.55	-0.65	-0.17	0.49	0.83	0.90	0.90	0.94	-0.29	-0.82	-0.02	0.42	0.37	0.19	0.98	0.99	0.99	-0.93	1.00			
T2	0.00	-0.51	-0.67	-0.51	-0.42	-0.52	-0.41	0.46	-0.08	-0.52	-0.53	-0.53	-0.78	-0.28	0.64	-0.87	0.55	-0.58	-0.70	-0.25	0.54	0.72	0.85	0.92	0.94	-0.39	-0.83	0.04	0.42	0.42	0.28	1.00	0.99	0.99	-0.92	0.99	1.00		
RH2	-0.04	0.63	0.74	0.58	0.40	0.56	0.41	-0.46	0.44	0.01	0.50	0.37	0.71	0.71	-0.84	0.80	-0.39	0.49	0.54	0.27	-0.42	-0.68	-0.79	-0.84	-0.96	0.45	0.84	-0.11	-0.47	-0.49	-0.34	-0.93	-0.94	-0.93	0.94	-0.93	-0.94	1.00	
THZ	0.08	-0.48	-0.63	-0.50	-0.41	-0.41	-0.29	0.46	-0.12	-0.66	-0.48	-0.62	-0.70	-0.38	0.68	-0.82	0.52	-0.50	-0.74	-0.20	0.58	0.61	0.76	0.84	0.85	-0.36	-0.74	-0.10	0.23	0.21	0.21	0.94	0.91	0.90	-0.81	0.91	0.94	-0.81	1.00

BCS-Body condition score, MP-Milk production, CMP-Change in milk production, MP-Milk production, PPH-Days of first post parturient heat, SP-Service period, SPC-Services per conception, X8-Parameter at 8.00 am, X2- Parameter at 2.00 pm

Table 5.3: Ranking of housing systems and dairy housing comfort index

Sl. No	Agro-climatic zone	Housing System based on roofing	Overall Rank	Rank in agro-climatic zone	Housing comfort index
1	North Eastern Zone	Tile	3	I	1.17537
2		Thatched	6	II	0.6393
3		Cement sheet	7	III	0.62732
4		Metal sheet	9	IV	-0.11986
5		Open	10	V	-0.01114
6	North Western Zone	Tile	8	I	0.38899
7		Metal sheet	9	II	0.37114
8		Cement sheet	12	III	-0.17183
9		Open	14	IV	-0.47193
10		Thatched	15	V	-0.74321
11	Western Zone	Cement sheet	16	I	-0.7844
12		Open	17	II	-0.93999
13		Thatched	18	III	-1.45285
14		Metal sheet	19	IV	-1.57605
15		Tile	20	V	-1.63564
16	Hilly Zone	Cement sheet	1	I	1.83584
17		Metal sheet	2	II	1.35513
18		Thatched	4	III	0.9113
19		Open	5	IV	0.90149
20		Tile	13	V	-0.29897

individual housing system. Hence, considering the trend in climate change, the dairy housing systems need further research in improving the comfort level of the animal for improving the production and ensuring the welfare of the animals.

References

Ahirwar, R.R., S. Nanavati, and N.K. Nayak, 2009. Studies on housing management of buffaloes under rural and urban areas of Indore district of M P. Indian J. Field Vet. 5 (2): 41-43.

Anon, 2010.Basic Animal Husbandry Statistics.Ministry of Agriculture.Department of Animal Husbandry, Dairying and Fisheries, Government of India.New Delhi. p 8, 53.

Balusamy, C., 2004. Productive and reproductive performance of buffaloes in northeastern zone of Tamil Nadu. Ph.D., Thesis submitted to Tamil Nadu Veterinary and Animal Sciences University, Chennai – 600 051.

Divekar, B.S., and L.H. Saiyed, 2010. Housing and breeding practices followed by professional Gir cattle owners of Anand district. Indian J. Field Vet.5 (4): 9-12.

Duncan, D.E., 1955. Multiple range and multiple F test. Biometrics.11:1-12.

Modi, R.J., K.B. Prajapati, N.B. Patel, and H.D. Chauhan, 2010.Dairy animal housing in rural area of milk shed of Gujarat. Indian J. Dairy Sci.63 (1): 46-49.

Sabapara, G.P., P. M.Desai, V. B. Kharadi, L. H. Saiyed, and R. R, Singh , 2010. Housing and feeding management practices of dairy animals in the tribal area of South Gujarat. Indian J. Anim. Sci.80(10): 1022-27.

Singh, P., M. Singh, M. L. Verma., and R. S. Jaiswal, 2004. Animal husbandry practices in Tarikhet block of Kumaon hill of Uttaranchal. Indian J. Anim. Sci.74(9): 997-999.

Sinha, R. R. K., D.Triveni, R. R. Singh, B. Bharat, S. Mukesh, and K. Sanjay, 2009. Feeding and housing management practices of dairy animals in Uttar Pradesh.Indian J. Anim. Sci.79(8): 829-833.

Whitaker, D.A., 2000. Use and interpretation of metabolic profiles. In: A.H.Andrews (ed.) The health of dairy cattle, Blackwell Science, Oxford, UK, 89-107.

6

Management of Lactating Dairy Cow During Period of Heat Stress

Dharmendra Kumar*, Asit Chakrabarti[1] and Prakash Kahate[2]

*Subject Matter Specialist (KVK, Kishanganj, Bihar Agricultural University) Bihar
[1]Senior Scientist, ICAR Research Complex for Eastern Region, Patna, Bihar
[2]Assistant Professor, Dr. PDKV, Akola, Maharashtra

Heat or thermal stress occurs when it becomes increasingly difficult for dairy cow to maintain a normal core body temperature of 101.5 to 102.8° F (rectal temp.). The thermo neutral or comfort zone (TNZ) for cow is at an environmental temperature of 41 to 77° F. Within the TNZ, heat production from normal metabolic functions is about equal to the heat loss from the body and maintaining a normal body temperature is relatively easy. Extent of heat stress can be measured by the Temperature Humidity Index (THI). A THI level above 75 is considered stressful for dairy animals. Two sources of heat impact the cow as the environmental temperature and the heat produced internally from basal nutrient metabolism. Heat produced from nutrient metabolism is a lesser factor than environmental heat sources. However, as milk production and feed intake increase, more heat from nutrient metabolism is produced aggravating any heat stress being incurred from environmental sources. Therefore, higher milk producing cow will be more affected by heat stress than low milk producing or dry cow. The primary sources of heat derived from the environment are solar radiation and elevated ambient air temperature. These are influenced by high relative humidity and a lack of air movement. Since cattle sweat at only 10 percent of the human rate, they are more susceptible to heat stress. This is the reason for which dairy cattle need mechanical means to reduce heat, such as body sprinkling to aid in evaporation and effective air movement systems to aid in cooling. Heat stress is one of the leading causes of decreased production and fertility in dairy cattle and buffaloes during summer months. Hot environment affect the performance of dairy cattle both directly and indirectly. Thermal factors consist of air temperature, humidity, air movement, and rate of radiation. In lactating Holstein cow, the comfortable temperature is within the range of 4-24°C (Hahn, 1981). The effects of heat stress on the cow begin to be observed above 24°C, and milk yield decreases markedly above 27°C (Johnson, 1965). A decline in milk yield, fertility and growth rate in hot environment is closely related to increase in body temperature. Body temperature results from the balance between heat production and heat loss. Since, humidity affects the Heat loss from an animal under high temperature condition, dairy cattle performance falls markedly in hot-humid summers. Moreover, heat production is associated with feed intake level, which in turn affects the production level. In high-producing cow, the heat production is higher, and the effect of hot environment is more pronounced.

6.1 What is Heat Stress?

- Heat stress occurs when a dairy cow produces heat by her own activities (such as exercise and heat produced from digestion) and the heat from the environment, exceeds the capacity of the cow to lose this heat and also referred as hyperthermia.

6.1.1 Primary causes of heat stress

- The most important factors affecting heat stress are air temperature, humidity, solar radiation and air movement.

- When air temperature is greater than about 23°C and relative humidity is greater than 80%, cow begin to experience heat-induced depression of feed intake, and lower productivity (Flamenbaum *et al.*, 1986).

- High relative humidity decreases evaporation and reduces the cow's ability to lose heat by sweating and breathing.

- Cow radiate heat during night to the cooler surroundings but high temperature, humidity and cloud cover at night influence heat stress.

6.1.2 Secondary causes of heat stress

- Individual cow respond differently to heat stress situations.

- High producing cow have greater metabolic activity and produce more body heat than low producing cow and consequently experience more severe heat stress.

- Different breeds of cow differ in their ability to cope up with heat stress, i.e. Friesian cow appear more affected than Jerseys.

- Black coats absorb more radiation than white coats, which means a greater heat load for black cow.

- Cow with thick coats also suffer more from heat stress than cow with sleek coats.

6.2 Physiological Responses to Heat Stress

Heat stress induces a number of physiological responses by the cow in an attempt to keep body temperatures within normal limits. The following are some of the physiological changes occurring in the cow as heat stress conditions are incurred:

- Respiration rate increase and may reach the stage of panting. In this attempt to increase evaporative cooling, increased amount of CO_2 are exhaled resulting a decrease in H_2CO_3 and an increase in blood pH. In response to the decrease in blood pH, the kidney increases absorption of H^+ and more HCO_3^- and cations, primarily sodium, are excreted in the urine.

- Heat stressed cow lose two thirds of their evaporative water loss by sweating and one third by panting. The maximum sweat loss at 95° F is estimated to be $150g/m^2$ of body surface per hour. Increased respiration rate can increased loss of sodium and potassium. This can shift the acid-base balance and result in a metabolic alkalosis.

- There can also be a decrease in the efficiency of nutrient utilization.

- Reticulo-rumen motility and overall rate of digesta passage is decreased during heat stress. There is also a change in rumen fermentation with less total volatile fatty acids produced and an increase in the molar percent of acetate.

- Blood flow to the digestive tract and other internal tissues is decreased and flow to the skin surface is increased.

- Urine volume generally increases.

6.3 Symptoms of Heat Stress

To cope up with a hot environment cow use a variety of strategies including:

- Elevated body temperature – Body temperatures > 102.5 °F (normal is 101.5 °F).

- Increased breathing rate > 70-80/minute.

- Increased water intake.

- Increased sweating.

- Dry matter intake decreases in dairy cows subjected to heat stress. This depression in dry matter intake can be either short term or long term depending on the length and duration of heat stress. Decreases of DMI 10 to 20% are common in commercial dairy herds.

- Decreased milk production.

- Change in milk composition, e.g. fat % and protein % declines.

- Change in blood hormone concentration, e.g. increased prolactin.

- Behavioural change.

- Seek shade.

- Crowd together in shed.

- Refuse to lie down.

- Change orientation to sun.

- Stand in water.

If these strategies fail and the cow's heat load exceeds the body's removal ability, her body temperature will increase. It has been estimated that with each 0.5°C increase in body temperature above 38.6°C, milk yield will decline by 1.8 kg/day (Johnson *et al.*, 1965). The critical cow body temperature is 42° C. Severe heat stress symptoms include extended head, protruding tongue, panting and drooling. Temperature effects on fertility depend on the duration and magnitude of heat stress. High environmental temperatures that persist for weeks may be especially detrimental to reproductive efficiency. Cool nights will help decrease the heat stress and may decrease the effects of high day time temperatures.

6.4 Effects of Heat Stress on Bulls

Signs of heat stress have been observed in bulls maintained in 90° F environmental temperature, 100° F can be dangerous and may produce advanced signs of heat stress. Experimental data indicate heat stress can damage semen quality, change in shape of the sperm cell head and tailpiece. Applying external insulation to the bull's scrotum can raise the scrotal skin temperature 1-4° F and yearling bulls for 24-72 hours has resulted in a decrease in normal sperm count. This decrease begins 1-2 weeks after the heat stress effect. When the heat stress is alleviated, semen quality continues to decline for an additional 1-4 weeks. Depending on the duration of heat stress, semen quality returns to pre-stress level approximately 4-8 weeks after the heat stress effect.

6.4.1 Effects of heat stress on reproduction in cow

Various studies (Jordan, 2003; Rensis, 2003; West, 2004) have shown heat stress challenges the reproductive performance of dairy cows through a variety of altered physiologic means, including:

- Altered follicular development
- Lowered estrus activity
- Impaired embryonic development

The first reproductive challenge in heat stressed cow is altered follicular development. Cow decrease feed intake causing less frequent pulses of the luteinizing hormone (LH) resulting in longer follicular waves. This lengthening of the follicular wave leads to the selection and ovulation of multiple, smaller dominant follicles (Sartori, 2002). Follicles are responsible for producing estrogen, a hormone that causes cow to show signs of heat. Smaller follicles will produce less estrogen than larger ones; therefore, resulting in less estrus activity. Estrus activity is also lowered due to the cows' reduced motor activity, a means of trying to decrease her endogenous heat output. Thus, the occurrence of silent ovulations or "silent heat" increases, which will ultimately reduce heat detection efficiency even in well-performing heat detection programmes. The high uterine temperature of the heat stressed cow can impair embryonic development, resulting in poor embryo implantation and increased embryo mortality (Jordan, 2003 and West, 2004).

Heat stress immediately after the time of breeding (7-10 days) may result in a lower conception rate. This is particularly true if there has been no or little adaptation to the high temperature. This change appears to increase the rate of early embryonic death, but not from a failure of the egg to be fertilized. With early embryonic death due to heat stress, the cow returns to estrus in 21 days. The developing embryo was lost before the maternal recognition of pregnancy, so the cow continues regular estrous cycle. The decreased fertility appears to be much more of a problem in lactating dairy cow, but some depression of fertility may be noted in heifers as well.

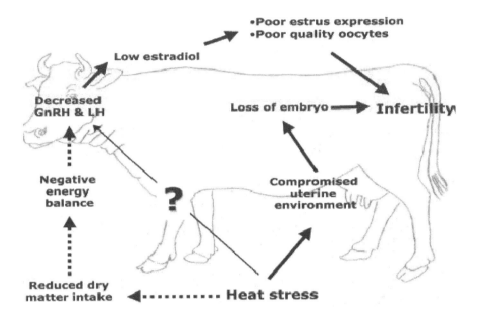

Fig. 1: Depiction of heat stress on dairy cows

6.5 Heat Stress Management

There are different ways to decrease the apparent effects of heat stress in dairy cattle.

Shade

Although shade trees are the best method for relieving heat stress, most trees don't survive intensive use. When not enough natural shade is available, artificial shades can provide needed shelter from the effects of solar radiation. If cow are to be confined under a shade structure, it should be oriented with a southeastern exposure of an open sidewall. Walls of free stall barns should be opened up to maximize air movement. Eliminate any wind block within 50 ft of the windward side of the building. Each cow should be provided with 60 sq ft of shade.

Ventilation

Have the shade in a location where there is a breeze and ensure that the sides are open for good ventilation. The shed roofs should be at least 10 feet tall. If the side wall is open it will allow the hot air to escape. If the roof is insulated, it will help reduce the radiant heat. Without proper ventilation in buildings and shaded areas, heat and moisture accumulate and the animals will be under stress.

Cooling

Increasing air flow is another important component. Be certain air moves freely in all sections of the barn. There are two main ways to increase air flow. One is to install fans so air movement is increased and the second is to open the sides of the barn. In many cases walls of the barn may be made partially of concrete. In this situation, opening the lower level of the barn to increase air flow is not an option. The addition of fans is essential. In barns where sheet metal is used for the walls it may be practical to remove the sides and install netting over these areas. The netting can be raised to increase air flow during the summer and lowered during the winter. Increasing the roof venting is another option. Research was conducted to study the feeding frequency of cattle from 8:00 a.m. to 8:00 p.m. with and without supplemental cooling. Results showed cattle shaded and cooled by sprinklers near the feed bunks ate between 63 percent to 100 percent of the time as compared to uncooled cattle. Therefore, offering a cool, shaded area for feeding during daylight hours will increase feed intake, thereby helping to maintain production.

Each area of the dairy facility must be considered when looking at cow cooling options. First, cool the holding pen near the milking parlor. Crowding cow into a small area restricts air flow and aggravates heat stress. Fans and sprinklers can reduce ambient temperature by 15° F, and cooled cow produce more milk than non-cooled cow. Also, it is important to minimize the time cow spend in the holding area.

1. Fan systems appear to be effective in areas of low or high humidity and cool the air while raising the relative humidity. Arizona studies found that fans and pads reduced the temperature by 20-24° F, which resulted in 7-13 lb more milk per cow (Shearer, 1996). Coolers are positioned every 20 ft in the roof and air is pulled through the cooler at very high rates. Although there is a substantial initial investment and operating expense, the cost is probably offset by increased milk yield and improved reproduction.

2. Other evaporative cooling methods include mist, fog, and sprinkling systems. A mist or fog system sprays small water droplets into the air and cools the air as the droplets evaporate. When an animal inhales the cooled air it can exchange heat with the air and remove heat from its body.

   a) High pressure foggers disperse a very fine water droplet which quickly evaporates and cools air while raising the RH. As fog droplets are emitted they are immediately dispersed into the fan's air stream where they soon evaporate. Foggers should operate during daylight hours only; humidity is too high at night but fans should operate continuously. They should not be used in low barns with side walls that restrict air flow, droplet evaporation, and reduce cooling, while making excessive wet conditions. Fog system is very efficient method of cooling air but also are more expensive than mist system and require more maintenance.

b) A mist droplet is larger than a fog droplet and animals are cooled primarily by inspiration of cooled air. Mist systems also cause respiratory and pneumonia problems when cow were exposed to mist particles in enclosed areas. A mist system probably is not advisable for most dairy operations, especially for free stalls bedded with sawdust.

c) An alternative to mist and fog systems is the sprinkling system. This method does not attempt to cool the air, but instead uses a large droplet size to wet the hair coat to the skin of the cow, and then water evaporates and cools the hair and skin. Sprinkling is most effective when combined with air movement. At least one 36-inch fan is needed for each 40 animals which will move air effectively for about 30 ft Cow are sprinkled for 1-2 minutes at 15-minute intervals. Concrete floors must be sloped to handle water runoff. The sprinkling system can be used in holding areas, shade structures, and feed barns.

3. Cooling ponds are man-made with approximate dimensions of 50 x 80 ft and 4-6 ft deep. The incidence of clinical mastitis for cow accessing ponds was half that of cow without ponds. These cow were cleaner and easier to milk, with lower somatic cell and bacteria counts; perhaps resistance was enhanced by lower heat stress.

6.6 Dry Matter Intake (DMI)

Voluntary DMI can decrease by 25-50% of that in the TNZ during heat stress. Much of the decrease in milk production observed during heat stress can be attributed to the decreased DMI. Cow decrease DMI in an attempt to reduce heat production from the digestion and metabolism of nutrients. Maintaining a normal body temperature is critical as 3 to 4 pounds decrease in milk production and TDN intake occur with each 1° F increase in body temperature above 101.5° F (West, 1996). At the same time DMI or nutrient intake is decreasing and nutrient requirements for active cooling processes like panting are increasing. Also, blood flow to internal organs like the mammary gland is reduced delivering fewer nutrients to these organs for metabolism. Thus, fewer nutrients are available and used for milk production during heat stress. Little is known about the effects of heat stress on DMI of the dry and particularly close-up dry cow. However, decrease in DMI during the dry period can lead to more health problems at parturition and potentially reduce milk production during the subsequent lactation. As much attention should be given to alleviating heat stress in dry cow during the last trimester of gestation through environmental and dietary changes as is given to lactating cow.

6.7 Nutrients

Water

Water is the most important nutrient for lactating cow and especially heat stressed cow. As milk production and DMI increase, water intake also increases. Dado and Allen (1994) reported a correlation between water intake and DMI of 0.94 for water intake and 0.96 milk production. The amount of water intake per liter of milk production will vary depending on DM, salt and protein content of the diet; however, cow generally consume between 3 to 5 liter of water per liter of milk produced and 2 to 4 kg of water for each kg of dry matter consumed. Water intake increases sharply as environmental temperature increases because of water loss from sweating and with more rapid respiratory rate (panting), both efforts aimed at increasing evaporative cooling for the cow. Water intake can increase by 10-20% in hot weather. Environmental temperature above body temperature, water intake decreases because of reduced DMI and inactivity.

Cow should have an unlimited quantity of fresh clean water in an easily accessible area. Water tanks should be located close to the feeding area to encourage both DMI and frequent drinking. Fans and sprinklers or at least shade should be located over the feeding and watering area to encourage consumption. Given a choice, cow will choose a shaded or cooled resting area over eating and drinking from an un-cooled and sunny area. Hence, keep feed and water available in cooled areas. Sterner, et al. (1986), Wilkes, et al. (1990) and Beede and Shearer (1996) research findings indicated that cooling of drinking water below well water temperature is unnecessary. Water cooled to below 59° F helped dissipate body heat during hot weather, but the small increase in milk production did not justify the cost of cooling the water. When given a choice, the cow preferred warm (70 to 80° F) over cooled water. Having easily accessible clean waterers with an ample source of clean fresh water should be the first and foremost consideration in water nutrition of dairy cattle. Never let water become stagnant. Be certain to routinely clean out waterer daily, to maximize intake. One important point to be considered is increasing the amount of available water and also provision cool water. Monitor temperature of water because cow prefer water at 70-86° F. Average milking animals take 50-60 liter water in winter and 70-80 liter /day in summer, its quantity depends upon milk production, body weight, breed and feed intake.

Protein

Both the quantity and form of protein in the diet need to be considered when feeding heat stressed cow (Huber, *et al.*, 1994). Feeding crude protein (CP) in either excess or less than required amounts will increase body heat production. Deficiency of CP reduce digestibility of the diet whereas feeding excess amount of CP increase energy requirements for the synthesis and excretion of urea from the body. Huber (1993) reported on Missouri research where cow were fed diets containing either 20 or 40% soluble protein in either TNZ or heat stressed conditions. Cow fed the low soluble protein diet had higher milk yields and DMI during both climatic conditions than cow fed the high soluble protein diet. Huber, *et al.* (1994) opined heat stressed cow fed a high CP (18.5%), high degradable protein diet (65% of CP) had 6 per cent lower DMI and 11 percent lower milk yield than cow fed either the high CP with lower degradable protein (59%) or a lower CP (16%) diet with either a high or low degradable protein content. Based on this research, it is suggested that during heat stress rumen degradable protein should not exceed 65 per cent of CP. A possible reason why high degradable protein diets are deleterious during heat stress is rumen motility and rate of passage decline allowing for a longer protein residence time in the rumen and more extensive degradation to ammonia. Any ammonia in excess must be detoxified to urea in liver through metabolism, which is very high in energy demands (1g urea=7.3 Kcal) in the body is an energy cost and increases heat production as it is metabolized to urea and excreted in the urine.

Fiber

As DMI decreases during heat stress, a concern about having adequate amounts of fiber (ADF, NDF and effective or forage NDF) in diets arises. However, fiber digestion results in a higher heat increment (sum of heat produced from rumen fermentation and nutrient metabolism) than digestion of fat or non-fiber carbohydrates (NFC). Acetate, the end product of fiber digestion, has a lower efficiency of utilization in the body than propionate or glucose from NFC digestion and both fiber and NFC end products are utilized less efficiently than fat (West, 1996). Thus, feeding high forage diets during summer months can add significantly to a cow's heat load. Decreasing the forage to concentrate ratio (feeding more concentrate) can result in more digestible rations that may be consumed in greater amount. However, many herds already fed the maximum amount of concentrate and more would cause problems with acid rumens and cow going off feed. However, never reduce the fiber level below 18 -19% ADF (acid detergent fiber) and 25% to 28 percent NDF. Fed Sodium bicarbonate or sesquicarbonate 0.75% of DMI can help buffer the rumen to accommodate higher levels of concentrate. High quality forages are the best source of digestible fiber that is effective and produces minimum heat when fermented in the rumen.

Feeding forages in a TMR is recommended during heat stress as cow will reduce forage intake relative to grain and concentrates when given a choice.

Depressions in milk yield during average daytime temperatures of 96° F were found to be less when ADF content of the diet was 14 compared to 17 or 21 percent (Cummins, 1992). Also, decreases in DMI were found to be more closely associated with the daily high minimum temperature than with the maximum temperature. At any given minimum temperature, DMI was highest in cow fed the lowest ADF diets. However, as the daily minimum temperature increased, the decline in DMI was greatest in the 14 percent ADF diet. Thus, heat loads in cow are associated as much with total energy or DMI as with fiber content of the diet. West, (1996) also found that as NDF content of the diets increased, DMI decreased during both cool and hot weather (Table 6.1). Amount of fiber consumed was not as different between diets as the percentage of NDF in the diets would suggest.

Table 6.1: Effect of increasing the percentage NDF in the diet on DMI and milk production under cool and hot, humid environmental conditions. West, (1996)

| | | Bermuda hay added to diets, % | | | |
| | | 0 | 7.6 | 15.2 | 22.8 |
		Dietary NDF, %			
Item	Environment	30.2	33.8	37.7	42.0
DMI, lb/day	Cool	51.4	48.1	45.4	41.9
	Hot	40.3	39.2	38.4	36.1
Milk, lb/day	Cool	71.2	71.9	69.2	63.7
	Hot	54.2	56.9	58.2	50.0

Fat

If cow reduce their intake during heat stress, more nutrients need to be packed into a smaller volume of feed. Remember that a cow's energy requirement for lactation is unchanged and her energy needs to remain cool actually increase. Therefore, maintaining adequate nutrient intake becomes critical to avoid undue milk production loss. Optional methods to increase dietary nutrient density include feeding high quality forage, feeding more grain, and use of supplemental fats. Total dietary fat should not exceed 7 percent and 2-3% fat already present in feed, so only 4-5% can extra fat fed to animals. The advantage to including fat in the diet during hot weather is improved efficiency of energy use and greater energy intake as fats are 2.25 times greater in energy than carbohydrates. Because fat is more efficiently utilized as a source of energy than other feeds, it produces less heat during digestion and utilization. However, research on feeding fat during heat stress periods has not consistently shown the improved milk yields expected from feeding an energy dense cool feed (Huber, *et al.*,1994 and West, 1996). Adding 2.8 percent fat to the diet only resulted in a small increase in milk production (average of 1.4 pounds per day) compared to evaporative cooling which increased milk production 3.5 pounds per day (Huber, 1993) . This data suggests modification of the diet to minimize heat stress is secondary to the potential benefits of providing cow with a cool comfortable environment.

Minerals and Vitamins

Minerals are also more easily depleted during hot summer months. The increase in respiration and perspiration will cause an excessive loss of water, thereby reducing mineral levels. Potassium can be increased to 1.3 percent to 1.5 percent of the total dietary dry matter, sodium to 0.5 percent, and magnesium levels increased to 0.3 percent. The electrolyte minerals, sodium (Na) and potassium (K), are important in the maintenance of water balance, ion balance and acid-base status of the heat stressed cow. Sweating by heat stressed cow results in a considerable loss of K. Beede and Shearer (1996) has shown cow in mid day during heat stress conditions without shade lost five times more K per day than cow under shade. Current Dairy NRC (1989) recommendations of 1 percent K in the dietary DM appear to be to low during periods of heat stress. Milk production increases of 3 to 9 percent and increased DMI intakes have been found when K level in the dietary DM has been 1.2 percent or higher. Increasing Na in diets to 0.45 percent or greater has improved milk production (7 to 18%) more than increasing K during heat stress periods (Sanchez, *et al.*, 1994).The ratio or balance of cations (Na and K) and anions (Cl and S) may be as important during heat stress periods as altering the concentrations of individual minerals in the diet. Heat stressed cow responded to increasing the dietary cation anion balance (DCAB, Na + K - Cl) from 120 to 464 milli equivalents per kilogram regardless of whether Na or K was used to increase the DCAB (West, *et al.*, 1992). Florida data indicates increasing Cl concentration in the diet decreased DMI and milk production resulting in a maximum recommendation of 0.35 percent of the DM (Beede and Shearer. 1996, Sanchez, *et al.*, 1994).

Some nutritionists have suggested raising levels of supplemented vitamins during heat stress. However, if you are supplementing 100,000 international units (IU) of vitamin A per day, 50,000 IU of vitamin D, and 500 IU of vitamin E, it would not appear that more would do any good. Cow can manufacture vitamin D with exposure to sunlight and summer is a time where we might need less supplementation. Also, cow receiving fresh cut plants or pasture will get high levels of vitamin A and E in the forage. Therefore, supplementing extra vitamins during summer is not usually warranted.

6.8 Feed Additives

Buffers

Sodium bicarbonate (bicarb) is a white crystalline compound derived from soda ash. The pH of a 1 percent solution is 8.4 and it buffers at a pH of 6.2. Bicarb is widely used as a buffer. Research has shown it increases rumen pH, produces a more desirable rumen fermentation, and increases rumen fluid outflow. Sodium sesquicarbonate is sold under the trade name S-Carb™. It contains a mixture of sodium bicarbonate and sodium carbonate and is an alkalizing agent. The pH of a one percent solution is 9.9.

Feeding buffers can be beneficial during heat stress periods for two reasons. First, if fiber content of the diet is minimized and/or cow are selecting against eating concentrate, buffers can help prevent a low rumen pH and rumen acidosis problems. Secondly, the most common macro-mineral in a buffer ($NaHCO_3$) is usually Na, exception of K in $KHCO_3$, which when increased in diets fed during heat stress has increased DMI and milk production. Sodium bicarbonate or sesquicarbonate can help buffer the rumen to accommodate higher levels of concentrate. Add 0.75% of the total ration dry matter as sodium bicarbonate or sodium sesquicarbonate (10 kg D.M. X .0075 = 75 grams/cow/day). Adjust intake based on dry matter consumption.

Fungal Cultures

In a series of experiments, Feeding *Aspergillus oryzae* reduced heat stress in cow through lowering rectal temperature. Milk production increased in some studies and was attributed to improved fiber digestion in the rumen (Huber *et al.*, 1994). Yeast maintains rumen health by a) Improving rumen pH hence reduce heat stress. b) Improve fibre digestion and nitrogen utilization hence increased feed efficiency. c) Rumen micro- flora stabilization.

Niacin

In a five-herd study, feeding niacin during the summer increased milk production across all cow by an average of about 2 pounds per day, but cow producing over 75 pounds per day increased over 5 pounds per day (Muller, *et al.*, 1986). No change in milk components occurred. Niacin helps to alleviate heat stress both by increasing evaporative heat loss from the body and also by reducing the effects of heat at the cell level (Landquist, 2008). Because niacin acts as subcutaneous vasodilator.

Antioxidants

Thermal stress generates disequilibrium of the oxidative balance. With important consequences over the vital function, life and death of affected cells. Treatment of buffaloes with antioxidants (Viteselen) before the beginning of months of heat stress period may increase the pregnancy rate, correct the infertility due to heat stress through the decrease in cortisol secretion and a decrease in the oxidative stress. (Megahed *et al.*, 2008). Supplementation of vitamin C and vitamin E have a negative effect on cortisol levels during heat stress, which relieved the severity of heat stress in goats (Sivakumar *et al.*, 2010). The reduction in cortisol levels by vitamin C and vitamin E is not yet fully understood but may be achieved by reducing the synthesis and/or excretion of cortisol or by breaking it down (Webel *et al.*, 1998). Supplemental beta-carotene may increase pregnancy rates for cow in the summer and can increase milk yield (Arechiga *et al.*, 1998)

6.9 Feeding Management and Strategies

When feeding lactating dairy cattle during hot weather are feeding frequency (an extra feeding or two), time of feeding (cooler time of day), adequate feed bunk space (all cow can eat together without crowding), plenty of cool water, and adequate air flow. Keeping cow comfortable is the key to keeping them eating which is critical in keeping them productive. During hot, humid weather, it is advisable to increase the number of times of feeding per day than one time feeding. Increasing the amount of feed available during the cooler period of the day, early morning or late evening, may be another alternative. Feeding 60 percent to 70 percent of the ration between 8:00 p.m. and 8:00 a.m. has successfully increased milk production during hot weather.

Increase the number of feedings offered per day has two advantages. First, the feed will be fresher encouraging more consumption; secondly, cow are curious and if the feeding area is comfortable, cow will be stimulated to come to the feed manger more frequently with increased feedings. Increasing the number enables one to feed less at any one time, thereby avoiding heating and spoilage of the feed in the bunk. Flies around the feed are also reduced, thereby reducing the insect population. The number of feedings to obtain benefits is not known, but practically it is probably a minimum of three per day.

Time of Feeding is also important. During hot days, cow will eat mostly during the night time and after milking. Having fresh feed in the mangers after milking, especially when cow have been cooled in holding areas, is a good way to encourage DMI. Feeding at sunset and then again about an hour before sunrise are good times for feeding.

Feeding a TMR is preferable to component or separate ingredient feeding during heat stress periods. A TMR with forages mixed in will help reduce the cow's tendency to selectively consume concentrates rather than forages. A well balanced TMR will allow diets to be formulated at minimum fiber levels encouraging DMI and minimizing rumen fermentation fluctuations and pH declines.

Adding Water to Diets may help DMI during summer months. Water will soften fiber feeds, and reduce dustiness and dryness of the diet increasing palatability and DMI. A three to five percent addition of water is recommended.

Also, be sure mangers or bunks are kept clean. Remove refused feed every day. Check and clean any moldy and/or heating feed from the corners and edges of feeding areas at least three times a week or often if animal protein and fats are fed. Feeding areas with a decaying feed smell reduce DMI even when fresh feed is offered.

6.10 Summary

A reduction in DMI is the primary reason milk production declines during heat stress periods. At the same time DMI decreases, maintenance cost of the cow increases in an attempt to maintain body temperature and thus, the overall availability of nutrients and energy for milk production is decreased. The most effective feeding management strategy to minimize production losses during heat stress periods is to provide a cool, comfortable environment by shading, sprinkling and/or forced air flow, always provide clean water which is simple and cheap strategy to keep cow cool. Modifying the environment will result in bigger gains, or fewer losses, during heat stress periods than any dietary manipulations. Diet changes will have only a small effect on productivity and should be considered supportive and an enhancement to environmental cooling. The concentration of all nutrients will need to be increased in diets as DMI decreases during heat stress. The CP in the diet should not exceed 18% while the level of RDP not exceed 65% of CP. More frequent but smaller meals should be offered in the cooler hours of day time to encourage DMI. Maintenance of proper mineral balance with special care of K and Na and positive DCAD is of important to minimize the effect of heat stress. Some feed additives like niacin, herbal preparation, fungal cultures, buffers etc. may be beneficial while the oxidative stress could be managed by supplementing antioxidants. Guidelines for nutrients that have been specifically shown to have an influence on DMI and milk production during heat stress are:

Nutrient	Change and dietary concentration (DM basis)
Energy	Increase to compensate for reduced DMI.
Fiber	ADF minimum - 18%. NDF minimum - 25%. NDF from forage or effective fiber - 21%.
Fat	Added amount should not exceed 5%.
Protein	Meet overall CP requirement by adjusting concentration as needed. Use a combination of rumen degradable and undegradable sources (RUP level of 36 to 40% of CP).
Sodium	Increase using buffers to 0.45 to 0.55%
Potassium	Increase to 1.2% or more. High quality Lucerne is a good source.
Salt	Feed 3 to 4 ounces per cow per day.
Chlorine	Minimum - 0.25%, maximum - 0.35%.
DCAB	Na + K - Cl ~ 35 to 45 meq/100g DM (Na + K) - (Cl + S) ~ 25 to 35 meq/100g DM
Magnesium	Increase between 0.30 and 0.35%.
Niacin	6 g per cow per day
Aspergillus oryzae	3 g per cow per day

References

Arechiga C F, Staples C R, McDowell L.R and Hansen P. J. 1998. Effects of timed insemination and supplementation beta carotene on reproduction and milk yield of dairy cow under heat stress. J. Dairy Sci. Feb; 81 (2): 390-402.

Beede, D. K., and J. K. Shearer 1996. Nutritional management of dairy cattle during hot weather. Prof. Dairy Management Seminar. Dubuque, IA. IA State Univ. p. 15.

Bluett, S. J.; Fisher, A. D.; Waugh, C. D. 2000. Heat challenge of dairy cow in the Waikat a comparison of spring and summer. Proceedings of the New Zealand Society of Animal Production 60: 226-229.

Cummins, K. A. 1992. Dietary acid detergent fiber for dairy cow during high environmental temperatures. J. Dairy Sci. 75:1465.

Flamenbaum, I.; Wolfenson, D.; Mamen, M.; Berman, A. 1986. Cooling dairy cattle by a combination of sprinkling and forced ventilation and its implementation in the shelter system. Journal of Dairy Science 69: 3140-3147.

Hahn G L. 1981. Housing and management to reduce climatic impacts on livestock. Jour. Animal Science 52: 175-186.

Huber, J. T., 1993. Feeding for high production during heat stress. Western Large Herd Management Conf. Las Vegas, NV. p. 183.

Huber, J. T., Higginbotham G, Gomez-Alarcon RA, Taylor RB, Chen SC and Wu,Z. 1994. Heat stress interactions with protein, supplemental fat and fungal cultures. J. Dairy Sci. 77:2080.

Jhonson H D. 1965. Environmental temperature and lactation (with special reference to cattle). International Journal of Biometerology 9: 103-116.

Jordan, E. R. 2003. Effects of heat stress on reproduction. J. Dairy Sci. 86: E. Suppl.: E104-E114.

Landquist, R. 2008. Rumen protected niacin has potential to reduce heat stress. Dairy Today, August:n 12-14.

Meghahed G A, Anwar M M, Wasfy S I and Hammadeh M E. 2008. Influence of heat stress on the cortisol and oxidant-antioxidants balance during oestrous phase in buffalo-cow: thermo protective role of antioxidants treatment: Report Domest. Anim. 43: 672-677

Muller, L. D. 1986. Supplemental niacin for lactating cow during summer feeding. J. Dairy Sci. 69:1416.

National Research Council. 1989. Nutrient requirements of dairy cattle. 6th ed. Nat'l Acad. Press. Washington, DC.

Rensis, Fabio D. and Rex.J. Scaramuzzi. 2003. Heat stress and seasonal effects on reproduction in the dairy cow—a review Theriogenology 60:1139-1151.

Sanchez, W.K., M.A. McGuire and D.K. Beede. 1994. Macromineral nutrition by heat stress interactions in dairy cattle: review and original research. J. Dairy Sci. 77:2051-2079.

Sartori, R., G.J. Rosa and M.C. Wiltbank. 2002. Ovarian structures and circulating steroids in heifers and lactating cows in summer and lactating and dry cows in winter. J. Dairy Sci. 85(11):2813-22.

Shearer, J. K. 1996. Utilizing shade and cooling to reduce heat stress. Prof. Dairy Management Seminar. Dubuque, IA. IA State Univ. p. 1.

Shearer, J.K., D.R. Bray, and R.A. Bucklin. 1999. The management of heat stress in dairy cattle : What we have learned in Florida. P. 60-71 in Proc. Feed and Nutritional Management Cow College, Virginia Tech.

Sivakumar A V N, Singh G. and Varshney V P. 2010. Atioxidants supplementation on acid base balance during heat stress in goats. Asian-Aust J.Anim. Sci. 11:1462-1468.

Webel D M, Mahan D C, Johnson R W. and Baker D H. 1998. Pretreatment of young pigs with vitamin E attenuates the elevation in plasma interleukin-6 and cortisol caused by a challenge dose of lipopolysaccharide. J.Anim. Sci. 77: 21-35.

West, J. W. 2004. Heat stress affects how dairy cows produce and reproduce. Pro. Southeast Dairy Herd Management Conference.

West, J. W. 1996. Dietary management of heat stress cow: Secrets of southern cooking. Proc., 1996 Heart of America Dairy Management Conf. Kansas State Univ. and Univ. of Missouri. p. 135.

West, J. W., Haydon KD., Mullinix BG and SAndifer TG. 1992. Dietary cation-anion balance and cation source effects on production and acid-base status of heat-stressed cow. J. Dairy Sci. 75:2776.

7

Thermoregulation and Resilient to Climatic Changes in Camel

Sajjan Singh and N.V. Patil*

National Research Centre on Camel, Bikaner, Rajasthan

The camel is an important animal component of the fragile desert eco-system. With its unique bio-physiological characteristics, the camel has become an icon of adaptation to challenging ways of living in arid and semi-arid regions. The proverbial Ship of Desert earned its epithet on account of its indispensability as a mode of transportation and draught power in desert but the utilities are many and are subject to continuous social and economic changes. The camel has played a significant role in civil law and order, defence and battles from the ancient times till date. The world famous Ganga-Risala of erstwhile Bikaner State was accepted as Imperial Service Troup and participated in World War I and II. The camel helped the engineers while constructing the Indira Gandhi Canal in Western part of Rajasthan. Presently, the camel corps constitutes an important wing of Border Security Force of Indian Para-Military Services. Considering the importance of camel in the socio-economic development of arid and semi-arid zones and its declining population due to change in shift in its utility is a serious concern.The arid and semi arid , the most fragile ecosystems, both together have 53.40% of total geographical area in the country (MoEF, 2001) have the climate parameters like extreme temperatures, high humidity during wet summet alers and winters, higher evapotranspiration and erratic rainfall.It causes much stress to the agricultural and livestock production systems prevalent in the areas. The most common effect of climate change has been increased desertification. Dregne (2000) estimated that approximately 3% area of the dry lands are irrigated croplands, 9% are rain fed croplands, and 88% rangelands and about 30% of the irrigated croplands, 47% of the rain-fed croplands and 73% of the rangelands are degraded. In these areas animal husbandry being the major vocation of a large segment of population providing drought proofness compared to crop husbandry has also been affected because of the dependence of livestock on rangelands for nutrition and their exposure to extreme climatic factors which affects the productivity.

The dromedary camel, an important animal of dry region livestock production system is a livelihood support to pastoral communities through provision of milk and draft power for transportation of goods, withstands harsh conditions of climate and scarcity of water, feed and fodder. But have also suffered a lot in last 4 to 5 decades due to aberrations resulting from the climate change. The most significant adverse impact of Climate change in camel inhabiting areas remain loss in its numbers during last 50 years from 1.0 to 0.52 million (Livestock census, 2007) which can be ascribed to increased pressure of abiotic stress factors /conditions in the form of increased temperature by $0.60^{0}C$, increased humidity in different seasons and more intense discomfort conditions due to higher THI values (70>) during day and nights of wet summer and during days in winter months(CAZRI,1962-2011 data, Fig.7.1).

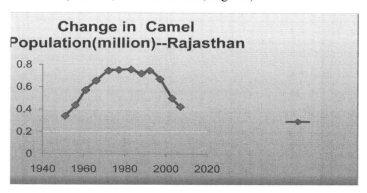

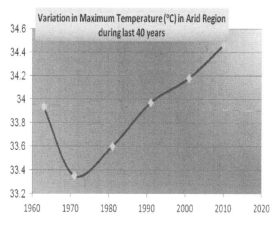

Fig. 7.1: Decline in camel population in last 50 years w.r.t. temp. change

The result of the climatic aberrations in the region is towards higher frequency years of recurrent droughts of moderate to severe nature (30-48% in last century) with its persistency of more than 2 to 3 years due to erratic rainfall patterns , higher evapotranspiration and higher THI values (Fig. 7.2) which has an indirect influence on the camel livestock production system due to low production from agriculture and also the resulting degraded rangelands which are important for the nutritional support to the camels. Similar situation remains true for the double humped camels living in the cold arid situations prevalent in northern parts of India in Leh and Ladakh of Jammu & Kashmir who have to suffer due exposure to extreme cold temperature conditions and also due to lack of sufficient and quality nutrition and even the water shortage is also experienced as a result of change in climate conditions. There are no studies undertaken to know the effect of climate change in areas of camel inhabitations on the quantity and quality of feed,fodder and water resources affected which will help to devise support policy for the traditional pastoral territory.

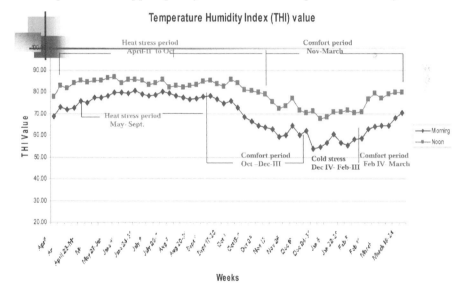

Fig. 7.2: THI values in arid region during 2010-11 indicating hot and cold stress periods

In the tropics the climate change conditions have although well demonstrated their negative impacts on the livestock production systems, the arid and semi-arid situations which are being largely exposed to these conditions of extreme thermal variations it affect the behavioral and physiological attributes most and thereby trigger the compensatory and adaptive mechanisms to re-establish the homoeothermic and homeostasis. These readjustments generally referred to as adaptations, may prove favorable or unfavorable to the productive functions of the camel but are essential for the survival of it.

The data analysis of female camel calf mortality (Bissa *et al.*, 2004) from NRC Camel from 1960 to 1990 for 30 years indicated that % calving occurred during October to Decemeber, January to February and March to May months were 17.57, 61.12 and 21.31%, respectively and maximum mortality (28.1%) was found in calves born during March to May months compared to calves born during October –December (22.3%) and January-February (12.8%). The respective figures for the calves died in the age group of 0 to 3 months age were 15.8,6.4 and 6.8% which were highest for March to May born calves. It may be because they suffered more from heat stroke due to less developed thermoregulatory system to combat the highest summer month temperatures of nearly more than 44^0C (Khanna *et al.*, 1992, Singh (1966), Bissa *et al.*, 2004) .The heat stress also had a negative effect on post natal growth performance. The birth weights of the female calves born during March-May which were significantly lower also continued to have lower post natal growth rate during various age groups and upto first calving compared to female calves born during October to December and January to February months therefore the culling rates for female calves of all age groups from 0 to age at first calving were also highest for March-May born calves (36.8%) compared to calves born during October to December (28.7%) and January to February (33.0) and the reasons for culling for different age group calves were poor growth, higher age at maturity, infertility and other factors indicating that summer stress in organized situations also affect the growth adversely and the calf mortality from the field may be still higher and appear to the major cause of loss of germplasm and continuous decline in camel population as about 23-25% of total calvings occur during March to May months. The studies of comparative basic physio-biochemical and hormonal differences in summer born calves than calves born in other season periods for higher susceptibility to various ailments causing higher mortality or low growth are lacking and need be undertaken to devise suitable management practices to reduce the calf mortality rates may be through proper shelter and feeding management systems.

Similarly there are very little researches conducted in camel in particular to understand the effect of changed climatic factors such as higher temperature and/or humidity and its combination on the physiological, biochemical and hormonal parameters of adaptations, so as to know and set comfort limits for camels of various other physiological states so that it is able to express optimum productivity in terms of lactation, reproduction and /or maintenance of health.

The available reports in dairy animals suggest that temperature increase and also higher humidity affect the animal productivity adversely causing decrease in milk production function to the tune of 10 to>25% (Upadhyay *et al.*, 2009, Singh and Upadhyay, 2009). Similarly studies indicated that the climate change /global warming has negatively impacted animal's growth and time to attain puberty is prolonged due to rise in temperature and it could be true for the livestock like camel maintained in extensive system of management. The Camel though it is reared by pastoralists for milk production no data is available to indicate level of decline in response to higher temperature or humidity or its combined effect. The studies in this regard are also lacking. Heat stress has also been reported to decrease reproductive performance in dairy animals. Hahn (1995) reported reduced conception rates in dairy cows by 4.6% for each unit change when THI reached above 70. And minimum temperature was also found to have greatest influence on the cows getting pregnant (Amundson et al., 2006). Upadhyay et al., (2009 b) reported effect of heat stress on reproduction by decreased length and intensity of estrus period, decreased conception, decreased growth, size and development of ovarian follicles and decreased foetal growth and calf size, increased risk of early embryonic mortality and higher number of services for conceptions. There are reports available in other ruminants (Beede and Collier, 1986, Kumar et al., 2009) indicate that the management factors like housing having proper designs, orientations, roof and wall materials can reduce the impact thermal factors by providing proper shade and ventilation to benefit the animals for physiological comfort and also to optimize the productive functions. Similar studies can be undertaken in camels utilizing different locally available agricultural materials to develop such type of shelters in which provision of the covered area and open area may be useful for free accessing of sunlight during winter and shed during summer. Under suitable shelter system, camel can be put in comfort zone which may lead to increase in BMR which may lead to increase feed intake and feed efficiency and as a result overall performance of animal may be improved. Further studies can be undertaken to understand whether and how and which intensity of these climatic factors affect the camel reproduction. In last few years the increased rainfall with higher temperatures in the arid and semiarid areas have resulted with increased risks and occurrences of animal diseases, as certain species of vectors such as biting flies and ticks have survived year round which might have contributed spread of disease and parasites into new regions which in turn, reduced animal productivity and possibly increased animal mortality (Baker and VIglizoo,1998). This may contribute to an increase in disease spread, including zoonotic diseases. Further the dromedary camels thought to be resistant to most of diseases commonly affecting domestic animals in the past but due to the climate change, new data conûrmed that camels are at risk to a large number of viral pathogens and are also incriminated in the transmission of many viruses (OIE report of AHG on Camelids, 2010). It is considered as a potential pathogen carrier in the transmission of Peste des Petits Ruminants (PPR) and several vector-borne infections such as Blue tongue (BT). Therefore the study is required of emergent infectious diseases in camelids as a result of climate change which also could be important from zoonotic point of view and need be prevented for developing sound bio-security measures to prevent the introduction and spread of disease in a population, herd, or group of animals. Based on which the Vaccination or Quarantine programs can be planned. The studies regarding the status of such diseases, its epidemiology and preventive methods for the safety of camels, the economic loss estimations to the camel herders and its significance to other livestock in the arid and semi arid region need be taken up urgently.

The milking dromedary camel during last decade is found to be affected by udder infection as mastitis, with heavy economic losses largely due to clinical and subclinical mastitis requiring immediate diagnosis as it reduce milk yield, alters milk properties, impairs preservation and processing and is a public health concern for consumers of camel milk but little work has been done on diagnosis of subclinical mastitis and the udder's response to bacterial invasion through conventional and also molecular approaches involving the production of recombinant cameline cytokines. Camel sustaining exposure to all climatic stress factors like extreme temperature ranges like 45^0 C to frosty -2 degree temperatures (and the double humped camels withstanding temperature conditions as low as - 40^0C), high speed winds and continued drought situations limiting the availability of water and feed has come out as an excellent example of adaptation and to know the Molecular basis of Adaptation of camels need be traced to genetic factors-like Hsp 70 genes for regulation of body mechanisms to adapt to thermal changes which also been known to be helpful for innate immunity modulation and stimulation of endogenous protective mechanisms. In recent years the importance of camel for human has been recognized in manifold ways like the health benefits of the camel milk as functional food to control the metabolic diseases like Type I diabetes, atherosclerosis, hypertension, hypercholestremia beside its exclusive immunological property for production of most effective vaccines and diagnostics for livestock and humans from its heavy chain nano-antibodies. The need is felt to rear this animal species for milk production and also take up effective measures to revert the declining trend of camel population. However the changed climate situations witnessed during last 4 to 5 decades in the arid and semi arid areas of camel inhabiting areas are posing challenges to maintain productivity and also to retain the desired numbers. Many scientists agree that climate change will alter destination habitat and increase vulnerability to invasion because of resource scarcity and increased competition among native fauna and flora. Animals in the desert must survive in a hostile environment. Intense heat, searing sun, and lack of water are just a few of the challenges facing desert animals.

Animals that live in the hot desert have many adaptations. Some animals never drink, but get their water from seeds (some can contain up to 50% water) and plants. Many animals are nocturnal, sleeping during the hot day and only coming out at night to eat and hunt. Some animals rarely spend any time above ground. It is pertinent to point out that desert inhibiting organisms share most genetic and physiological traits with closely related species living in identical environments. The suites of adaptive physiological traits of desert species are those that allow the organism to remain within the survivable thermal energy balance and water balance. It is important to understand that many species of desert plants and animals are living close to their limits of tolerance for one or more environment variables. Many desert animals are able to behaviourally seek out benign microclimates. Animals that are capable of maintaining temperature close to optimum for longer periods of time have an advantage in obtaining necessary resources. High temperature and dry air increases water needs thus requiring behavioural and/or physiological mechanisms for reducing water loses. There are several general categories of adaptations of animals that contribute to success of many species in hot arid environment. These include avoidance of extremes, a physiological and morphological characteristic that reduces water loses and thermal stress, rapid response to availability of water and nutrients, reduced metabolic rates and mechanisms to conserve essential nutrients. Camel is one such species that have all these physiological and morphological characteristics that reduces water loses and thermal stress.

7.1 Morphological Characteristics

1. **Feet suitable for sandy soils-**Padded feet are convenient for sandy soils and also for other difficult terrains. Toenails protect the feet damage from a bump. Calloused structures on feet protect the animals from injuries and extreme hot grounds during hot summer. Padded feet helps camel walk in sandy soils without sinking.

2. **Chest pad-**Also called as sternal pad keep the body elevated from hot ground while in sitting position and protect from hot grounds. It helps in copulation and acts as an external lock with prominences of pelvic girdle during mating and avoids slipping.

3. **Eyelashes-**The eyes of camels have two eyelash layers. The eyelashes interlock like a trap and protect the eyes of the animal from harsh sand storms.

4. **Nostrils-** Nostrils are naturally designed in such a way that air enters but keep out sand. A feature of their nostrils is that a large amount of water vapor in their exhalations is trapped and returned to their body fluids, thereby reducing the amount of water lost through respiration

5. **Split upper lip:** The camel's split upper lip allows it to grip and draw food into its mouth. Its lips are also very tough to protect against thorny desert plants.

6. **Hump:** Camels are best known for their impressive fat-storing humps that are prominent on the camels back. This fat can be used in emergency as energy and water source.

7. **Skin:** Camels rarely sweat, even in desert temperatures that reach 120°F (49°C), so when they do take in fluids they can conserve them for long periods of time. In winter, even desert plants may hold enough moisture to allow a camel to live without water for several weeks.

8. **Fur:** The dense wool like hair layer reduces insensible evaporative water loses and reduces the rate of heat gain of camel exposed to the intense thermal environment of a mid day summer. All camels lose their fur in spring and grow a new coat.

9. **Long legs:** When a camel walks, it moves both legs on one side and the both legs on the other, rocking side-to-side. This is why camels are nicknamed "The ships of the desert." Camel legs are incredibly strong, which allows them to carry up to 1000 pounds. They also can walk 100 miles per day and sprint at 12 miles per hour. Legs keep camel body sufficiently raised from ground to prevent heating from ground reflected radiations.

10. **Small pinna (Ear):** Camels have small rounded ears located far back on the head covered with hairs to help keep out sand and dust

7.2 Physiological Adaptations

1. Plasma volume is maintained at the expense of tissue fluid, so that circulation is not impaired.

2. The small oval erythrocyte of the camel can continue to circulate in situations of increased blood viscosity.

3. Camels can take in a very large amount of water at one session to make up for previous fluid loss. In other animals, this would result in severe osmotic problems. Camels can do this because water is absorbed very slowly from their stomach and intestines, allowing time for equilibration. Furthermore, their erythrocytes can swell to 240% of normal size without bursting, (Other species can only go to 150 %).

4. Their kidneys are capable of concentrating their urine markedly to reduce water loss. The urine can become as thick as syrup and have twice the salt content of sea water.

5. They can extract water from their fecal pellets so much that these can be used immediately for fuel upon voiding. Water loss via urine, feces and evaporation is reduced by a relatively efficient kidney, a capacity to recycle urea and an ability to maintain plasma volume.

6. A further adaptation solely for heat is involved in the camel's ability to have a large fluctuation in body temperature (from 97.7 to 107.6 degrees F). During the day, its body acts as a heat sink, and during the cool night of the desert, excess body heat is dissipated by conduction.

7. One of the important physiological characteristics of camel is associated with its ability to continue lactation on low water inputs and on low quality diets.

8. Camels conserve nitrogen by recycling urea back into the fermentation structure of the gut and use high C:N ratio and maintain protein balance.

Camels are incredibly resilient to the desert climates and the arid conditions that would easily kill another animal. Camels also possess the incredible ability to lose nearly 40% of their body weight as water and be unharmed.

7.3 Opportunities

Some of the steps for climate change adaptation described above involve more significant changes to current practice than others - some will require a major change to the thought processes involved with livestock management. However, there are a number of basic measures that require little change to best practice and should be considered as no-regret options: Species mixtures will provide some insurance against climate change - not all will be affected to the same extent. Provenance mixtures will provide insurance. Climate change predictions should be considered in the choice of livestock.

References

Abdoun K.A., Samara E.M., Okab A.B. and Al-Haidary A.A., 2013. Therelationship between coat color and thermoregulation in dromedary camels (*Camelus dromedarius*). J. Camel Pract. Res., 20(2): 251-255.

Al Jassim, R and Sejian, V 2015. *Review paper*: Climate change and camel production: impact and contribution. Journal of Camelid Science, 8: 1-17.

Elrobh M.S., Alanazi M.S., Khan W., Abduljaleel Z., Al-Amri A. and Bazzi M.D. 2011. Molecular cloning and characterization of cDNA encoding a putative stress-induced heat-shock protein from camelus dromedarius. Int. J. Mol. Sci., 12(7): 4214-4236.

Faye B., Chaibou M. and Vias G. 2012. Integrated Impact of Climate Change and Socioeconomic Development on the Evolution of Camel Farming Systems. Br. J. Environ. Clim. Change 2(3): 227-244.

Faye B., 2014. The Camel today: assets and potentials. Anthropozoologica 49 (2): 167-176.

Garbuz D.G., Astakhova L.N., Zatsepina O.G., Arkhipova I.R. and Nudler E., 2011. Functional Organization of hsp70 Cluster in Camel (*Camelus dromedarius*) and Other Mammals. PLoS ONE, 6(11): e27205.

Guerouali A. and Laabouri F.Z., 2013. Estimates of methane emission from the camel (*Camelus dromedarius*) compared to dairy cattle (*Bos taurus*). Adv. Anim. Biosci. 4: 286 (Abstract).

Musaad A., Faye B., Al-Mutairi S., 2013. Seasonal and physiological variation of gross composition of camel milk in Saudi Arabia. Emir. J. Food Agric., 25(8): 618-624.

Nolan J.V., Hegarty R.S., Hegarty J., Godwin I. R. and Woodgate R., 2010. Effects of dietary nitrate on fermentation, methane production and digesta kinetics in sheep. Anim. Prod. Sci., 50: 801-806.

Samara E.M., Abdoun K.A., Okab A.B. and Al-Haidary A.A., 2012. A comparative thermophysiological study on water-deprived goats and camels, J. App. Anim. Res., 40(4): 316-322.

Turnbull K.L., Smith R.P., Benoit St-Pierre B. and Wright A-D.G. 2011. Molecular diversity of methanogens in fecal samples from Bactrian camels (*Camelus bactrianus*) at two zoos. Res. Vet. Sci., 93: 246-249.

8

Physio-Genomic Responses of Pigs to Heat Stress: Strategies for Mitigation of Climatic Stress

Mohan N.H.

National Research Centre on Pig, Indian Council of Agricultural Research Guwahati, Assam – 781 131

The Fourth Assessment Report of the United Nations Intergovernmental Panel on Climate Change (IPCC), after analysis of more than 6000 peer reviewed publications has assessed potential effects of changing climate on various spheres of life (Solomon *et al.*, 2007). Under different scenarios, the estimated temperature rise of 1.1 to 6.4 °C in the global surface temperature is predicted at the end of 21st century (period 2090-2099 relative to 1980-1999). Global average sea level in the last interglacial period (about 125,000 years ago) was likely 4 to 6 m higher than during the 20th century, mainly due to the melting of ice in the polar regions. It has been detected in changes of surface and atmospheric temperatures in the upper several hundred metres of the ocean, and in contributions to sea level rise (Solomon *et al.*, 2007). The change in climate is concurrent with the changes in global levels of carbondioxide, methane, oxides of nitrogen etc, loss of biodiversity, natural habitat and other natural resources. The IPCC report also estimates a confidence level of 90% that there will be more frequent warm spells, heat waves and heavy rainfall and a confidence level of 66% that there will be an increase in drought, tropical cyclones and extreme high tides. The magnitude of the events will vary depending on the geographic zones of the World (Solomon *et al.*, 2007).

The International Fund for Agricultural Development (IFAD) acknowledges climate change as one of the factors affecting rural poverty and as one of the key challenges needs to be addressed (www.ifad.org/sf/). While climate change is a global phenomenon, its negative impacts are more severely felt by poor people in developing countries who rely heavily on the natural resource base for their livelihoods. Agriculture and livestock keeping are amongst the most climate-sensitive economic sectors and rural poor communities are more exposed to the effects of climate change. The IPCC predicts that by 2100 the increase in global average surface temperature may be between 1.8 and 4.0 °C. With global average temperature increases of only 1.5 – 2.5 °C, approximately 20-30 percent of plant and animal species are expected to be at risk of extinction (Solomon *et al.*, 2007, FAO, 2007).

Indian economy is closely tied to its natural resource base and climate-sensitive sectors such as agriculture, water and forestry. India may face a major threat because of the projected changes in climate and its impact on agriculture. Climate change may alter the distribution and quality of natural resources and adversely affect the livelihood of people of India (NPCC, 2012). It is assumed that due to higher demographic pressure on natural resources and weak mechanisms to offset adverse effects, the impact of climate change on agriculture in India will have a much higher impact of food security in comparison to developed countries of the world (DARE/ICAR Report, 2011). It is expected that in coming days one of the key concerns will be to assess and develop strategies for combating adverse effects of climate change on production of agricultural and animal produces. On the other hand, the animals contribute towards climate change through emissions of 18% of total greenhouse gas (GHG) emissions (9 % CO_2, 37 % methane and 65 % N_2O) (Steinfeld *et al.*, 2006). India's contribution to global CO_2 emissions is about 4.7%. Hence the livestock can be considered as contributors as well as victims of the increasing global temperature.

It is apprehended that the global warming may reduce agricultural production severely and may increase the gap between demand and supply of animal feed materials. Keeping in mind that there is more than 60% gap in demand and supply of concentrate feed, the major effect of climate on animal production can be summarized as 1. Reduction in production (milk, meat, egg, wool etc.) and fertility due to impact on normal physiological functions of the body, 2. Higher ambient temperature and extended range of temperature and increasing incidence of animal diseases, 3. Reduction in the availability of animal feed components, which is further compounded by reduction in cropping area due to increasing population. The twin major factors of reduction in the production with increasing population will place a severe economic burden, calls for a definite policy for combating impact of climate change and implementation of multiple measures to reduce adverse impact on animal production, keeping in the long term interests of the country.

8.1 Impact of Climate Change on Pig Production

According to the FAO statistics more than 50% of world pig production occurs in tropical and subtropical regions (FAO, 2007). In India, pig rearing is one of the most important occupations of rural society especially the tribal masses. The pig is one of the most efficient food converting animals among domesticated livestock, and can play an important role in improving the socio-economic status of the weaker sections of the society, particularly in the North-Eastern Hill Region of the country. However, compared to other species of farm animals, pigs are more sensitive to high environmental temperatures because of limted capabality of seating and panting. Economic losses due to heat stress on animal is large, calculations and predictions have been made keeping in view of impending climate change by several authors (Gerber *et al*., 2010; Pelletier and Tyedmers, 2010; Thornton, 2010; Lake *et al*., 2012; Hoffmann, 2013; Moore and Ghahramani, 2013).

Despite many challenges faced by pig industries in these developing countries including the fluctuations in the price of raw feed components, economical crisis, environmental problems, it is still predicted that these areas will continue to support future growth of pig production (Delgado *et al*., 1999). Among the multiple factors, climate becomes the prime limiting factor for livestock production in general and pig farming in general throughout the world. The heat stress is mainly an occasional challenge during summer heat waves in temperate regions, it is a constant problem in many tropical areas. In tropics regions, the effects of high temperature are often aggravated by a high relative humidity and by the fact that pigs are generally housed in semi-open buildings and directly exposed to outdoor climatic conditions. In the present paper, the understanding on physiological responses of pigs towards increasing ambient temperature, which is the most pronounced climatic change anticipated as a result of global warming is summarised.

8.2 Biological Responses to Heat Stress

The influence of meterorological parameteres on the various aspects of biological responses and swine production has been studied by several authors over a span of more than 3 decades. A substantial amount of data is also available on the factors affecting heat production in pigs (Bruck *et al*., 1969; Kaciuba-Uscilko and Poczopko, 1973; Farkas and Donhoffer, 1976; Rafai and Papp, 1976; McCracken, 1980; Labussiere *et al*., 2013; Pearce *et al*., 2013a; Weller *et al*., 2013). The principal climatic parametres (ambient temperature, humidity, and wind speed) can affect pig production directly as well as indirectly. The extent of indirect effect of thermal load is determined by climatic parametres themselves, besides the management and availablility of quality feed (Ingram, 1970; Poetschke, 1985a, b; Poetschke and Muller, 1991; Huynh *et al*., 2005).

8.3 Physiological Responses, Behaviour and Growth Rate

The average body temperature in pig is 39.1°C with a normal range between 38.7 to 39.8 °C. An environmental temperature range of 18 to 21°C has generally been found to support optimal productive performance of growing - finishing pigs. Pigs have a limited capacity to lose heat by water evaporation from the skin (Heath and Ingram, 1983). Another response ton increased ambient temperature is increase in respiration rate (RR). The effects of increase thermal load on pigs have been shown to be more pronounced at high levels of relative humidity. In confined housing, the pigs eliminate their heat load mainly by respiratory evaporation. Heitman and Hughes (1949) found that at ambient temperature of 32 °C and with a rise in relative humidity from 30% to 95% the respiration rate increased as an acute response. It has been observed that when the room temperature increases, sensible heat loss decreases rapidly, and evaporative heat loss becomes a vital route for pigs to eliminate heat load (Huynh et al., 2005). Above an inflection point temperatures (IPt) of approximately 22.4°C the pigs increases RR pronouncedly with increasing ambient temperatures (Huynh et al., 2005). The rectal temperature is found to increase with the ambient temperature rise. A recent study has reconfirmed the effect of heat stress in increasing rectal temperature by 1.5°C and 100% increase in respiration rate (from 54 to107 breaths per minute) (Pearce et al., 2013a). The animals with increased thermal load is found to consume more water and excreted more urine as a mechanism to disspiate heat. If the temperature does not exceed the critical temperature, where the balance between heat production and heat loss can be maintained, evaporative heat loss is at its maximum and respiratory evaporation will be inadequate for sufficient heat loss to keep the temperature homeostasis. Beyond this ambient temperature levels the body temperatures rise and followed by an adaptive depression of the heat production (Quiniou et al., 2001). The reduction in the heat increment associated with feeding is considered as an efficient mechanism to reduce heat load (Verstegen and Hoogerbrugge, 1987). Previous studies on short-term heat stress showed that at each degree C above a daily mean temperature of 21 °C, pigs gained 36-60 g/day less body weight (Steinbach, 1987). A decrease in 30 g/day in daily gain for each 1 °C above the optimum temperature was observed by (Mount 1979). Serres (1992) found that in the temperature range from 21 to 32°C, the growth rate of 70 kg pigs was reduced by 46 %. The body weight gain in growing pigs and/or milk output in lactating sows are directly related to the amount of feed intake (Renaudeau et al., 2005; Gourdine et al., 2006a, b; Renaudeau et al., 2006). Therefore, an increase of ambient temperature above the upper limit of the thermoneutral zone (25°C and 18°C in growing pigs and in lactating sows, respectively) negatively affects performance.

The increasing ambient temperature also alters the behavior of pigs to lose more heat, hence reducing the thermal load on them. Mount (1979) reported that the pigs could modify their postures in relation to ambient conditions, to either increase or decrease heat loss. Hahn (1985) reported that the several behavioural changes in farm animals, including pigs in response to hot environments, as animals attempted to maintain homeostasis by postural adjustment. For example, when lying down, the pigs will avoid contact with other pigs and will seek out cooler places in the pen. It is reported that fattening pigs preferred to lie on slatted floor at high ambient temperature and the animals shifted their excreting area to the solid floor and smeared themselves with dung and urine in order to cool them by facilitating evaporative cooling (Aarnink et al., 1996). Physiological responses to thermal stress includes the activation of heat shock transcription factor 1 (HSF1), increased expression of heat shock proteins (HSP) and decreased expression and synthesis of other proteins, increased glucose and amino acid oxidation and reduced fatty acid metabolism, activation of endocrine system to combat stress, and immune system activation via extracellular secretion of HSP (Collier et al., 2008). The heat acclimation changes are reported to switch in various cellular and molecular mechanisms causing plasticity in the central nervous system, leading to drop in temperature thresholds for activation of heat-dissipation mechanisms (Horowitz and Robinson, 2007).

8.4 Feed Intake and Metabolism

The primary consequence of heat stress is that animals reduce feed intake progressively with increased temperature (Kemp and Verstegen, 1987). Progressive reduction in the feed intake directly reduces the performance parameters in pig such as growth and reproduction. Heat stress significantly decreased the average daily feed intake in pigs with body weight range of 15-60kg. However when the pigs with higher body weight (30- 60kg) when fed with a diet containing 0.306% phosphorous had best performance in both thermoneutral (22°C and RH at 77%) as well as heat stress (32°C and RH at 73%) environments (Weller et al., 2013). The reduction in feed intake (metabolic energy intake) by pigs with increased thermal load is found to be between 38 (Renaudeau et al., 2013) to 47% (Pearce et al., 2013a). As another mechanism for thermal acclimation, besides reduction in metabolic energy intake, the pigs are found to reduce heat production, measured through indirect calorimetry, rather than changes in the heat dissipation mechanism (Renaudeau et al., 2013). It was observed that (Huynh et al., 2005) for each degree C rise in ambient temperature above inflection temperature (IT), the voluntary feed intake (VFI) declined steadily by an average of 95.5g. Nienaber et al. (1996; 1999) reported that for grower and finisher pigs an increasing ambient temperature by 10 °C (from 20 to 30°C) reduced voluntary feed intake from 65 to 74g/day/°C. The panting increased their total oxygen consumption at maintenance or constant feed intake and therefore an increase in physical activity thus increasing in total heat production in pigs (Ingram 1965, cited by Huynh *et al*, 2005). For each degree Celsius increase, pigs at high RH as compared to low RH levels RH consumed more water than feed, thus widening water:feed ratio (Huynh *et al*., 2005). Recent studies indicate that both heat stress and reduced feed intake decreases intestinal integrity and increase endotoxin permeability. It is hypothesized by the authors that that these may lead to increased inflammation, contributing to reduced pig performance during warm summer months (Pearce *et al*., 2013b).

8.5 Adaptation, Breed Variability and Thermotolerence

The gene pools of mammals contain allelic variants of specific genes that control body temperature regulation and cellular responsiveness to increase in thermal load. Therefore it is anticipated that the, genetic selection, both natural and artificial, can modulate the impact of heat stress on animal physiology, including reproductive function (Hansen, 2009). It is well documented that there is breed and stain variability in thermo-tolerence among pigs. The animals with different genotypes are found to have widely ranging physiological responses to thermal load. Studies on French Large White pigs selected for residual feed intake (Renaudeau *et al*., 2013), Creole and Large White (Renaudeau *et al*., 2007a). Detailed studies on pigs with an experiment thermal challenge have shown evidence of a biphasic pattern of heat acclimation with a short term (STHA) and a long term (LTHA) component (Renaudeau, 2005; Renaudeau *et al*., 2007b; Renaudeau *et al*., 2008). The use of an experimental challenge has allowed characterizing the different thermoregulatory responses occurring during these two phases and the underlying physiological mechanisms. The STHA adaptations allow the animals to compensate for increased heat stress before more permanent adaptation can be made. The LTHA adaptation occurs when changes during STHA are complete and result in a decrease of metabolic heat production and an increase of the animal capacity to dissipate heat from the environment.

8.6 Haematology, Blood Biochemistry and Endocrine Responses

The red blood cells exposed to single thermal stress showed an increase of osmotic resistance with increasing temperature. However, the authors concluded that the erythrocytes adapt to thermal stress on prolonged exposure (Jozwiak *et al*., 1991). Insignificant alterations in the bood rheology in response to heat stress has been observed, with a variation in response between different breed of pigs (Waltz *et al*., 2013). One of the studies (Hicks *et al*., 1998) have documented effect of acute stress on behavioural, immune and blood biochemical responses for 132 animals. The authors studied heat, cold and shipping stress on pig weanlings and concluded that the social status had large effects on plasma cortisol, globulin, acute-phase proteins, body weight, and weight changes. Acute shipping stress resulted in weight loss. Many immune and blood measures were not changed among acutely stressed pigs; however, the relationship between social status and mitogen-induced lymphocyte proliferation and natural killer cell cytotoxicity was disrupted during acute stress. Pig behavior was significantly changed by each stress treatment in a unique manner conlcuding that during acute stress, behavioral changes seem to be the most consistent and reliable indicators (Hicks *et al.*, 1998). Another previous study (Morrow-Tesch *et al.*, 1994) has also indicated intimate relationship between social order and heat stress in pigs. The socially dominant and submissive pigs had acute alterations in the immunity as compared to socially intermediate ones.

The endocrine responses to pig is well established (Kattesh *et al.*, 1980; Larsson *et al.*, 1983; Becker *et al.*, 1985; Salak-Johnson and McGlone, 2007; Yu *et al.*, 2007; Malmkvist *et al.*, 2009), with stress related hormones, with changes in hypothalamo hypophyseal endocrine axis. It is observed that during short-term exposure to cold, and in exercise, there is a rapid catecholamine response, producing concentrations in the blood which could be high enough to stimulate thermogenesis. On the other hand, longer-term cold exposure, does not increase catecholamine output in a fashion similar to those in acute exposure (Barrand *et al.*, 1981), indicating difference between short and long term physiological response. One of the significant features of heat stressed pigs is increase in plasma insulin levels (Pearce *et al.*, 2011) and has been observed in several other species (Hall *et al.*, 1980; Torlinska *et al.*, 1987; Itoh *et al.*, 1998; Wheelock *et al.*, 2010).

8.7 Reproduction

Heat stress can have profound effects on almost all aspects of reproduction in pigs. The reproductive events that can be affected negatively includes spertmatogenesis, oogenesis, maturation of oocyte, ovulation, responses to endocrine stimuli by reproductive organs, fetal and embryonic growth and post natal events such as lactation and parental care. The effect of incresing thermal load in mammals have been reviewed (Hansen, 2009). In pigs the problem of seasons infertility is documented (Love, 1978; Hennessy and Williamson, 1984; Peltoniemi *et al.*, 2000; Tast *et al.*, 2002). One of the early records (Love, 1978) indicate that the period of infertility correlates with the period of summer heat stress and is charactersied by a delayed return of sows to oestrus after mating and an increase in the number of non-pregnant sows in an organised herd. However, the litter size was apparently not adversely affected and the gilts and first parity sows were most frequently involved. However, countering the argument of summer infertility, (Hennessy and Williamson, 1984) conclude that stressful factors other than summer heat stress can cause delayed or failure to return to oestrus, and perhaps the phenomenon is wrongly called 'summer or seasonal infertility' and when given the appropriate combination and/or intensity of stressful stimuli, it can be manifest at any time of the year. Compounding to effects of heat stress in neonatal pigs, it has been observed that despite higher nursing frequency (39 vs 34 sucklings per day), milk production in sows from 10.43 to 7.35 kg/d when temperature increased from 20 to 29°C (Renaudeau and Noblet, 2001). The high ambient temperature is found to increase embryonic mortality (Wildt et al., 1975).It is understood that when maternal core body temperature increases of about 2°C above normal for extended periods of time, 2-2.5°C above normal for 0.5-1 h, or $\geq$4°C above normal for 15 min have resulted in developmental abnormalities (Ziskin and Morrissey, 2011). The same authors concluded that corresponding specific absorption rate (SAR) values that would be necessary to cause such temperature elevations in a healthy adult female would be in the range of $\geq$15 W/kg (whole body average or WBA), with <"4 W/kg required to increase core temperature 1°C. However, smaller levels of thermal stress in the mother that are asymptomatic might theoretically result in increased shunting of blood volume to the periphery as a heat dissipation mechanism resulitng in altered placental and umbilical blood perfusion and reduce heat exchange with the foetus. Exposure of boars to elevated ambient temperatures compounded with high humidity causes reductions in sperm morphology, semen quality, sperm output and fertility (Wettemann and Bazer, 1985; Kunavongkrit et al., 2005; Suriyasomboon et al., 2005). The increased temperature has an inhibitory effect on spermatid maturation and on testicular androgen biosynthesis. The authors (Wettemann and Bazer, 1985) concluded that improvements in reproductive performance can be achieved by increasing evaporative cooling of boars and about 5 weeks is required for boars to recover from the detrimental effects of heat stress and to produce semen with potential for maximal fertility. Evaporative cooling systems built into boar accommodation are often used to reduce fluctuations in both temperature and humidity during the hot and humid months along with other management factors, such as housing comfort, social contact, mating conditions and the frequency of mating, were considered very important boar management aids that assist good quality semen production (Kunavongkrit et al., 2005). Exposure of male and female pigs to elevated ambient temperatures can result in reduced reproductive efficiency and it has been observed that during early pregnancy, gilts are especially susceptible to heat stress. Decreased conception rates and reduced litter size occur when gilts are exposed to elevated ambient temperature during Days 0 to 16 after mating (Wettemann and Bazer, 1985).

8.8 Cellular Metabolism

The heat stress is found to induce changes in the rates of proliferation, protein turnover, and abundance of heat shock protein mRNA and its protein in porcine muscle satellite cells cultures, indicating that the physiology of porcine muscle satellite cells is affected in ways that could affect muscle growth in swine (Kamanga-Sollo *et al.*, 2011). It has been observed that the porcine pathenotes when subjected to heat stress undergo apoptosis and overall development of the cells is reduced (Tseng *et al.*, 2006).

8.9 Thermal Stress and Gene Expression Pattern

The thermal stress is found to trigger a complex program of gene expression and biochemical adaptive responses in animals (Georgopoulos and Welch, 1993; Sieck, 2002; Van Breukelen and Martin, 2002). However, the most of the information available on the gene expression profiles are based on *in vitro* studies and rodent models and data under *in vivo* conditions and pigs are sparse. Among various proteins whose expressions are affected by increased thermal load at cellular level that can be used as a marker for heat stress, the most prominent ones are Heat Shock Proteins (HSPs). HSPs are molecular chaperones that control the correct folding of newly synthesized or damaged proteins. Among HSPs, Hsp90, Hsp70 and Hsp27 are constitutive and Hsp72 and Hsp86 are inducible and are expressed in several tissues including skeletal muscle, heart, liver and kidney. Traditionally a variety of HSPs have been found to be modulated by heat stress such as small HSPs, HSP 40, 47, 56, 70, 90, 110 and ubiquitins. Presently more than 50 genes other than HSPs are reported to be upregulated as response to heat stress. The list includes C-*fos,* C-*jun,* C-*myc, egr*-1, CAAT enhancer binding proteins, peroxisome proliferators activated receptors (PPAR), nuclear factor kappa, C reactive protein, acid glycoprotein, orsomucoid protein (Bukh *et al.*, 1990; Yiangou *et al.*, 1998), Integrin beta binding protein 4, Ras-related nuclear proteins (Balcer-Kubiczek *et al.*, 2000), annexin-1(Rhee *et al.*, 2000), iNOS (Sharma *et al.*, 2000), superoxide dismutase (Balcer-Kubiczek *et al.*, 2000), Bradykinin receptor B_1, NMDA receptors (Le Greves *et al.*, 1997), amyloid precursor proteins, interferons (Singh and Hasday, 2013), interleukins, bcl-2 and related genes (Tavernier *et al.*, 2012), netrin-1 (Son *et al.*, 2013), osteopontin (Chung and Rylander, 2012) and a plethora of protein kinases (Chang *et al.*, 2013; Lollo *et al.*, 2013; Matsushima-Nishiwaki *et al.*, 2013). Heat stress caused significant damage to the pig small intestine and altered gene expression in the pig jejunum (Yu *et al.*, 2010). Since the intestinal epithelium is crucial for nutrient uptake, theses damages could be responsible, atleast partially, for the impairment of growth performance of pigs under heat stress (Liu *et al.*, 2009).

8.10 Responses to Cold Stress in Pigs

The neonatal pigs with limited capability to produce sufficent heat, adipose tissue reserves as the thermoregulatory system is not fully mature and are highly vulnerable to cold stress. The physiological and endocrine responses of neonatal pigs to cold stress is well known and forms one of the important reasons for neonatal mortality (Curtis, 1970; Carroll et al., 2012). The thermoneutral zone drops from 30-34 degree C to 25-30 degree C over a period of 3-4 weeks (Simmons, 1976). The lower critical temperature for piglets have been reported as 34.6 degree C at 2hrs of age (Herpin et al., 2002). Maintaining optimal ambient temperature close to 34.6degree C thorugh use of insulating bedding material such as straw, heating lamps and wodden boxes in piglet house ensures piglet survival, proper growth rate and development of immunity.

8.11 Strategies for Mitigating Thermal Stress in Pigs

The following section provides mitigating strategies for reducing thermal stress briefly. There are three principal methods to improve animal performance during the period of thermal stress.

8.11.1 Modification of ambient environment through optimal housing conditions

The heat and moisture produced by the metabolism of pigs must be removed by providing optimal ventilation. Generally in pig houses in India are consturted as semi-open with a wallowing tank in the open area. The semi-open construction ensures proper ventillation and wallow allows surface cooling of pigs during hotter parts of the day. Providing water sprinklers also form a method to reduce ambinent temperature, eventhough it increases humidity inside the animal house. Providing fans to increase evaporative cooling has also been used successfully in many farms. Several studies provide comparative analysis of various methods to allevieate thermal stress in animals such as use of water drips, application of different flooring material and type etc. (Hahn, 1981; McGlone et al., 1988; Malmkvist et al., 2009). One study concluded that water drip was an effective cooling technique for heat-stressed sows, especially when floors are plastic. The snout coolers, partial concrete slots and high energy-density diets provided only minor benefits to heat-stressed sows and were not of benefit to piglets nursing heat-stressed sows (McGlone et al., 1988).During winter months, the covering of windows and open areas with jute material such as sacks or plastic sheets reduces wind circulation inside the house, thus preventing excess heat loss from the skin surface.

8.11.2 Nutritional management

The commonest nutritional strategy to reduce heat stress is to increase the nutrient density in general and specifically energy density to compensate for the reduction in the feed intake by animal.Addition of extral protein, fat and reduction in fibre content of the feed are some important points in the pig feed preparation during hotter months of the year. The importance of clean *ad libitum* drinking water cannot be over emphasized and must be available at all times of the day.The feeding of probiotics and other supplements including mineral and vitamins has been experimented. However, providing optimal nutrition throughout the animals' lifetime is key to ensure maximum performance and successful pig farming as feeding contributes to about 70% of animal raising costs.

8.11.3 Genetic improvement and development of heat tolerant animals

It has been established that the differences in thermal tolerance exist between various animals with wide variation between breeds or strains of a species and within each breed. It has been observed that there is variation in the thermoal tolerence among sow line (Bloemhof *et al.*, 2008). The existance of variation opens the opportunity to select and develop animals with improved heat tolerence. The use of next generation technologies with large data sets provide valuable data for identification of animals with better thermal tolerance. However, presently the role of epigenetic regulation of gene expression and thermal imprinting of the genome is less known. Pigs with shorter generation intervels will be one of the suitable livestock to develop an understanding the role of epigenetics on animal adaptation. Based on experiences at NRC on Pig, purebred exotic pigs of Hampshire, Large White Yorkshire and Landrace, eventhough they are of temperate origin, performs quite well in the agro-climatic conditions of Assam.

8.12 Conclusions

It can be seen from the literature availble, that the basic data effect of thermal stress in pigs is available, which needs to be extrapolated to Indian context. However, with the advancement of technologies and knowledge, especially with respect to gene function, genome oraganisation and its association with various traits and epigenetics, there is still paucity in information with respect to thermal tolerence, responses to stress and long term adaptation including thermal imprinting of the genome. In the future, this information will be essentially required for development of heat resistant breeds of animals and devising strategies for mitigation of excess thermal load on the animal to maintain or improve performance. In nutshell, the impact of increasing temperature is multidimensional and can challenge the physiological resilience and induces various responses for adaptability along with reduction of natural resources. The some of the key measures to combat adverse impact of climate on animal production will include adopting optimal management options including housing, modified managemental practices, genetic selection of animals with increased stress tolerance, improved biosecurity for prevention, early diagnosis and treatment of animal diseases, development of alternate feed resources, contingency plans and emergency preparedness for addressing natural disasters including development of resource inventory and locator. Above all, the creation of awareness among stakeholders (Veterinarians, farmers, researchers, building engineers and policy makers) through print and electronic media will be essential along with the definite policy backup, long term planning, research and implementation of steps towards climate resilient agriculture.

References

Aarnink, A. J. A., van den Berg A. J., Keen A., Hoeksma P. and Verstegen M. W. A. 1996. Effect of slatted floor area on ammonia emission and on the excretory and lying behaviour of growing pigs. Journal of Agricultural Engineering Research 64: 299-310.

Balcer-Kubiczek, E.K., Harrison, G.H., Davis, C.C., Haas, M.L., Koffman, B.H., 2000. Expression analysis of human HL60 cells exposed to 60 Hz square- or sine-wave magnetic fields. Radiat Res 153, 670-678.

Barrand, M.A., Dauncey, M.J., Ingram, D.L., 1981. Changes in plasma noradrenaline and adrenaline associated with central and peripheral thermal stimuli in the pig. J Physiol 316: 139-152.

Becker, B.A., Nienaber, J.A., Christenson, R.K., Manak, R.C., DeShazer, J.A., Hahn, G.L., 1985. Peripheral concentrations of cortisol as an indicator of stress in the pig. Am J Vet Res 46: 1034-1038.

Bloemhof, S., van der Waaij, E.H., Merks, J.W., Knol, E.F., 2008. Sow line differences in heat stress tolerance expressed in reproductive performance traits. J Anim Sci 86: 3330-3337.

Bruck, K., Wunnenberg, W., Zeisberger, E., 1969. Comparison of cold-adaptive metabolic modifications in different species, with special reference to the miniature pig. Fed Proc 28: 1035-1041.

Bukh, A., Martinez-Valdez, H., Freedman, S.J., Freedman, M.H., Cohen, A., 1990. The expression of c-fos, c-jun, and c-myc genes is regulated by heat shock in human lymphoid cells. J Immunol 144, 4835-4840.

Carroll, J.A., Burdick, N.C., Chase, C.C., Jr., Coleman, S.W., Spiers, D.E., 2012. Influence of environmental temperature on the physiological, endocrine, and immune responses in livestock exposed to a provocative immune challenge. Domest Anim Endocrinol 43: 146-153.

Chang, C.C., Chen, S.D., Lin, T.K., Chang, W.N., Liou, C.W., Chang, A.Y., Chan, S.H., Chuang, Y.C., 2013. Heat shock protein 70 protects against seizure-induced neuronal cell death in the hippocampus following experimental status epilepticus via inhibition of nuclear factor-kappaB activation-induced nitric oxide synthase II expression. Neurobiol Dis 62C: 241-249.

Chung, E., Rylander, M.N., 2012. Response of preosteoblasts to thermal stress conditioning and osteoinductive growth factors. Cell Stress Chaperones 17: 203-214.

Collier, R.J., Collier, J.L., Rhoads, R.P., Baumgard, L.H., 2008. Invited review: genes involved in the bovine heat stress response. J Dairy Sci 91: 445-454.

Curtis, S.E., 1970. Environmental—thermoregulatory interactions and neonatal piglet survival. J Anim Sci 31: 576-587.

DARE/ICAR Annual Report. 2011. Government of India.

Delgado, C., Rosegrant, M., Steinfeld, H., Siméon, E. and Courbois, C. 1999. Livestock to 2020: The Next Food Revolution. Vision initiative food, agriculture, and the environment discussion Paper 28. International FoodPolicy Research Institute, Washington D.C., pp. 1-88.

FAO, 2007. The state of the world's animal genetic resources for food and agriculture.

Farkas, M. and Donhoffer, S., 1976. The relationship between heat production and body temperature in the new-born and the adult guinea pig after termination of exposure to hypoxia and hypercapnia: with notes on body weight vs. body surface as the frame of reference. Acta Physiol Acad Sci Hung 47: 1-14.

Georgopoulos, C., Welch, W.J., 1993. Role of the major heat shock proteins as molecular chaperones. Annu Rev Cell Biol 9: 601-634.

Gerber, P.J., Vellinga, T.V., Steinfeld, H., 2010. Issues and options in addressing the environmental consequences of livestock sector's growth. Meat Sci 84: 244-247.

Gourdine, J.L., Bidanel, J.P., Noblet, J., Renaudeau, D., 2006a. Effects of breed and season on performance of lactating sows in a tropical humid climate. Journal of Animal Science 84: 360-369.

Hahn, G.L., 1981. Housing and management to reduce climatic impacts on livestock. J Anim Sci 52: 175-186.

Hall, G.M., Lucke, J.N., Lovell, R., Lister, D., 1980. Porcine malignant hyperthermia. VII: Hepatic metabolism. Br J Anaesth 52: 11-17.

Hansen, P.J., 2009. Effects of heat stress on mammalian reproduction. Philos Trans R Soc Lond B Biol Sci 364, 3341-3350.

Heath, M., Ingram, D.L., 1983. Thermoregulatory heat production in cold-reared and warm-reared pigs. Am J Physiol 244: R273-278.

Heitman, H. and E.H. Hughes. 1949. The effects of air temperature on relative humidity on the physiological well being of swine. J. Anim. Sci. 8: 171-181.

Hennessy, D.P. and Williamson, P.E., 1984. Stress and summer infertility in pigs. Aust Vet J 61: 212-215.

Herpin P, Damon M and Le Dividich J. 2002. Development of thermo-regulation and neonatal survival in pigs. Live Prod Sci 2002; 78:25–45.

Hicks, T.A., McGlone, J.J., Whisnant, C.S., Kattesh, H.G. and Norman, R.L., 1998. Behavioral, endocrine, immune, and performance measures for pigs exposed to acute stress. J Anim Sci 76: 474-483.

Hoffmann, I., 2013. Adaptation to climate change—exploring the potential of locally adapted breeds. Animal 7 Suppl 2: 346-362.

Horowitz M. 2007. Progress in Brain Research 162: 373-92.

Horowitz, M., Robinson, S.D., 2007. Heat shock proteins and the heat shock response during hyperthermia and its modulation by altered physiological conditions. Prog Brain Res 162: 433-446.

Huynh, T.T., Aarnink, A.J., Verstegen, M.W., Gerrits, W.J., Heetkamp, M.J., Kemp, B., Canh, T.T., 2005. Effects of increasing temperatures on physiological changes in pigs at different relative humidities. J Anim Sci 83: 1385-1396.

Ingram, D.L., 1970. Meteorological effects on pigs. Prog Biometeorol 1: 71-79 passim.

Itoh, F., Obara, Y., Rose, M.T., Fuse, H., Hashimoto, H., 1998. Insulin and glucagon secretion in lactating cows during heat exposure. J Anim Sci 76: 2182-2189.

Jozwiak, Z., Palecz, D., Leyko, W., 1991. The response of pig erythrocytes to thermal stress. Int J Radiat Biol 59: 479-487.

Kaciuba-Uscilko, H., Poczopko, P., 1973. The effect of noradrenaline on heat production in the new-born pig. Experientia 29: 108-109.

Kamanga-Sollo, E., Pampusch, M.S., White, M.E., Hathaway, M.R., Dayton, W.R., 2011. Effects of heat stress on proliferation, protein turnover, and abundance of heat shock protein messenger ribonucleic acid in cultured porcine muscle satellite cells. J Anim Sci 89: 3473-3480.

Kattesh, H.G., Kornegay, E.T., Knight, J.W., Gwazdauskas, F.G., Thomas, H.R., Notter, D.R., 1980. Glucocorticoid concentrations, corticosteroid binding protein characteristics and reproduction performance of sows and gilts subjected to applied stress during mid-gestation. J Anim Sci 50: 897-905.

Kemp, B., and M. W. A. Verstegen. 1987. The influence of climatic environment on sows. In: M.W.A.Verstegen and A.M. Henken (eds.) Energy metabolism in farm animals effects of housing, stress and disease. p 115. Martinus Nijhoff, Dordrecht.

Kunavongkrit, A., Suriyasomboon, A., Lundeheim, N., Heard, T.W., Einarsson, S., 2005. Management and sperm production of boars under differing environmental conditions. Theriogenology 63: 657-667.

Labussiere, E., Dubois, S., van Milgen, J., Noblet, J., 2013. Partitioning of heat production in growing pigs as a tool to improve the determination of efficiency of energy utilization. Front Physiol 4, 146.

Lake, I.R., Hooper, L., Abdelhamid, A., Bentham, G., Boxall, A.B., Draper, A., Fairweather-Tait, S., Hulme, M., Hunter, P.R., Nichols, G., Waldron, K.W., 2012. Climate change and food security: health impacts in developed countries. Environ Health Perspect 120: 1520-1526.

Larsson, K., Einarsson, S., Lundstrom, K. and Hakkarainen, J., 1983. Endocrine effects of heat stress in boars. Acta Vet Scand 24: 305-314.

Le Greves, P., Sharma, H.S., Westman, J., Alm, P., and Nyberg, F. 1997. Acute heat stress induces edema and nitric oxide synthase upregulation and down-regulates mRNA levels of the NMDAR1, NMDAR2A and NMDAR2B subunits in the rat hippocampus. Acta Neurochir Suppl 70: 275-278.

Liu, F., Yin, J., Du, M., Yan, P., Xu, J., Zhu, X. and Yu, J. 2009. Heat-stress-induced damage to porcine small intestinal epithelium associated with downregulation of epithelial growth factor signaling. J Anim Sci 87: 1941-1949.

Lollo, P.C., Moura, C.S., Morato, P.N. and Amaya-Farfan, J. 2013. Differential response of heat shock proteins to uphill and downhill exercise in heart, skeletal muscle, lung and kidney tissues. J Sports Sci Med 12: 461-466.

Love, R.J., 1978. Definition of a Seasonal Infertility Problem in Pigs. Vet Rec 103: 443-446.

Malmkvist, J., Damgaard, B.M., Pedersen, L.J., Jorgensen, E., Thodberg, K., Chaloupkova, H., Bruckmaier, R.M., 2009. Effects of thermal environment on hypothalamic-pituitary-adrenal axis hormones, oxytocin, and behavioral activity in periparturient sows. J Anim Sci 87: 2796-2805.

Matsushima-Nishiwaki, R., Kumada, T., Nagasawa, T., Suzuki, M., Yasuda, E., Okuda, S., Maeda, A., Kaneoka, Y., Toyoda, H. and Kozawa, O. 2013. Direct Association of Heat Shock Protein 20 (HSPB6) with Phosphoinositide 3-kinase (PI3K) in Human Hepatocellular Carcinoma: Regulation of the PI3K Activity. PLoS One 8: e78440.

McCracken, K.J., 1980. Fasting heat production and maintenance requirement of early-weaned pigs. Proc Nutr Soc 39: 14A.

McGlone, J.J., Stansbury, W.F., Tribble, L.F., 1988. Management of lactating sows during heat stress: effects of water drip, snout coolers, floor type and a high energy-density diet. J Anim Sci 66: 885-891.

Moore, A.D., Ghahramani, A., 2013. Climate change and broadacre livestock production across southern Australia. 1. Impacts of climate change on pasture and livestock productivity, and on sustainable levels of profitability. Glob Chang Biol 19: 1440-1455.

Morrow-Tesch, J.L., McGlone, J.J. and Salak-Johnson, J.L. 1994. Heat and social stress effects on pig immune measures. J Anim Sci 72:2599-2609.

Mount, L. E. 1979. Adaptation to thermal environment: Man and his productive animals. Edward Arnold Limited, Thomson Litho Ltd, East Kilbride, Scotland.

Nienaber, J. A., G. L. Hahn, and R. A. Eigenberg. 1999. Quantifying livestock responses for heat stress management: A review. International Journal of Biometeorology. Springer-Verlag Heidelberg 42: 183 - 188.

Nienaber, J. A., G. L. Hahn, T. P. McDonald, and R. L. Korthals. 1996. Feeding patterns and swine performance in hot environments. Transactions of the ASAE 39: 195-202.

NPCC (2012). National Plan for Climate Change, Prime Ministers Council on Climate Change, Government of India.

Pearce S.C., Upah N.C., Harris A., Gabler N.K., Ross J.W., Rhoads R.P. and Baumgard L.H.. 2011. Effects of heat stress on energetic metabolism in growing pigs. FASEB J. 25:1052.5.

Pearce, S.C., Gabler, N.K., Ross, J.W., Escobar, J., Patience, J.F., Rhoads, R.P. and Baumgard, L.H. 2013a. The effects of heat stress and plane of nutrition on metabolism in growing pigs. J Anim Sci 91: 2108-2118.

Pearce, S.C., Mani, V., Weber, T.E., Rhoads, R.P., Patience, J.F., Baumgard, L.H. and Gabler, N.K. 2013b. Heat stress and reduced plane of nutrition decreases intestinal integrity and function in pigs. J Anim Sci 91: 5183-5193.

Pelletier, N., Tyedmers, P., 2010. Forecasting potential global environmental costs of livestock production 2000-2050. Proc Natl Acad Sci U S A 107: 18371-18374.

Peltoniemi, O.A., Tast, A. and Love, R.J., 2000. Factors effecting reproduction in the pig: seasonal effects and restricted feeding of the pregnant gilt and sow. Anim Reprod Sci 60-61:173-184.

Poetschke, J., 1985. [Adaptation of breeding swine to high environmental temperatures and delimitation of tolerance ranges of thermically stressed swine populations]. Beitr Trop Landwirtsch Veterinarmed 23: 325-338.

Poetschke, J. and Muller, I. 1991. [Bioclimatic assessment of a tropical site—a criterion for decisions concerning breeding and technology for successful animal production in the tropics]. Beitr Trop Landwirtsch Veterinarmed 29: 475-485.

Quiniou, N., Noblet, J., van Milgen, J., Dubois, S., 2001. Modelling heat production and energy balance in group-housed growing pigs exposed to low or high ambient temperatures. Br J Nutr 85: 97-106.

Rafai, P. and Papp, Z. 1976. Effect of the moistening of body surface on heat production and heat sensation in fattening pigs. Acta Vet Acad Sci Hung 26: 95-104.

Renaudeau, D., 2005. Effects of short-term exposure to high ambient temperature and relative humidity on thermoregulatory responses of European (Large White) and Carribbean (Creole) restrictively fed growing pigs. Animal Research 54: 81-93.

Renaudeau, D., Frances, G., Dubois, S., Gilbert, H. and Noblet, J. 2013. Effect of thermal heat stress on energy utilization in two lines of pigs divergently selected for residual feed intake. J Anim Sci 91: 1162-1175.

Renaudeau, D., Giorgi, M., Silou, F., Weisbecker, J.L., 2006. Effect of breed (lean or fat pigs) and sex on performance and feeding behaviour of group housed growing pigs in a tropical climate. Asian-Australasian Journal of Animal Sciences 19: 593-601.

Renaudeau, D., Gourdine, J.L., Quiniou, N. and Noblet, J., 2005. Feeding behaviour of lactating sows in hot conditions. Pig News and Information 26: 17N-22N.

Renaudeau, D., Huc, E. and Noblet, J. 2007a. Acclimation to high ambient temperature in Large White and Caribbean Creole growing pigs. J Anim Sci 85: 779-790.

Renaudeau, D., Huc, E., Noblett, J., 2007b. Acclimation to high ambient temperature in Large White and Caribbean Creole growing pigs. Journal of Animal Science 85: 779-790.

Renaudeau, D., Kerdoncuff, M., Anais, C., Gourdine, J.L., 2008. Effect of temperature level on thermal acclimation in Large White growing pigs. Animal 2: 1619-1626.

Renaudeau, D. and Noblet, J., 2001. Effects of exposure to high ambient temperature and dietary protein level on sow milk production and performance of piglets. J Anim Sci 79: 1540-1548.

Rhee, H.J., Kim, G.Y., Huh, J.W., Kim, S.W. and Na, D.S., 2000. Annexin I is a stress protein induced by heat, oxidative stress and a sulfhydryl-reactive agent. Eur J Biochem 267: 3220-3225.

Salak-Johnson, J.L., McGlone, J.J., 2007. Making sense of apparently conflicting data: stress and immunity in swine and cattle. J Anim Sci 85: E81-88.

Serres, H. 1992. Manual of pig production in the tropics. 2 ed. CAB International, Cedex, France.

Sharma, H.S., Drieu, K., Alm, P. and Westman, J. 2000. Role of nitric oxide in blood-brain barrier permeability, brain edema and cell damage following hyperthermic brain injury. An experimental study using EGB-761 and Gingkolide B pretreatment in the rat. Acta Neurochir Suppl 76:81-86.

Sieck, G.C., 2002. Molecular biology of thermoregulation. J Appl Physiol (1985) 92, 1365-1366.

Simmons, J.R., 1976. Keeping piglets warm. Vet Rec 98:381-382.

Singh, I.S. and Hasday, J.D. 2013. Fever, hyperthermia and the heat shock response. Int J Hyperthermia 29:423-435.

Solomon, S., Intergovernmental Panel on Climate Change., Intergovernmental Panel on Climate Change. Working Group I., 2007. Climate change 2007 : the Physical Science basis : contribution of Working Group I to the Fourth Assessment Report of the Intergovernmental Panel on Climate Change. Cambridge University Press, Cambridge ; New York.

Son, T.W., Yun, S.P., Yong, M.S., Seo, B.N., Ryu, J.M., Youn, H.Y., Oh, Y.M., Han, H.J., 2013. Netrin-1 protects hypoxia-induced mitochondrial apoptosis through HSP27 expression via DCC- and integrin alpha6beta4-dependent Akt, GSK-3beta, and HSF-1 in mesenchymal stem cells. Cell Death Dis 4: e563.

Steinbach, J. 1987. Swine. Effects of the tropical climate on the physiology and productivity of the pig. Elsevier: 181:199.

Steinfeld H, Gerber P, Wassenaar T, Castel V, Rosales M & de Haan C. 2006. Livestock's long shadow: environmental issues and options. Food and Agriculture Organisation, Rome.

Suriyasomboon, A., Lundeheim, N., Kunavongkrit, A., Einarsson, S., 2005. Effect of temperature and humidity on sperm morphology in duroc boars under different housing systems in Thailand. J Vet Med Sci 67: 777-785.

Tast, A., Peltoniemi, O.A., Virolainen, J.V. and Love, R.J. 2002. Early disruption of pregnancy as a manifestation of seasonal infertility in pigs. Anim Reprod Sci 74: 75-86.

Tavernier, E., Flandrin-Gresta, P., Solly, F., Rigollet, L., Cornillon, J., Augeul-Meunier, K., Stephan, J.L., Montmartin, A., Viallet, A., Guyotat, D. and Campos, L. 2012. HSP90 inhibition results in apoptosis of Philadelphia acute lymphoblastic leukaemia cells: an attractive prospect of new targeted agents. J Cancer Res Clin Oncol 138: 1753-1758.

Thornton, P.K., 2010. Livestock production: recent trends, future prospects. Philos Trans R Soc Lond B Biol Sci 365: 2853-2867.

Torlinska, T., Banach, R., Paluszak, J. and Gryczka-Dziadecka, A. 1987. Hyperthermia effect on lipolytic processes in rat blood and adipose tissue. Acta Physiol Pol 38: 361-366.

Tseng, J.K., Tang, P.C., Ju, J.C., 2006. In vitro thermal stress induces apoptosis and reduces development of porcine parthenotes. Theriogenology 66: 1073-1082.

Van Breukelen, F. and Martin, S.L., 2002. Invited review: molecular adaptations in mammalian hibernators: unique adaptations or generalized responses? J Appl Physiol (1985) 92: 2640-2647.

Verstegen, M.W., Hoogerbrugge, A., 1987. [Health and fertility in relation to production. Climate and health in young farm animals]. Tijdschr Diergeneeskd 112: 1062-1068.

Waltz, X., Baillot, M., Connes, P., Gourdine, J.L., Philibert, L., Beltan, E., Chalabi, T., Renaudeau, D., 2013. Effect of heat stress on blood rheology in different pigs breeds. Clin Hemorheol Microcirc.

Weller, M.M., Alebrante, L., Campos, P.H., Saraiva, A., Silva, B.A., Donzele, J.L., Oliveira, R.F., Silva, F.F., Gasparino, E., Lopes, P.S., Guimaraes, S.E., 2013. Effect of heat stress and feeding phosphorus levels on pig electron transport chain gene expression. Animal 7: 1985-1993.

Wettemann, R.P., Bazer, F.W., 1985. Influence of environmental temperature on prolificacy of pigs. J Reprod Fertil Suppl 33: 199-208.

Wheelock, J.B., Rhoads, R.P., Vanbaale, M.J., Sanders, S.R. and Baumgard, L.H., 2010. Effects of heat stress on energetic metabolism in lactating Holstein cows. J Dairy Sci 93: 644-655.

Wildt, D.E., Riegle, G.D. and Dukelow, W.R. 1975. Physiological temperature response and embryonic mortality in stressed swine. Am J. Physiol 229: 1471-1475.

Yiangou, M., Paraskeva, E., Hsieh, C.C., Markou, E., Victoratos, P., Scouras, Z. and Papaconstantinou, J., 1998. Induction of a subgroup of acute phase protein genes in mouse liver by hyperthermia. Biochim Biophys Acta 1396: 191-206.

Yu, H., Bao, E.D., Zhao, R.Q. and Lv, Q.X. 2007. Effect of transportation stress on heat shock protein 70 concentration and mRNA expression in heart and kidney tissues and serum enzyme activities and hormone concentrations of pigs. Am. J. Vet. Res. 68:1145-1150.

Yu, J., Yin, P., Liu, F., Cheng, G., Guo, K., Lu, A., Zhu, X., Luan, W. and Xu, J., 2010. Effect of heat stress on the porcine small intestine: a morphological and gene expression study. Comp. Biochem. Physiol. A. Mol. Integr. Physiol. 156: 119-128.

Ziskin, M.C. and Morrissey, J. 2011. Thermal thresholds for teratogenicity, reproduction, and development. Int J Hyperthermia 27:374-387.

9

Impending Climate Change and Thermal Adaptability of Yaks

G. Krishnan[1,], V. Paul[1], P.J. Das[1], S.S. Hanah[1], T.K. Biswas[1]*
M. Bagath[2] and V. Sejian[2]

[1]*ICAR-National Research Centre on Yak, Dirang – 790101, Kameng Arunachal Pradesh*
[2]*ICAR-National Institute of Animal Nutrition and Physiology, Bangalore – 560030*

Yak (*Poephagus grunniens*)is the most remarkable and multipurpose domestic animal living at high altitudes of the Himalayas. Features common to the environment in which yak live are extreme cold, mountainous terrain, high altitudes with reduced oxygen in the air, high solar radiation and short growing seasons for herbageanda variable assortment of herbage, sparse insome areas. Generally, the yak habitat ranges from 3000-6000 m abovemean sea level (msl), where the alpine pastures are found and there is no absolutely frost-free period during any part of the year. Wilted herbage provides some sustenance for the yak at other times of year, but not insufficient quantity for their requirements (Fig. 9.1). Many of thecharacteristics of the yak can be regarded as adaptations to these conditions, in which cattle of other species have difficulty in surviving (Wiener *et al.*, 2003).The yaks are considered as the life line of the highlanders in remote terrain because yak is the only sustainable livelihood due to non-availability of arable land for major agriculture. Yak husbandry is one of the important and indispensable aspects of the Indianstates, namely, Arunachal Pradesh, Himachal Pradesh, Jammu and Kashmir, Sikkim and Garhwal hills of Uttarakhand where other livestock species can hardly live but yaks can survive, reproduce and produce milk, meat, wool and other by-products. Yak milk and meat are the only source of protein for the isolated highlanders who have little access to modern life. The yaks are inextricably linked to the lifestyle, culture, and religion of highlanders and are the fundamental means of subsistence for

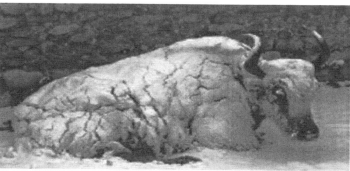

Fig. 9.1: Pictorial representation of Yak survival in extreme climatic condition at high altitude

the highlanders in harsh areas. Yaks in different forms and shapes are widely depicted in oriental cultures. The highland people in the Indian Himalayan states, as in neighboring countries, profess Buddhism and consider the yak of celestial origin. The yak appears symbolically in the Buddhist scriptures in different forms and is extant in the mythology. Yaks also feature in dances, festivities and pantomime. Yak keeping is a prestigious issue among the yak farmers by the number of yaks owned.

The yaks are commonly reared by the local tribes called Brokpas under semi-migratory free range system without any shelter even in the worst time of the blizzard.The association of yak as a pack animal (Fig.2) in the transportation of essential goods as well as in the tourism sector is a vital element to carry the loads of the tourist in the Himalayan region. As the effect of global warming, the atmospheric temperature of earth has increased by $0.74 \pm 0.18°C$ in the 20th century and is predicted to be increased by 1.8 to 4°C by the end of the 21st century (IPCC, 2007b). The gradual increase of environmental temperature in yak tracts may affect the themoneutral zone of yak which ranges from 5°C to 13°C (Wiener *et al.*, 2003) with an average of 10°C (Zhang, 2000 and Krishnan *et al.*, 2009a). Therefore, the increasing trend of ambient temperature may absolutely cause heat stress to the yaks at high altitudes.

9.1 Adaptive Characteristics of Yak to High Altitude

9.1.1 Body confirmation

The yak's body is compact with short neck, short limbs, no dewlap, small ears and a short tail. The scrotum of the male is small, compact and hairy, and the udder of the female is small and hairy (Wiener *et al.*, 2003). The skin has fewer wrinkles, and the surface area of the yak is relatively small per unit of bodyweight (0.016 m^2 per kg). Sweat glands are distributed in the skin over the whole body which is of the apocrine type. However, the sweat glands are nonfunctional and sweating is restricted to muzzle region of the yak and not on other parts of the body. The absence of sweating in the yak assists in cold tolerance by minimizing dissipation of body heat (Li Shihong,1984) but the same enforces the yak intolerant to heat stress (Ouyang and Qianfei, 1984).

9.1.2 Hair fibers of the yak

Heat conservation is enhanced by a thick fleece on the whole body, composed of an outer coat of long hair and an undercoat of a dense layer of fine down fibres that appear in the winter. In general, the coat of the yak seems well suited to insulating the animal from cold and repelling moisture. All these factors are important to survive in the prevailing harsh climate at high altitudes. Since, generation of extra heat would ultimately require additional feed, which is in short supply throughout the winter (Yousef, 1997).

The hair coat consists of three types of fiber; coarse, long fibres with a diameter above 52μ, down fiber with a diameter below 25μ and mid-type hairs with diameters between these two values. The down fiber provides an additional insulation to yak during winter, whereas the mixed type of fibres is able to maintain a constant air temperature within the coat (Ouyang and Qianfei, 1984). The hair coat helps yak to survive in very cold and wet conditions by its enhanced quality of low water absorption (Xue Jiying and Yu Zhengfeng, 1981). However, there is a seasonal change in hair coat growth and its composition, as air temperature falls with the approach of winter, down fibres grow densely among the coarser hairs, especially on the shoulder, back and rump (Ouyang, 1983). When the air temperature rises with the onset of the warm season, down fibres begin to be shed from the fleece(Wiener *et al.*, 2003).

9.1.3 Traditional system of yak rearing

The strategies practiced by the yak rearers to adjust and adapt their natural surroundings have resulted in aunique system of natural resource management. The transhumance pastoralism is the most common and popular adaptive method in the high altitudes to utilize the seasonal pastures and also to minimize heat stress during summer (Fig.9.2). Traditionally, the yak farmers are practicing the transhumance system of rearing where during the summer months they used to migrate to higher altitudes and return to lower altitudes called winter pastures of the low altitudes at around 3000 m above msl during winter months. The yak keepers practice two-pasture utilization strategy, the summer pasture extends for about 190 days (May to October) and the winter pasture extends for about 140-150 days (November to April). The rest of period are spent ontransit from winter pasture to summer pasture.

This transhumance pastoralism provides almost same ambient temperature to yaks round the year, which is also one of the important practices to minimize the heat stress in yak at high altitudes. Though yaks can withstand lower than freezing temperatures, but are more sensitive to high temperatures. When the temperature on lower points of Himalayas climbs, herds also need to climb

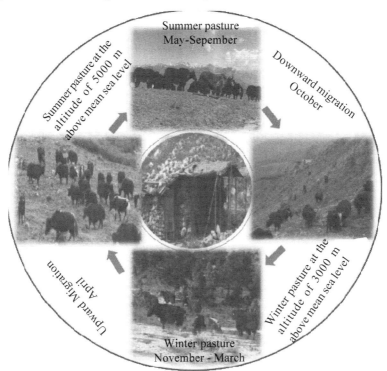

Fig. 9.2: Migratory system pattern followed at high altitude

elevated parts and they used to practice this on schedule, according to the temperature and availability of forage. However, global warming may significantly alter thenormal schedule. And when there's less forage at the elevated points as a result of climate change, yaks won't go down because of increased environmental temperature and fodder scarcity at lower altitudes forced the herds to move further higher altitudes.

9.2 Climate Change at High Altitude

Global warming may cause a variety of risks to mountain habitats and affects thedistribution of plant and animal communities (Beckage *et al.*, 2008). Under the expected climate scenarios in the final perspective results in the loss of rare species of alpine habitats (Dirnbock *et al.*, 2011) which may result in dearth of pasture lands for high altitude animal species. Parmesan and Yohe (2003) reported that a significant range shift toward the poles or toward higher altitudes for many organisms may be attributed to the increased global temperatures. Since many observations have proven that high altitude animals are under the hot spots of global warming. The impact of climate change on high altitude animals and their habitats to sustain their production has become big challenge in the 21st century where few places in the world have more affected by the climate change like yak rearing areas. The climate change is already happening in the high altitudes with an average rise of 0.01 to 0.04°C per year; the highest rates of warming are in winter with the lowest or even cooling in summer. The past trends and climate change projections suggest that temperatures will continue to rise and rainfall patterns will become more variable, with both localized increases and decreases. Annual mean temperatures are projected to increase an average of 2.9°C by the middle of the century with a range (places/ seasons) of 2.9 to 4.3°C by the end of the century (Sharma *et al.*, 2009). The expected current trend of increase in temperature of 2-3°C due to climate change have potentially catastrophic for Greater Himalayan animals and ecosystems (Hansen, 2008; Solomon, 2009; Xu *et al.*, 2009). The data available for temperature in Himalayas indicate that warming during last 3-4 decades has been more than the global average of 0.75% over the last century (Du *et al.*, 2004). Temperature increases are more during winter and autumn than during summer, and they clearly increase with the altitudinal rise (Lin and Chen, 2003).

9.2.1 Impact of climate change on rise of temperature with altitude

The seasonal temperature fluctuations are also changing in both timing and extent and the rate of change of temperature with altitude is becoming less that is higher altitude areas are warming faster than lower altitude areas. Hence, the difference in temperature between different altitudes is becoming less and gradually more increase in temperature with elevation or altitude, with areas above 4000 m experiencing the greatest warming rates up to 0.06°C (Shrestha *et al.*, 1999; Sharma *et al.*, 2009). Whereas, Lin and Chen (2003) reported that the decadal temperature rise remains 0.2°C up to 2000 m above msl and it often exceeds 0.3°C when the altitude is above 2000 m of msl.

9.2.2 Effect of climate change on high altitude alpine pasture

In the relatively mesic and productive Himalaya, the primary habitat types are alpine meadow and shrub vegetation; vegetation in this region provides relatively favorable forage for yak and sheep, the pastoralists' primary livestock. One phenomenon that may be affecting the Himalayan grasslands is a warming and drying trend. Direct field measurements (French and Wang, 1994), ice core data (Thompson *et al.*, 2003), and interviews with pastoralists over 25 years in the Himalayas, all provide evidence of a warming trend on Himalayan alpine pasture (Yao *et al.*, 1997). The altitude at which the alpine habitat is found decreases with increasing latitude and studies have shown that the mean temperatures in the Himalayan alpine zones have increased by 0.6 to 1.3°C between 1975 and 2006 (Dimri and Dash, 2011). Being located near the mountaintops, alpine species are particularly vulnerable to global warming (Beniston, 2003). Many species of the alpine grasslands are able to start their growth with the supply of snow melt water, well before the commencement of monsoon (Singh*et al.*, 2010). Obviously, their growths and life cycles are to be disrupted because of the lack of snow-melt water once the glaciers are disappearing. Several organisms and ecosystem process requiring a flow of water are going to be affected once snow reserves are vanished. The species composition and structure and functioning of alpine meadows are going to change both because of increased temperatures and loss of snow cover. The shift of alpine pasture to higher elevations may affect the high altitude animal rearing and also cause fodder scarcity. However, alpine pasture has little scope to march upward as the temperature rises. Because, the species that depend on snow cover for protection would be exposed to frost, and others that require winter chilling for bud-break may not get sufficiently low temperatures over a long enough period to survive (Cannone *et al.*, 2007). The effect of climate change or global warming and grazing on high altitude vegetation aboveground biomass decreased in total and composition (Klein *et al.*, 2004).

9.2.3 Effects of climate change on yak production

9.2.3.1 Effect of Climate Change on the Distribution Pattern of Yak

The temperature is one of the most important environmental variables that can affect the health, welfare and the production efficiency of animals (Kuczynski *et al.*, 2011). The mean environmental temperature of the yak habitat at the altitude of 3000 m above msl varies from -1.16 to 11.06 and 7.88 to 19.69 during winter and summer respectively, which absolutely cause thermal stress to the yak during summer. The physiological and metabolic adjustment resulting from the climate change may have negative consequences on yak production. The climate change has complicated impacts on animals affecting distribution, growth,incidence of diseases, availability of prey, productivity and even extinction of species in extreme cases due to habitat loss (Nardone *et al.,* 2010). In domestic animals, the climate change has multifacetedimpactsofthe production system affecting feed supply, challenging thermoregulatory mechanism resulting thermal stress, emerging new diseases due to change in the epidemiology of diseases and causing many other indirect impacts (Thornton *et al.,* 2008). Thermal stress is one of the greatest climatic challenges faced by the domestic animals (West, 2003) and global climate warming may further aggravate the condition and even provoke new episodes of thermal stress condition.

Global warminghasan adverse effect on yakproduction at high altitudes; since the rising temperature is disturbing their natural habit and creating conditions that they may not live in. In the yak's native regions, at present, the stocking density of yaks declines with the increase in average annual temperature. As a result, the yaks are moving upwards through the Himalayas and quickly running out of comfortable living conditions. As the global warming continues, changing climatic patterns could possibly affect high altitude indigenous species like yak by creating a shortage of water, fodder and an increase in pests and diseases (Zhang, 2000).

9.2.3.2 *Prediction of Heat Stress in Yak*

As the environmental temperature increases, certain thermoregulatory responses are initiated, including reduced feed intake, decreased activity and increased water intake, shade or wind seeking, increasedperipheral blood flow, sweating and panting. However, these thermoregulatory activities may not be sufficient to maintain a normal body temperature during the periods when ambient air temperature and humidity are particularly high.A variety of indices were used to estimate the degree of heat stress affecting cattle and other animals (Dikmen and Hansen, 2009). However, thermoregulation in cattle is affected largely by air temperature and relative humidity, the temperature-humidity index (THI), which combines the effects of these parameters into one value, which is one of the most commonly usedparametersto assess the impact of heat stress on dairy cows (Bouraoui *et al.,* 2002). Mader *et al.* (2004) reported that the upper comfortable limit of cattle is THI 72 and when it exceeds 72 cattle start experiencing the heat stress. Whereas, the modified THI for yak indicates that the THI 52 is comfortable and when it exceeds 52 the yaks started to experience the heat stress at higher altitudes (Krishnan *et al.,* 2009b).Factors other than air temperature and relative humidity can also impact heat stress. The presence of sunshine can add several degrees to the THI, whereas wind can lower it by a few points because it brings cooler air to the animal and carries away excess heat. However, the THI is the simplest and best tool to evaluate the level of the heat load in animals with species specific and their thermoneutral zone.

9.2.3.3 Responses of Yak to Heat Stress

The various adaptation characteristicsof yak to cold,harsh environmental conditions are more accountable to heat stress at high altitudes.When animals are experiencing heat stress, the first response observed is an increase in respiration rate (Beatty *et al.*, 2006). As the ambient temperature increases above 13°C the respiration rate of the yak starts to rise (Wiener *et al.*, 2003) whereas Krishnan *et al.* (2009a and 2010a) reported that the respiration rate has increased when the ambient temperature exceeds 10°C at the altitude of 3000 m above msl.It has been suggested that yak alter their respiration rate not only in response to a changing need for oxygen at high altitude, but also to regulate the body temperature (Fig. 9.3). Since, the yak has thick skin with few functional sweat glands and a heavy coat, has few means at its disposal for heat dissipation, other than the respiratory system. The heart rate and body temperature start to rise when the ambient temperature reaches 16°C.When

Fig. 9.3: Yak showing symptoms of heat stress

environmental temperatures reach 20°C, yak will start panting, stand nearby water sources or wading into streams and ponds, laying in the shade of trees, ifavailable, without moving, grazing, drinking or ruminating (Li Shihong, 1984). At the other extreme of environment where, yak canfeed and move normally on grasslands with air temperatures ranging as low as -30°C to -40°C or even lower(Cai and Wiener, 1995). In addition, seasonal variation in adaptive capability is also been reported in yaks (Krishnan *et al.*, 2009b; Sarkar *et al.*, 2009).

9.2.3.4 Effect of Climate Change on Growth Performance of Yak

The yak is a multi-purpose animal, providing milk, meat, hair and wool, hide, work as a draught animal (packing, riding, ploughing) and faeces-important as fuel in the absence of trees. The air temperature has been reported as the most important environmental factor influencing the growth and body size of yak (Chen Zhihua, 2000; Krishnan *et al.* 2010b). As a consequence of the abundant grazing in summer and early autumn, yak is normally able to develop a layer of subcutaneous fat as an energy reserve for the period of nutritional deprivation over winter and spring. Scanty feed availability in the long winter period is a major nutritional factor limiting the performance of yaks in terms of production and reproduction. Generally, yaks gain their body weight during summer and lose 25-30 per cent of their body weight during winter (Pourouchottamane, 2008). Due to cyclic nutrient deficiency or unbalanced nutrient supply within short growing seasons for herbage growth, annual cycle of weight loss in cold seasons and weight gain or compensatory growth in warm seasons in yak is a common phenomenon (Yongqiang *et al.*, 2000).The yaks are able to gain their body weight during summer inspite of thermal stress, if scientific management/ ameliorative measures have been adapted more body weight can be achieved.

9.2.3.5 Effect of Climate Change on Milk Yield of Yak

In general, lactating yaks irrespective of age, parity and even location, tend to peak in yield in July and August when the grass is at its best in terms of quality and quantity. Before July, though the grass has started to turn green and to grow, the amount of grass available is not high. After August, air temperature falls, the nutritive value declines - as the grass produces seeds and then wilts, and the content of crude fibre of the grass is high. One of the important factors that influencing milk yield is pasture production, which includes the quantity, growth status and nutritive value of the herbage or alpine pastures.It is clear that the climate change has already affected the quantity and quality of alpine pasture,which in turn affects the milk production of yak. In the warm season, variable temperature affects the milk yield of yak grazing on pastures. Yak produces less milk at too high temperature with strong solar radiation on clear days, but more milk within short periods of cloudy or rainy time. This tendency can be explained that high temperature might prevent cows from producing heat, thus the milk yield dropped with the decreased metabolism. Inversely, low temperature might cool down the cow, so the milk yield goes up with the increased metabolism (Dong et al., 2012). Hence, day to day fluctuations of weather is tend to affect the milk yield of the yak (Xu Guilin et al., 1983) and the optimum temperature for high production of yak is around 5°C-13°C (Zeng Wenqun and Chen Yishi, 1980). The milk production becomes less when the ambient temperature is exceeding the upper limit of its thermoneutral zone, especially when this is associated with strong solar radiation and a lack of wind on a clear day.

9.2.3.6 Effect of Climate Change on Reproductive Performance of Yak

Yaks are multipurpose, domesticated strong and hardy animals, which survive in a very harsh high altitude environment, including seasonal nutritional deficiencies (Zhang, 2000). They are considered as seasonal breeders with mating and conception restricted to the warm part of the year. The onset and end of the breeding season are affected by climatic factors such as ambient temperature, relative humidity, availability of feed and fodder, latitude and altitude (Shrestha et al., 1999). The severity and duration of malnutrition are primarily related to body condition of yak which is related to the availability of feed and fodder. The quality and quantity of fodder are decreasing as a result of global warming at high altitude. The detrimental effects of malnutrition appear to be manifested as reduced fertility; the greater the loss of body condition, the greater the reduction in pregnancy rate. Yaks exhibited estrous almost round the year, but mating and conception restricted to warm part of the year. The breeding season, normally starts in July and reaches its peak from September to November at an altitude of 2750 m above msl on 94°4' East longitude and 27° North latitude (Krishnan et al., 2010a). However, very few animals show estrous in July and August as a result of global warming at high altitudes with increased temperature of 7.88 to 19.69°C during summer, which absolutely cause heat stress to the yaks. The estrus is most common in the cool morning or evening and bred during overcast days in the breeding season. The duration of estrus is difficult to quantify because signs of estrus may be weak or vague during summer. However, the duration of estrus is longer during cool weather comparedto warm weather and age (average 23.8 and 36.2 hours in young versus mature yaks, respectively) (Zhang, 2000). The sexual activity of yak cows is suppressed by heat, lower altitudes (or a higher oxygen content in the ambient air), low-growing grasses in the pasture and other unfavorable factors. The conception rate of yak cows at high-altitude or summer pastures is 75.8 percent when compared with yaks at lower altitudes that is 66.7 per cent. However, during hot days, heart and respiratory rates were elevated, grass intake was decreased, and conception rates were low (Zhang, 1996).

9.2.3.7 Effects of Climate Change on Yak Diseases

The heavy coat and other specialized thermoregulatory mechanism of yak keep them alive in extreme conditions of high altitude, but are responsible for heat load in warm climatic condition (Brower, 1991). Moreover, the absence of immunity to lowland cattle diseases makes them more prone to parasitic and vector borne diseases. Yaks can survive in the summer pasture in high altitude even during the severe winter provided that there is sufficient fodder, but they cannot withstand the warmer climate of the lower settlement in the sub alpine region with more ticks and flies which force the yaks to migrate higher altitudes. The rising ambient temperature due to climate change will result in an increase in pests as well as diseases, especially in the lower permanent settlements in the sub-alpine region thereby making these areas unsuitable for yaks.

9.3 Strategies to Ameliorate Heat Stress in the Yak

In high altitude climate, the ecosystems are very fragile and sensitive to global climate change. As the global warming continues, changing climatic patterns could possibly affect high altitude indigenous animals like yak by creating a shortage of water, fodder and an increase in pests and diseases. Hence, different strategies have to be developed to ameliorate heat stress specific to yaks to minimize the heat stress and to augment the productivity of high altitude animals. Yaks are more susceptible to heat stress; hence it is extremely necessary to reduce the heat stress to obtain optimum productive and reproductive performance in the yak. Mitigation of the effect of climate change on yak possibly can be done through three major ways viz., physical modification of the environment; improved nutritional management; and genetic development of strains that would be less sensitive to heat stress (Krishnan *et al.*, 2008).

9.3.1 Physical modification of the environment

Generally, yaks are reared in a free range system without any shelter at alpine pasture lands. It is necessary to constructsuitable shed in the grazing areas for free-range yaks to protect them from adverse climatic conditions. Moreover, at organized farms, suitable housing with necessary air circulation and cooling facilities should be provided for the animals maintained under semi-intensive conditions. Housing of milking yaks during hot days improves the milk production than the free range. The perspective of yak milk production is bright if all or most of these options are adopted timely and properly (Dong *et al.*, 2012).

9.3.2 Improved nutritional management

The yaks must be maintained on quality feed with low fibre, optimum protein and energy during the period of heat stress. However, yaks are not fed with any supplementary feed where they are solely depending on the alpine pastures of grasslands at high altitude. Hence, grazing management remains an important consideration to improve the fodder quality at high altitude.Energy requirements of yaks may be standardized by a scientific approach to determine the optimum time for supplemental feeding to prevent loss of bodyweight in the winter season as well as to augment the production during summer.Supplementation of electrolytes is also one of the heat ameliorative measures where it helps in maintaining the physiological responses and homeostasis of animals during heat stress (Krishnan *et al.*, 2009b).

9.3.3 Genetic development of strains with less sensitive to heat stress

Systematic crossing of yak with other cattle is popular in the traditional herd farming system of yak husbandry and practiced for many years. From the earliest times, the name "PianNiu" (and other variants) has been used to describe these hybrids. These crosses find a special niche with herdsmen, usually at a somewhat lower altitude than typical yak. Crossbred females (Dzomo) are an important source of milk and dairy products. Since crossbred males are infertile, they cannot be used for breeding, which are used as draught animals or are slaughtered for meat. These hybrids are very suitable for work as they are easily tamed and have better heat tolerance than pure yak. The cross breeding/hybridization, can benefit from the higher performance of hybrid vigor (more milk and meat production) and the crosses are better adapted than the parents to various ranges of altitudes and ecological zones.

9.4 Conclusion

Yak husbandry is declining at an alarming rate due to lower economic benefits in addition to global warming. The younger generations of highlanders are unwilling to remain in traditional yak husbandry because of more lucrative opportunities in the industries in cities. Many aspects of the environment cannot be altered, although their worst effects can be lessened by, for example, the provision of shelter and, if possible, supplementary feed - especially in the winter. However, the cost effectiveness of such measures has been always an over-riding consideration. It is possible, of course, that global warming may alter the challenges to be faced by yak herders. Developing genetically superior animals through selective breeding / hybridization that would be less sensitive to heat stress is another possible option in to minimize the heat stress. The approach should be made to develop animals that would maintain desired productive and reproductive performance specific to the high altitude environmental conditions. Creating suitable temporary sheds in the grazing areas for free-range yaks are necessary to protect them from adverse climatic conditions. Climate change is one of the possible causes of rangeland degradation and of desertification of some areas. Hopefully, however, this consideration will not lessen recognition of the dangers posed to the grazing resources by over-grazing. Alpine pasture restoration programme, sustainable yak stocking rate along with control grazing scheme based on people's participation should be launched in the yak rearing areas. More detailed study of yak physiology and nutrition are needed to determine whether specific nutrients may be limiting factors and whether specific supplements at critical times like winter would be cost-effective. Research and development on yak husbandry should be accomplished in the light of highlander's perspective so that conservation of yak as well as heat ameliorative measures could be achieved in the near future.

Climate change has an adverse effect on yak production at high altitudes since the rising temperature is disturbing their natural habit and creating conditions which may not be favourable for their survival. The body confirmation, absence of sweat glands and thick fleece are important adaptive characteristics of yaks to survive at high altitudes. The climate change at high altitudes may prove detrimental for the productive performance of yaks. The temperature is one of the most important environmental variables that can affect the health, welfare and the production efficiency of yaks. Further, the shift of alpine pasture to higher elevations as a result of changing climate may affect yak production by causing fodder scarcity. The physiological and metabolic adjustment resulting from the climate change may have negative consequences on yak production. There are also seasonal variations in adaptive capability of yak to changing climate. The increased temperature as a result of climate change affected their productive parameters such as growth, milk production and reproductive performance. Further, the sudden outbreak of diseases as a result of climate change also leads to reduced production in yaks. Hence, different strategies have to be developed to minimize the heat stress effect on yak and to augment their productivity. Proper housing management and nutritional interventions are required to improve yak production under the changing climate scenario. In addition, there are also efforts pertaining to developing thermo-tolerant yaks to counter environmental extremes. Further intensified research efforts are needed in developing yak husbandry so that conservation of yak as well as heat ameliorative measures could be achieved in the near future.

Reference

Beatty, D.T., A. Barnes, E. Taylor, D. Pethick, M. McCarthy and S.K. Maloney, 2006. Physiological responses of *Bostaurus* and *Bosindicus* cattle to prolonged, continuous heat and humidity. Journal of Animal Science, 84: 972-85.

Beckage, B., B. Osborne, D. G. Gavin, C. Pucko, T. Siccama and T. Perkins, 2008. A rapid upward shift of a forest ecotone during 40 years of warming in the Green Mountains of Vermont. Proceedings of the National Academy of Sciences of the USA, 105 (11): 4197-4202.

Beniston, M. 2003. Climatic change in mountain regions: a review of possible impacts. Climatic Change, 59:5-31.

Bouraoui, R., M. Lahmar, A. Majdoub, M. Djemali and R. Belyea, 2002.The relationship of temperature-humidity index with milk production of dairy cows in a Mediterranean climate.Animal Research, 51: 479-491.

Brower, B. 1991. Sherpa of Khumbu: People, Livestock, and Landscape. Oxford University Press.

Cai, L. and G. Wiener, 1995. The yak.FAO (Food and Agricultural Organization of the United Nations) Regional Office for Asia and the Pacific, Bangkok, Thailand.

Cannone, N., S. Sgrobati and M. Guglielmin, 2007. Unexpected impacts of climate change on alpine vegetation. Frontiers in Ecology and the Environment, 5(7): 360-364.

Chen Zhihua, 2000. The Relationship between yak body size and ecological factors. Journal of Southwest Nationalities College, 24(4): 403-406.

Dikmen, S. and P. J. Hansen, 2009. Is the temperature-humidity index the best indicator of heat stress in lactating dairy cows in a subtropical environment? Journal of Dairy Science, 92:109-116.

Dimri, A. P. and S. K. Dash, 2011.Wintertime climate trends in the Western Himalayas. Climatic Change: doi: 10.1007/A10584011-0201-y.

Dirnbock, T., F. Essl and W. Rabitsch, 2011.Disproportional risk for habitat loss of high altitude endemic species under climate change.Global Change Biology, 17(2): 990-996.

Dong, Q. M., X. Q. Zhao, G. L. Wu, J. J. Shi, Y. L. Wang, L. Sheng, 2012. Response of soil properties to yak grazing intensity in a *Kobresiaparva*-meadow on the Qinghai-Tibetan Plateau, China. Journal of Soil Science and Plant Nutrition, 12 (3): 535-546.

French, H. M. and B. Wang, 1994. Climate controls on high altitude permafrost, Qinghai-Xizang (Tibet) Plateau, China. Permafrost Periglacial Processes, 5:87-100.

Hansen, J., 2008. Target atmospheric CO_2: where should humans aim? The Open Atmospheric Science Journal, 2:217-231.

IPCC (2007b).Summery for policymakers.*In:* Solomon, S., D. Qin, M. Manning, M. Marquis, K. Averyt, M. M. B. Tignor, H. L. Miller and Z. Chen, (eds.) Climate change 2007: The physical science basis. Contribution of working group I to the fourth assessment report of the intergovernmental panel on climate change. Cambridge University Press, Cambridge, United Kingdom and New York, NY, USA.

Klein, J. A., J. Harte and X. Q. Zhao, 2004. Experimental warming causes large and rapid species loss, dampened by simulated grazing, on the Tibetan Plateau. *Ecology Letters*, **7**:1170-1179.

Krishnan, G., Ramesha, K.P., Chakravarty, P., Kataktalware, M.A.andSarvanan, B.C. (2009a).Modified temperature humidity index for yaks (*Poephagusgrunniens* L.).Indian Journal of Animal Sciences, 79 (8): 788-790.

Krishnan, G., Ramesha, K.P., Chakravarty, P., Chouhan, V.S and Jayakumar, S. (2009b). Diurnal variations in the physiological responses of yaks (*Poephagusgrunniens* L.).Indian Journal of Animal Sciences, 79 (11): 1132-1133.

Krishnan, G., K. P. Ramesha, G. Kandeepan, V. S. Chouhan, and S. Jayakumar, 2010a.Effect of seasonal variations on primary physiological responses of yak.Indian Journal of Animal Sciences 80 (3): 271-272.

Krishnan, G., Ramesha, K.P., Chakravarty, P., Chouhan, V.S and Jayakumar, S. (2010b). Effect of environment on reproductive traits in yaks.Indian Journal of animal sciences, 80 (2): 123-124.

Krishnan, G., K. P. Ramesha, P. Chakravarty, M. A. Kataktalware, and B. C. Sarvanan, 2009.Modified temperature humidity index for yaks (*Poephagusgrunniens* L.).Indian Journal of animal sciences, 79 (8): 788-790.

Krishnan, G., M. Sarkar, M. A.Kataktalware, B. C. Saravanan, B. S.Sanjeeth and K. P. Ramesha, 2008. Envronmental stress and its amelioration in yaks. *Livestock International*, 12(2): 4-5

Kuczynski, T., V. Blanes-Vidal, B. M. Li, R. S. Gates, I. A. Naas, D. J. Moura, D. Berckmans and T. M. Banhazi, 2011. Impact of global climate change on the health, welfare and productivity of intensively housed livestock.International Journal of Agricultural Biology and Engineering, 4(2): 1-23.

Li Shihong, 1984. The observation on yak's heat resistance.A research on the utilization and exploitation of grassland in the northwestern part of Sichuan province, Sichuan National Publishing House, pp. 171-174.

Lin, C. W. R. and H. Y. S. Chen, 2003. Dynamic allocation of uncertain supply for the perishable commodity supply chain. *International Journal of Production Research*, 41:3119-3138.

Mader, T. L., M. S. Davis and J. B.Gaughan, 2004. Wind speed and solar radiation adjustment for the temperature humidity index. 16th *conference on Biometeorology and Aerobiology*, 23-27 August 2004, Vancouver, Canada.Pub.American Meteorology Society.

Nardone, A., B. Ronchi, N. Lacetera, M. S. Ranieri and U. Bernabucci, 2010.Effects of climate changes on animal production and sustainability of livestock systems.*Live stock Science*, 130: 57-69.

Ouyang, X. and W. Qianfei, 1984.An observation on adaptation of calf yak.A research on utilization and exploitation of grassland in the northwestern part of Sichuan province, Sichuan National Publishing House, pp. 159-161.

Ouyang, X., 1983.Effects of seasonal change of natural ecological conditions on yak's hair-coat.*Journal of Southwest Nationalities*, 4: 1-5.

Parmesan, C. and G. Yohe, 2003. A globally coherent fingerprint of climate change impacts across natural systems. Nature, 421 (6918): 37-42.

Pourouchottamane, R., G. Krishnan, M. Sarkar and K. P. Ramesha, 2008.Adaptation of yak to harsh environment of high altitude. Yak moving treasure of the Himalayas (eds. K. P. Ramesha) published by NRC on Yak, Dirang-790101, Arunachal Pradesh, India, pp. 53-59.

Sarkar, M., Bandyopadhyay, S., Krishnan, G. and Prakash, B.S. (2009).Seasonal variations in plasma glucocordicoid levels in female yaks (*Poephagusgrunniens. L)*. Tropical Animal Health Production, DOI 10.1007/s11250-009-9437-1.

Sharma, E., N. Chettri, K. Tse-ring, A. B. Shrestha, F. Jing, P. Mool and M. Eriksson, 2009. Climate change impacts and vulnerability in the Eastern Himalayas. Kathmandu: ICIMOD

Shrestha, A. B., C. P. Wake, P. A. Mayewski and J. E. Dibb, 1999. Maximum temperature trends in the Himalaya and its vicinity: an analysis based on temperature records from Nepal for the period 1971-94. Journal of Climate, 12: 2775-2786.

Singh, S., V. Singh and M. Skutsch, 2010. Rapid warming in the Himalayas: Ecosystem responses and development options. Climate and Development, 2: 221-232.

Solomon, S., 2009.Irreversible climate change due to carbon dioxide emissions. Proceedings of the National Academy of Sciences of the United States of America 106:1704-1709.

Thompson, L.G., E. M. Thompson, M. E. Davis, P. N. Lin, K. Henderson and T. A. Mashiotta, 2003. Tropical glacier and ice core evidence of climate change on annual to millennial time scales. Climatic Change, 59:137-155.

Thornton, P., Herrero, M., Freeman, A., Mwai, O., Rege, E., Jones, P. and McDermott, J., 2008. Vulnerability, climate change and livestock-Research opportunities and challenges for poverty alleviation. ILRI, Kenya.

West, J. W., 2003. Effects of heat-stress on production in dairy cattle.Journal of Dairy Science, 86: 2131-2144.

Wiener, G., H. Jianlin and L. Ruijun, 2003.The yak, 2ndedn.Published by FAO Regional Office for Asia and the Pacific.

Xu Guilin, 1983. Analysis of factors concerned with the milking performance and milk quality of yak. Journal of China Yak, 1: 21-29.

XueJiying and Yu Zhengfeng, 1981. The property and utilization of yak's down hair. *Journal of China Yak*, 1: 1-5.

Yao T., Y. Shi and L. G. Thompson, 1997.High resolution record of paleo-climate since the little ice age from the Tibetan ice cores.Quaternary International, 37:19-23.

Yongqiang, T. Zh. Xingxu, W. Minqang, L. Zhonglin and Zh. Rongchang, 2000. Endocrine changes and their relationships with bodyweight of growing yak. Proceedings of the Third International Congress on Yak held in Lhasa, P.R. China, 4–9 September 2000, pp. 380-387.

Yousef, M. K., 1997. Physiological adaptations of less well-known types of livestock in cold regions: yak and reindeer. In: *Stress Physiology in Livestock*. Vol. II. Ungulates (ed. M. K. Yousef).Boca Raton, Florida, CRC Press Inc. pp. 142-148.

Zeng Wenqun and Chen Yishi, 1980.Yak in ancient China. *Journal of China Yak*, 1: 71-74.

Zhang, R. C., 1996. Development of yak production in China in the past, present and future. In: Proceedings Res ProgAnimIndAnim Prod Process, Beijing, China. China AgricSci Press, pp.43-50.

Zhang, R. C., 2000. Ecology and Biology of Yak Living in Qinghai-Tibetan Plateau. In: Recent Advances in Yak Reproduction, Zhao, X. X. and Zhang, R. C. (Eds.) Publisher: International Veterinary Information Service (www.ivis.org), Ithaca, New York, USA.

10

Adaptation and Mitigation in Poultry Production against climate change/variability

A. Natarajan

Professor and Head and Nodal Officer, Gramin Krishi Mausam Seva, Animal Feed Analytical and Quality Control Laboratory, Veterinary College and Research Institute, Namakkal, Tamil Nadu

Mammals are homeothermic, meaning that they have the ability to control their body temperature within a narrow range in an environment whose temperature may change over a wide range.

India is experiencing a phenomenal spurt in the growth of livestock and poultry sectors through various timely and quality measures scientifically over a few decades. Milk production rose by a large leap from about 17.0 million tones in 1950 to about nearly 100 million tones in 2005 standing first in the world for maximum quantum of milk production. From about a 'naught' during 1950, the commercial layers have increased to 220 millions in 2005 producing 4500 crores of eggs every yearranking 5th in the world. The meat production is absolutely fascinating with 20 million birds in 1960 to about 2100 million in 2005 (Table 10.1).

This was made possible through

1. Improved breeds and strains

2. High quality dense diet

3. Effective disease management through vaccination and bio-security

Table 10.1: Per capita availability of milk, egg and chicken in India

Year	Milk Production - Million tones/year (Per capita availability)	No of improved layers (millions)	Egg Production –(billions per year) and per capita availability	No. of broiler birds (millions/year)	Per capita chicken availability (grams/year)	Human Population (crores)
1950s	17.0 (124ml)	Nil	5.0(10)	Nil	Nil	50
1971-1989	53.9(176 ml)	100	28.0(30)	190	266	85
1990-2000	84.5 (225 ml)	155	33.0(33)	700	700	100
2000-2005	100 ?(240 ml)	220	45.0(42)	2100	2100	107

Source: All India Poultry Business Year Book (2003-04) and Dairy Year Book (2005)

While these were handled and modified or improved from time to time efficiently, the climate or weather change is a natural occurrence that often brings about subtle to strong interference in the health, growth and production in livestock and poultry. These changes are seasonal in a year and sometimes sudden during the given season. They often require meticulous feed, farm, disease managemental practices and housing designs. To plan these activities effectively, the basics of animal/poultry micrometeorology or biometeorology needs to be understood.

10.1 Animal –Environment Interactions

Animals live within an environment complicated by multitude of factors encompassing both physical and psychological aspects of animal's surroundings. The thermal environment has a strong influence on farm animals with air temperature having the primary effect, but altered by wind, precipitation, humidity, and radiation. Animals compensate within limits for variations in effective ambient temperature by altering food intake, metabolism, and heat dissipation, which in turn alter the partition of dietary energy by the animal. Animals are homoeothermic in nature. They have to maintain a relatively constant core (body) temperature by balancing the heat gained from their body metabolism against the heat gained through concerted effects of physiological, morphological, and behavioral thermoregulatory mechanisms. Too rapid a rate of heat loss leads to *hypothermia*; too slow a loss to *hyperthermia*. Neither can be tolerated for an extended time. Under most conditions, there is a continual net loss of *sensible* heat from the body surface by conduction, convection, and radiation and under all conditions there is a continual *insensible* (evaporative) heat from the respiratory tract and skin surface. The net rate of heat loss depends upon the demand of the environment and the resistance to heat flow of tissue, skin and hair or feather. This thermal demand is a function of meteorological factors and hence reflects the cooling power of the surroundings. Because the animals are always exposed to and affected by several components of climatic environment, an index called *Effective Ambient Temperature* (EAT) was developed to evaluate the responses of animals. This index is the temperature of an isothermal environment without appreciable air movement or radiation gain that results in the same heat demand as the environment in question. Quantification of EAT differs from one species to another and is highly influenced by the animal's variable efficiency to combat thermal stress by physiological and behavioral reactions. The EAT is further influenced by:

1. Thermal radiation

2. Humidity

3. Air movement

4. Contact surfaces and

5. Precipitation

10.2 Thermal Zones

Evaluation of the relationship between animals and their environment begins with the thermo neutral zone (TNZ). Thermo neutral zone is the range of ambient (EAT) temperatures over which

- metabolic heat production remains basal,

- body temperature remains normal,

- sweating and panting do not occur,

- heat production remains at minimum,

- there is a sensation of maximum comfort and

- there is maximum performance and least stress.

Hence, the TNZ is defined as the range of effective ambient temperatures within which the heat from normal maintenance and productive functions of the animal in nonstressful situations offsets the heat loss to the environment without requiring an increase in rate of metabolic heat production.

LCT: At temperatures immediately below optimum, but still within TNZ, there is a cool zone (Fig 10.1) where animals invoke mechanisms to conserve body heat. These are mainly postural adjustments, changes in hair or feathers, and vasoconstriction of peripheral blood vessels. The effectiveness of various insulative and behavioural responses to cold stress is maximal at the lower boundary of the TNZ, a point called the lower critical temperature (LCT). Below this point is the cold zone where the animal must increase its rate of metabolic heat production to maintain homeothermy. Initial responses to cold stress rely more on increasing metabolic heat production, but long-term exposure to cold gradually results in adaptive responses through physiological and morphological change. Increased tissue insulation (fat, skin) and external insulation (hair coat, wool, and feathers) are type of insulation animal develops. As an animal's insulation changes, so do the limits of its thermal zones described in Fig.10.1&10.2.

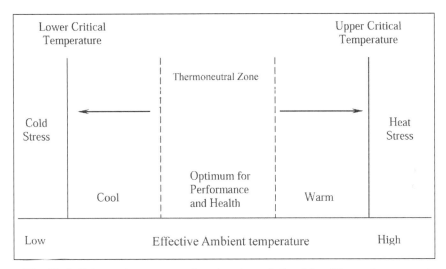

Fig. 10. 1: Schematic representation showing relationship of thermal zones and temperatures

Fig.10.2 shows expected zones of thermo neutrality for several species; however, it should be noted that shifts in the TNZ occur as a result of acclimatization by the animal to cold or hot environments as for the cow which is able to shift the TNZ as much as 15°C through cold acclimatization during a winter season.

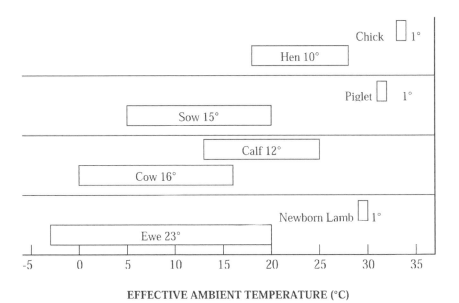

EFFECTIVE AMBIENT TEMPERATURE (°C)

Fig. 10.2: Estimated range in thermo-neutral temperature for newborn and mature animals of different species

Lower critical temperature may vary considerably depending upon specific housing and pen conditions, age, breed type, lactational state, nutrition, time after feeding, history of thermal acclimatization, hair or wool or feather coat, and behaviour. For example, a group of pigs has an LCT several degrees less than a single pig, because huddling behaviour of the pig in a cold temperature reduces the exposed surface and thus heat loss to the environment.

Estimates of Lower Critical Temperature (LCT)

Species or its type	LCT in °C
Cattle – fasting	18
Cattle- maintenance	7
Cattle full feed	- 1
New born calf	9
One month old calf	0
Dairy cow	- 14 to - 24
Swine	14 - 20
Chick	34
Five week chick	32
Adult	18

UCT: As EAT rises above optimum, the animal is in the warm zone where thermoregulatory reactions are limited. Insulation is decreased by vasodilatation (expansion of peripheral blood vessels to enable more heat exchange) and increasing surface area by changing posture. When EAT exceeds UCT, animals must employ evaporative heat loss mechanisms such as sweating and panting. The animals are considered heat stressed.

In a hot environment, animals are faced with dissipating metabolic heat in a situation where there is a reduced thermal gradient between the body core and environment, resulting in a reduced capacity for sensible heat loss. The immediate response of the animals to heat stress is reduced feed intake to attempt bringing metabolic heat production in line with heat dissipation capabilities. The higher producing animals with greater metabolic heat (from product synthesis) tend to be more susceptible to heat stress. Evaporation of moisture from skin or respiratory tract is used by animals to lose excess body heat in a hot environment but is limited by air vapor pressure but enhanced by air movement.

Key aspects of effects of heat stress and alleviation to improve the health and production with respect to the major livestock species dairy cattle and poultry merit attention in tropical countries and hence covered in this discussion.

10.3 Effects of Heat Stress in Dairy Cattle

Normal body temperature in cattle ranges from 101.1 to 102.2°F (38.4 – 39.0°C). During heat stress, cattle pant with their mouths closed so heat exchange occurs from the nasal mucosa of the upper respiratory tract near the turbinate bones and the cooled blood reduces the heat of the arterial blood near skull which supplies to brain and brain is kept cool that way.

Disregarding humidity, the TNZ of cattle is 41-77°F (16-25°C). When temperatures exceed the upper limits of the TNZ, the milk production, reproduction and overall health suffer.

The most commonly used formula in arriving at a figure that is used in deciding heat stress (HS) is Temperature Humidity Index (THI) and is calculated as follows.

THI = T + (0.36) Tdp + 41.5 where T – Temperature, Tdp - dew point temperature and when THI is > 72, HS occurs. This is around 21°C at moderate humidity.

10.4 Responses to Heat Stress

Behavioural Changes: Seeking out shade, reducing feed, increased water intake, standing rather than lying down, and bathing in ponds or mud.

Physiological Changes: Increased respiration and sweating to aid evaporative loss, drooling, increasing blood flow to peripheral blood vessels, vasodilatation, changes in important hormonal secretions leading to altered electrolyte balance, reproductive disturbance, reduced milk and immunity.

10.4.1 Effects of heat stress

1. Water is considered the most important nutrient during periods of HS. Milk contains 87 % water and will thus suffer for want of water. Approximately, water intake increases by 33 % durig HS.

2. Reduction in dry matter intake also occurs, which starts at a THI of 72 and above. For each unit of increase in THI thereafter, there is a decrease in feed intake by 0.23 kg of hay/day. Decreased DMI and negative energy balance has pronounced effect on milk production and reproduction. The decrease starts at THI of 76 and the decrease is in the order of 0.26 kg / day/animal.

3. Decrease in the cycling cows showing poor estrus signs and number of inseminated cows that establish pregnancy due to ovarian dysfunction as result of HS, change in uterine micro-environment and early embryonic death.

10.4.2 Heat Abatement

Currently used methods:

1. Shades, fans, sprinklers and evaporative cooling methods.

2. Nutritional management programs such as total mixed rations (TMR), low quantity-high quality fiber rations and rations supplemented with necessary amino acids.

Potentially Useful methods:

1. Strategic cooling.

2. Embryo transfer/*in-vitro* fertilization.

3. Administration of pharmacological thermo-protective agents for developing embryos.

4. Development of heat sensitive cross breeds by genetics, records and modern reproductive techniques.

Effect of heat stress in poultry is illustrated in Fig. 10.3.

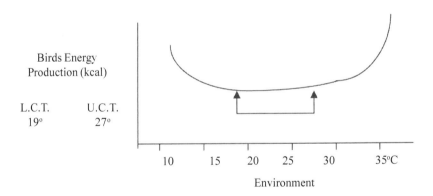

Fig. 10.3: Schematic representation of heat stress in birds

Poultry generally thrive well in warm conditions. Minimal body heat production (and hence the most efficient situation) is seen at around 23°C. Above 27°C, birds start to use more energy in an attempt to stay cool. But sudden raise of 2-3° C & above in day temperature from the previous day maximum temperature puts the birds in heat stress. Heat stress is induced in two ways. Environmental temperature increases and remains high for extended period at and above 35°C. Or, sudden heat waves elevate ambient temperature for short periods. Though the birds use evaporative cooling system (panting) to lose heat at higher temperature, humidity of inhaled air becomes critical. Thus higher temperature with high humidity, continued high temperature, wide difference in day and night temperature are much more stressful to birds than high temperature alone. This becomes more critical when the air speed and air movement inside the shed become small. Under these conditions, following changes take place. As house temperature goes up, the birds require less heat to maintain body temperature (more easily observed by the characteristic panting and wing drooping). Up to upper critical temperature, birds respond positively since the energy available in the feed is managed well. Above critical temperature, the energy available for production becomes gradually reduced and the egg production starts to decline. The upper critical temperature is 27°C in cooler places and in warmer countries, it is probably 33°C. During periods of elevated temperature, egg production is maintained at acceptable level but feed intake and egg weights are found to reduce. Mortality is high when birds are suddenly exposed to heat wave conditions (temperature touching mercury mark of 38° C and above) but this may not be the case when acclimatization to higher temperature is gradual or when protective measures are adopted anticipating the heat wave. To overcome the effects of high temperature with other related meteorological parameters, the following are attempted.

10.4.3 Correct the energy balance

a) **Increase Diet Energy**: This can be done by increasing the inclusion level of cereals in the diet but this is not perfect in practice since increasing the energy level through cereals reduces the feed intake further. Inclusion of fats or oils in the diet is another way to increase the energy density but most of the times it is prohibitively costly. Recent practical experience has shown that rice polish, high in oil content, added up to a level of 5-7 parts/100 parts, is useful in overcoming the reduced energy intake.

b) **Stimulate Feed Intake:** Birds can be stimulated to feed more by either encouraging them to feed during the cooler parts of the day or increase the palatability of the feed.

Body Energy Reserve: Correct the pullets nearing summer to have extra energy reserve in their body, but again care must be taken not to over-fattening them.

Wet Mash Feeding: Twenty five percent of the hen's feed can be made wet and top dressed on the dry feed in the morning to encourage the bird's feed intake and the rest be supplied in the evening i.e. cooler part of the day.

i) **Correct the Protein Level:** The protein level should be increased or adjusted so that loss through reduced protein intake is compensated. However, it is always beneficial to increase the essential amino acid contents of the diet by direct supplementation of methionine (intake 350 mg/day) and lysine (720 mg/day) since protein utilization dissipates more heat.

ii) **Minerals & Vitamins**: Similarly, the calcium, available phosphorus and vitamin D_3 levels should be increased to compensate the loss. Extra calcium and D_3 are essential to maintain the shell quality.

iii) **Electrolyte Balance**: During the extreme temperature conditions, there is an impairment of electrolyte balance, which should be corrected. The final anion and cation balance should be around 250 meq/kg of the diet.

	Mild $\rightarrow$	severe $\rightarrow$	loss of $\rightarrow$	affects +	general
	alkali	alkali	bicarbions	shell	bird
(pH) -	7.5	7.7		quality	health

Supplementation with part of sodium bicarbonate and part of common salt to maintain 250 meq/kg diet for layers.

iv) **Water**: Cooling the water is found to be very advantageous during the hottest times of the day and this helps to reduce the heat dissipation in the birds. Provision of cool water in the modern poultry farms with nipple type is beneficial. Nutrient supplementation is not beneficial.

Though there are many corrective measures followed to overcome the effects of high temperature, most of them are tentative and do not give satisfactory results. Cooling the inner side of the shed with foggers through separate pipeline is often a so-far-most effective method to reduce the mortality among the caged birds. This method cools off the shed and improves the atmosphere inside the sheds but not efficient in curtailing the mortality to a great extent. Birds are very often not inclined to drink water through the pipeline that is provided over their heads as the water is found hot. Rather they refrain from drinking and this condition results in deficient water intake, reduced feed intake, more stress, heat stroke and death.

It was for long time contemplated that birds need be given cooled water through addition of ice in the water tank, frequent flushing of water, pressurizing through pipeline to evacuate the warm or hot water that flows as a result of conduction of heat from environment. But all worked out to be partially ineffective in controlling the increase in bird mortality. An effective technique was found to be elusive.

10.5 A New Management Technique Found Effective

During this summer, a sudden and sharp increase in day time temperature was experienced as early as during last week of February, 2009. In a poultry farm of 184,000 birds where fogger was operated from 1030 hours to 1730 hours, mortality was not controllable and found to be alarmingly high.

Birds were not drinking sufficient water and feed intake thus went down significantly. The water running inside the pipeline was found to be more than warm and in late after noon, hot. The usual method of flushing out the water once every hour for about 4 – 5 minutes did not result in desirable results. It was decided to keep the water flowing continuously at the end of the pipeline in an effort to maintain the temperature of the water cool.

The following observations were recorded.

1. Continuous water-flow though the drinking water pipeline was found to control the mortality in a large and significant numbers.

2. Mortality pattern after continuous water-flow technique was similar to the period that was before summer.

3. Feed intake was found to be within reasonable limits.

4. Water that flew out was recycled for other purposes.

5. More research on this technique is the need of the hour.

Nearly 7902 (4.32%) birds were saved from a possible mortality with this new technique in this particular farm alone and this number is equal to a saving of Rs.15, 80,000 for a summer period of 90 days. Keeping in mind the population of layer birds in Tamil Nadu at a given time is 298,00,000 for the last 12 months, it can easily be extrapolated that any where near to 12,87,000 birds can be saved this way in 90 days of scorching summer effect. This is equal to net saving of Rs.25.75 crores in Tamil Nadu alone. Similar operation was followed in many farms of Namakkal zone where water could be spared and results were appreciably good.

In Broilers

The continuous water-flow technique was attempted in a broiler farm in Vellore district in Tamil Nadu where mortality and temperature of waters were recorded.

Period : From 13.05.2009 to 06.06.2009 (17 days)

Age: 9th day to 33rd day of broiler (40-41 days are usually market age)

Numbers: 9461 (Placed in two sheds with almost equal numbers)

Outside temperature: 96.8-112°F (12 days recording more than 100°F)

10.6 Diet Change

Sudden change in the diet composition or change on the day of extreme heat stress is not recommenced but instead recommended well in advance or earlier when the birds are under moderately stressful conditions.

10.7 Extreme Chill

It is observed mostly in northern parts of India for an extended period of time during the winter. In south, chill temperature remains actually for a very short time only. Birds respond to chillness by taking more feed to compensate the need for extra energy to maintain its body temperature at 41° C.

During the period of chill (Min. temperature 18 ° C) below the energy density of the feed should be increased using the equation $W^{.75}(173-1.95T)+5.5 DW + 2.07 EE$ where W = body weight of the bird, kg; T= °C; W = Weight gain/day, g; and EE = daily egg mass. In this equation, T is environmental temperature, which is the average of maximum and minimum temperature of the day.

10.8 High Air Velocity

At some periods during the course of the southwest and northeast monsoons, wind velocity would be moderately high and at times may well reach more than 10 km per hour on an average. The average day's wind speed would be around three times the average of the year's normal wind speed. During such high wind speed, there always exists an increase in feed intake. But this increase is entirely not due to feed intake. High wind causes weightless and smaller feed and medicinal particles to get carried away in the wind and part of the feed is actually wasted unnoticed in this way. The wastage due to wind velocity is calculated to be 2-3 g/bird in a day of high wind, approximately causing a feed loss of 200 – 300 kgs worth of Rs.1400 – 2100/day in a farm having 1,00,000 birds in raised type. Farmers can net extra profit if such loss is avoided. Proper covering of sides of the shed during such high wind periods should prevent the loss of feed in this manner. Loss may also be avoided to some extent by making the feed much stickier in nature by addition of molasses. Care must also be taken when feeding is done manually.

10.9 High Humidity, Low Temperature and Dampness Due to Continuous Rain

This weather phenomenon is common during the monsoon when it is active/ vigorous. This is conducive for the fungal growth and release of different mycotoxins in the feed ingredient and feed. Ingredient management becomes more critical during this type of weather especially cereals which normally have higher moisture (above 11 %). Rains during harvest, insufficient drying, and inappropriate storage make them vulnerable for fungal spoilage and ultimately spoil the feed. Sufficient stocking of the grains, good drying facilities and quality control measures (mycotoxin estimation) will not only certainly safeguard the birds from a possible out break of mycotoxins but also minimize the loss through reduced egg production, poor feed conversion etc.

Acknowledgement

The permission extended by Tamil Nadu Veterinary and Animal Sciences University, Chennai, to attend and present this topic, is greatly acknowledged.

References

Effect of Environment on Nutrient Requirements of Domestic Animals. 1981. The National Academy of Sciences, National Academy Press, Washington, D.C.

Fang, W 2003. In 'Environmental Engineering to reduce Heat Stress in Dairy Cattle', Technical Bulletin 164, Food and Fertilizer Technology Center, Republic of China on Taiwan.

Kurihara, M and S.Shioya 1993. In 'Dairy Cattle Management in a Hot Environment', Extension Bulletin 529, Food and Fertilizer Technology Center, Republic of China on Taiwan.

Leeson and Summers 2005. Commercial Poultry Nutrition, Third Edition, University Books, PO Box 1326, Guelph, Ontario.

Poultry House Ventilation Basics. A guide to Ventilation Management for Better Bird Performance, Hired-Hand, Breman, Alabama, USA.

Shan-Nan Lee 2003. In 'Feeding Management and Strategies for Lactating Dairy Cows Under Heat Stress' Extension Bulletin 530, Food and Fertilizer Technology Center, Republic of China on Taiwan.

11

Livestock Production and Health – Strategies for Sustainable Production under Climate Change Scenario

Deepa Ananth

Directorate of Entrepreneurship, Kerala Veterinary and Animal Sciences University, Pookode, Wayanad

Climate change is one of the most serious and long standing challenges facing livestock around the world and has moved up to the top of agenda among scientific community, civil society, business and government. Majority of scientific experts around the world believe that the climate change is already occurring by human activities, by use of fossil fuel, deforestation and agricultural practices, livestock rearing and that the developing countries in particular would be more vulnerable to the continuously changing climate. Rise in temperature due to climate change is likely to affect livestock production and health. Large deltas and low lying coastal areas would be inundated by a rise in sea level along with increased precipitation intensity. (DFID, 2004).Major contributors to the climate change are the man-made greenhouse gases, such as carbon dioxide, methane and nitrous oxide and hydroflurocarbons in the atmosphere. The accumulation of gas is causing the climatic change globally, as evident from the increased frequency of floods, droughts, cyclones and torrential rains in the recent past. The green house gases (GHGs) emission from agriculture sector alone accounts for 25.5% of global emission and over 60% from anthropogenic sources (FAO,2009)

11.1 Impact of Climate Change on Livestock Production

The climate has great influence on the type and distribution of livestock farming at large and definite influence on animal health and welfare. The varying conditions in physical environment, like poor rainfall, change in salinity of water, availability of water resources, humus content of soil, soil pH, humidity, temperature, wind, parasitic conditions, vector surveillance and disease outbreak are having huge impact on animal-agriculture practices, type of farming thus influencing the economics of livestock production. The impact of climate change on animal production has been categorized by Rotter and Van de Geijn (1999) as: a) availability of feed grain, b) pasture and forage crop production and quality, c) health, growth and reproduction and, d) disease and their spread.

11.1.1 Availability of feed

Any increase in the frequency and intensity of extreme events in climate such as storm, flood, drought, and cyclone would harm the health in many ways. Grain production will be severely affected. This causes the reduced feed availability and increased cost of grain. Hence, more land will be diverted for food production reducing the fodder availability. The extreme events also affect forest and trees as this would uproot trees, increases fungal growth and make trees more susceptible to diseases. Food safety hazards arise at various stages of food chain from primary production to consumption due to poor shelf life and more susceptibility to bacterial and fungal contaminations.

11.1.2 Effect on production and quality of forages

The production and productivity will be greatly affected as the adaptability of plants widely varies with the species and varieties. The performance of livestock fed on tropical forages is usually lower due to poorer feeding value. The feeding value depends on the chemical and physical composition of the forages which is related to soil condition, forage species, stage of growth and the plant part being eaten. Plant produces highly digestible sugars from photosynthesis in warmer condition. The rate of plant development is greater than in cool condition which reduces leaf/stem ratio, their digestibility and lower forage quality (Nelson and Moser, 1994). Each degree increase of temperature can decrease digestibility by 0.3 to 0.7% units. Temperature also affects the yield of forages. Elevated temperature depresses digesti bility, is associated with higher indigestible cell-wall (NDF) concentrations (Buxton and Fales, 1994).Severe drought may stop growth and kill all above-ground herbage and limit livestock production due to dead herbage and insufficient supply. Moderate moisture stress has either no effect or increases the digestibility due to slow growth of stem and resulting leafier sward (Wilson, 1981). Phosphorus is often at a low concentration in water-stressed forage (Rahman *et al.*, 1971) and could be a limitation to animal production where soil phosphorus levels are low. Increase of alkali or hydrocyanic acid and tannin contents may arise during stress and affect palatability of forages (Hoveland and Monson, 1980). In situation of excessive rainfall, lowland area grasses in over wet conditions may contain low CP and high call wall content (Pate and Snyder, 1979). The nutritive value will be decreased due to greater lignifications usually occurs at high growth temperature, reducing the digestibility (Ford *et al.*, 1979). Intake of herbage is restricted when CP levels are below 7% (Minson, 1980); this limitation occurs mostly in tropical grasses and not for legume. Pasture grown in soil deficient in minerals such as P, K, Ca and S may lead to low contents of these minerals. Both incidence of higher lignin, reduced digestibility with low yield simultaneously increasing the methanogenicity of diet threaten to reduce milk produced per unit of intake.

11.1.3 Effect on animal health and production

Apart from direct effects like injuries and death, epidemics such as diarrhea and respiratory diseases, population displacement, contamination of water and other natural resources also caused by seasonal change. Animal health may be affected in four ways: heat-related diseases and stress, extreme weather events, adaptation of animal production systems to new environments, and emergence or re-emergence of infectious diseases, especially vector borne diseases which are critically dependent on environmental and climatic conditions.Heat stress represents serious problem in dairying and other livestock enterprises. Livestock growth, reproductive ability, feed intake, lactation, age of attaining puberty would also be affected due to warming. It prolongs the oestrus cycle, weakening of signs of estrus with short oestrus period, low gestation rate and high fetal death rates. In India, direct heat stress on lactating cows and buffaloes cause a production loss of more than 1.8 million tons of milk. The increase in thermal stress days due to temperature rise has been estimated to cause an additional loss in milk production of 1.6 million tons in 2020, accounting about Rs.2365.8 Crores (Upadhyay *et al.*, 2007) . Maurya (2010) concluded that the length of service period and dry period of all dairy animals was increased from normal during drought. Singh *et al.* (1996) reported that higher incidence of clinical mastitis in dairy animals during hot and humid weather due to increased heat stress and greater fly population associated with hot–humid conditions. The outbreak of the disease was observed to be correlated with the mass movement of animals which in turn is dependent on the climatic factors (Sharma *et al.*, 1991). The epidemiology of many diseases are based on the transmission through vectors such as ticks, mosquitoes and flies and the developmental stages of which are often heavily dependent on temperature and humidity. The abundance of midges and mosquitoes may be increased through generation of vector breeding sites by periods of drought followed by heavy rainfall. The hot–humid weather conditions were found to aggravate the infestation of cattle ticks like: *Boophilusmicroplus, Haemaphysalisbispinosa and Hyalommaanatolicum*

(Basu and Bandhyopadhyay, 2004; Kumar *et al.,* 2004). Transmission of wind born diseases such as foot and mouth disease, infections transmitted by ticks, flies, mosquitoes, midges and other arthropods may be great concern with respect to changing climate. (FAO, 2004; Pattanaik and Sharma, 2010).Biological effects of climate change on animal and plant life include effects on physiology, metabolism or development rate, effects on distribution, effects on timing of life cycle, adaptation particularity of species with short generation times and rapid population growth rate (Huges, 2000). Heavy rainfall events can transport terrestrial microbiological agents into drinking water sources resulting in outbreaks of cryptosporidiosis, amoebiasis and other infection. Ocean warming and climate change related acidification and changes in ocean salinity and precipitation affect the biochemical properties of water, water microflora fisheries distribution persistence and patterns of occurrence of pathogenic *vibrio* and harmful algal blooms and chemical contaminants in fish and shell fish (Tirado and Clarke, 2010). Rodent born diseases associated with flooding including leptospirosis, tularaemia and viral haemorrhagic diseases, plague, Lyme disease are likely to arise. The assessment anticipates that for at least half of twenty first century additional UV exposure will augment the severity of sunburn and incidence of skin cancer. The immunity will also be suppressed. The exposure to UV- B may diminish the hosts response to certain vaccinations, Zoonotic diseases which causes important threat to human health. Zoonotic diseases likely to be affected by climate change are pasteurellosis, listeriosis/cirling diseases, anthrax infectious necrotic hepatitis/fasioliasis and Rift valley fever. (Singh,2010).Climate change and variability pose a challenge to pest and disease control measures such as Good agriculture and good veterinary practices with potential implications for the presence of chemical residues in the food chain.

11.2 Current Strategies Adapted by Farmers

- Farmers own perception and local traditional knowledge help them in evolving measures and technique to deal with situations arising due to climatic vagaries. These measures and techniques are locale specific, require no external help and are inherently scientific.

- Plantation fodder tree lines around animal shed/ house to reduce effects of cold/ heat waves was an important coping strategy adopted by most of the farmers.

- Provide cold water during hot and humid climate and provision of shade to reduce heat stress.

- Provide fresh air/ fan/cooler during extreme hot condition.

- Selling livestock during the hot climate or exchanging with more indigenous varieties which are more adaptable and disease resistant.

11.2.1 Strategies to combat climate change in fodder based livestock system

• Integrated approach towards mitigation of green house gases that are responsible for drastic change in climate, in management of soil, water, fodder and animal has to be adopted.

• Adoption of cost effective technologies to improve the production by reducing the energy loss in GHG emission through sustainable environment conservation efforts like mulching the crop residue instead of burning, ensiling the excess nutrient rich fodder, ammonia treatment of fibrous materials, soaking of straws, chopping of fodder, inclusion of C4 plants such as maize fodder etc. Nutritional strategies to reduce enteric methane production like balanced feeding, use of probiotics, plant extracts and use of Biotechnological methods by manipulating rumen fermentation and techniques for delignification should promoted.

• Managemental practices like breeding for improved varieties of fodder which reduce the methane production in rumen and reduced ADF content of the fodder on farm practises for better utilisation of ligniferous materials like early harvesting of fodder, soaking, ensiling, urea treatment.

• Strict water management should be enforced. The wastage of water will not only reduce the resource but also increase the water vapour in the atmosphere which is a major cause for global warming. Draught tolerant varieties may be developed for arid areas. Coastal areas may be planted with varieties that are adaptable to water logging, saline tolerant and to check soil erosion (Paragrass, setaria and other new varieties). Permitting sufficient humus in the surface of soil to retain water and other nutrients.

• Planting of Legume trees should be encouraged. This will improve the organic nitrogen content of the soil. This will also improve better carbondioxide assimilation and provide shade to farm reducing the stress to the animals. Legumes have potential to sustainably boost yields through self fertilizing with N, and usually have enhanced protein and energy content relative to grass forages. Also, flavenoids, condensed tannins, and other substances that are known to occur in significant concentrations in legumes have potential to enhance efficiency of N use, reduce parasite loads, and reduce methane emissions. Selection of grass and legume fodders and mixes should be based on potential for yield, feed quality, and desirable probiotic traits.

• Precision nutrition management of farm with application of climate information system will reduce the cost of production and prevent the wastage of resources. Optimizing fodder production based on the soil testing and precise application of fertilizers and other nutrients based on weather forecast will improve assimilation of nutrients and reduce the loss of CO_2, methane, N_2O and ammonia. Quality of fodder impacts both methane emissions and quantities of N excreted from livestock. Excreted N may increase N_2O emissions.

- Application of Phosphorus and Potassium to the soil has to be monitored. Excess Phosphorus would contaminate the surface water and promote the growth of harmful bacteria. Lowering of Phosphorus in soil would reduce the flowering and assimilation of nutrients.

- Unethical construction would hamper the normal water pathways resulting in waterlogging and impaired drainage. This would increase the accumulation of antinutritional factors such as oxalates in the plants and harbor parasites and their vectors.

- Surveillance and monitoring of infectious diseases during different climatic condition has to studied in detail. This allows to be prepared with the right strategy for control in place of requirement. The Veterinary services including medication and vaccination of animals, epidemiological research and veterinary public health should be given priority in the areas where diseases are emerging. Policies at Government level has to be taken for control of emerging diseases and should be strictly implemented.

- Geographic information system which monitor the climate change an be used to bring alertness regarding the spread of disease, draught, rainfall and warn for other environmental hazards so that the strategies can be designed to prepare for the emergencies. Strategic cost effective planning of the inventory can be made depends on the carbon credit so that effective utlisation of resources can be ensured.

- Effective knowledge dissemination methods should be utilized for early information to farmers and also line departments so that integrated strategy for effective management could be adopted. Different mobile channels such as voice and text messages and on-demand videos, a Farmer Helpline, including multiparty teleconferences can be used to update the weather forecast, disease forecast and feed and fodder availability and economics. Biweekly information to veterinary institutions and District veterinary centres about the weather and possible outbreak and surveillance reports would help to improve the preparedness and plan the control measures accordingly.

Hence, we need to create much better national, state wise and regional structures and systems to adapt to climate change and reduce the effect. Short, medium and long term strategies are to be developed to mitigate the sources of climate change, climate adaptation and use of Information and communication technologies for timely updates of climate changes, disaster management and disease surveillance has to be developed. Productive innovations are to be incubated and commercialized for sustainable production, growth and productivity in this respect.

11.2.2 Conclusion

The greenhouse gas of carbon dioxide, methane and nitrous oxide, being constantly added to the atmosphere by human activity, are the potential source of global warming. Environmental degradation caused by gradual increase of global temperature, drought and depletion of soil composition affecting plant production, their yield and quality. The important environmental influence on forage yield and nutritive quality which affects the feed and fodder resources of animals. The environmental temperature affects the growth, reproduction, productivity and economics of animal production systems. Improvements in the management practices for better adaptability to changing environment better nutrition of animals to reduce stress and mitigate the green house gases, precision farming techniques for optimizing the use or resources, improving the information system to counteract the health hazards and risk of diseases are the future tasks for sustainable livestock production.

References

Basu, A. K., Bandhyopadhyay, P. K. 2004. The effect of season on the incidence of ticks, Bull Anim Health Prod Afr 52(1): 39–42.

Buxton, D.R. and Fales, S.L. 1994.Plant environment and quality. IN: Forage quality, evaluation, and utilization. Eds G.C. Fahey Jr. et al. American Society of Agronomy, Madison, WI, USA.pp155-199.

DFID. 2004. The Department for International Development climate change in Asia Key sheet DFiD United kingdom available at:http//www.dfid.gov.uk/pubs/files/climate change/ IIasia.pdf.

FAO. 2009. Submission to UNFCCC AWG LCA, Enabling Agriculture to Contribute to Climate Change. Retrieved from: http://unfccc.int/resource/ docs/2008/smsn/igo/036.pdf.

Ford, C.W. and Wilson, J. R. 1981 Changes in levels of solutes during osmotic adjustment to water stress in leaves of four tropical pasture species. Australian Journal of Plant Physiology, 8:77-91.

Hoveland, C. S. and Monson, W. G. 1980.Genetic and environmental effects on forage quality. In: Crop quality, storage and utilization. Madison, Wisconsin, American Society of Agronomy, pp139-168.

Hughes, L. 2000. Biological consequences of global warming are the signal already there? Trends Eol.evol. 5: 56-61.

Kumar, S., Prasad, K. D. and Deb, A. R. 2004 Seasonal prevalence of different ectoparasites infecting cattle and buffaloes. BAU J Res 16(1):159–163.

Maurya, R. K 2010. Alternate dairy management practices in draught prone areas of Bundelkhand region of Uttar Pradesh, M.V.Sc. Thesis, Deemed University Indian Veterinary Research Institute, Izatnagar, India.

Minson, D. J. 1980. Nutritional differences between tropical and temperate pastures. In: Grazing Animals. F.H.W. Morley (Ed.). Elsevier Scientific Publishing Co.Amsterdam, pp143-157.

Nelson, D.J. and Moser, L. E. 1994. Plant factors affecting forage quality. In: Forage quality, evaluation, and utilization. Eds G.C. Fahey Jr. et al. American Society of Agronomy, Madison, WI, USA.pp115-154.

Pate, F. M.,Synder, G. H. 1979. Effect of high water table in organic soil on yield and quality of forage grasses-Lysimeter study. Proceedings Soil and Crop Science Society of Florida, 38:72-75.

Pattanaik, B and Sharma, G. K. 2010. Impact of climate change on animal health and Performance IN Proc.: National Symposium on Climate change and Livestock Productivity in India (Eds: R.C. Upadhyay, S.V. Singh, Ashuthosh, ManjuAshuthosh and Anjali Aggarwal) held at N.D.R.I. Karnal from October 7-8, 200pp:47-55.

Rahman, A. A. Abdel., Shalaby, A.F., Monayeri, M.O. El. 1971. Effect of moisture stress on metabolic products and ions accumulation. Plant and Soil, 34:65-90.

Rotter, R. and S.C. Van de Geijn, 1999.Climate change effects on plant growth, crop yield and livestock. Climate Change, 43: 651-681.

Smith, D. 1973. The non-structural carbohydrates. In: Chemistry and biochemistry of herbage. Editors G.W. Butler and R.W. Bailey. London, Academic Press, Vol. 1,pp.106-155.

Sharma, S. K., Singh, G. R., Pathak R. C. 1991. Seasonal contours of foot-and-mouth disease in India, Indian Journal of Animal Sciences 61(12):1259–1261.

Singh, K. B., Nauriyal, D.C., Oberoi, M. S. and Baxi, K.K. 1996. Studies on occurrence of clinical mastitis in relation to climatic factors, Indian Journal Dairy Science 49(8):534–536.

Singh, R.K. 2010. Impact of climate change on pattern of zoonotic disease occurance and spread in livestock. In Proc.:"National Symposium on Climate change and Livestock Productivity in India (Eds: R.C. Upadhyay, S .V. Singh, Ashuthosh, ManjuAshuthosh and Anjali Aggarwal) held at N.D.R.I. Karnal from October 7-8: 200Pp:38-46.

Tirado, M.C. and Clark, R.2010. Climate change and food safety: A review. Food research International.

Upadhyay, R.C., Singh, S.V., Kumar, A., Gupta, S.K. andAshutosh, A. 2007. Impact of Climate change on Milk production of Murrah buffaloes. Italian Journal of Animal Science 6 (2s): 1329-1332.

Wilson, J. R. 1981. The effects of water stress on herbage quality. Proceedings of the 14th International Grassland Congress, Lexington, U.S.A.

12

Advances in Nutrition and Production Management in Ruminants to Mitigate Climate Change

Dharmendra Kumar*, Asit Chakrabarti[1], Prakash Kahate[2]
Suraj P.T[3] and C. Balusami[4]

**Subject Matter Specialist,Krishi Vigyan Kendra, Kishanganj, Bihar
Agricultural University, Bihar
[1]Senior Scientist,ICAR Research Complex for Eastern Region, Patna, Bihar
[2]Assistant Professor,Dr.PDKV,Akola [3]&[4]Assistant Professor, KVASU Mannuthy
Bihar*

The greenhouse effect is thought to be due to the absorption of solar infrared (IR) radiation by gases and the earth's surfaces, which, as a result, is heated and then reemit IR radiation at low frequency with a high absorptive power. In fact greenhouse gases in the atmosphere are essential for maintaining life on earth, as without them the planet would be permanently frozen because all of the incoming heat from the sun would be radiated back into space by the earth's surface (Moss, 1993). The threshold concentration of these gases at which their greenhouse effect would be minimized is not known, but it is accepted that their concentrations in the atmosphere should not be allowed to continue to rise. In recent years, there has been an increase in public concern over environmental damage originating from animal feeding operations. The increased concentration of greenhouse gases (CO_2, CH_4, and N_2O) in the troposphere has been implicated in the consistent increase in atmospheric temperature and global warming over the last few decades (Intergovernmental Panel on Climate Change, 2001). The concentration of methane (CH_4) has increased at a rate of 10 nL/L per yr (1 nL = 10^{-9} L) since the preindustrial revolution (Moss *et al.*, 2000). Domesticated ruminants are estimated to produce about 80 Tg of CH_4

annually (1 Tg = 1 million metric tons), accounting for about 22% of CH_4 emissions from human-related activities (NRC, 2002). Methane is produced in ruminants as an unavoidable by-product of OM fermentation in the rumen, and it represents a significant energy loss to the host animal. Johnson and Johnson (1995) estimated that the energy loss as CH_4 in ruminants can range from 2 to 12% of the gross energy intake.

Methane (CH_4) has 21 times more global warming potential (GWP) than carbon dioxide (CO_2). Enteric fermentation is an important source of CH_4 emissions from anthropogenic activities, contributing nearly 22 % of the anthropogenic CH_4 emissions. Ruminant animals are unique due to their four-compartment stomachs and ability to digest otherwise indigestible, highly fibrous feedstuffs. But in this process of digestion, methane, a potential greenhouse gas is produced as a product of enteric fermentation, making cattle and other ruminants contributing factors in global warming. India has largest livestock population in the world and is estimated to emit about 10.8 Tg of CH_4 annually from enteric fermentation (Singhal et al., 2006). Of the various livestock enterprises, dairying is most popular in the country and dairy animals, which comprise of the majority of the livestock, account for nearly 60% of these enteric emissions. Crossbred cattle, indigenous cattle, buffaloes, goats and sheep and other livestock (mule, yak, camel, donkey, pig, mithun, horse and pony) emitted CH_4 about 4.6,48.5, 39, 4.7,1.8 and 1.4%, respectively (Singhal et al., 2005). Rectal emissions account for about 2 to 3 percent of the total CH_4 emissions in sheep or dairy cows. The soil sink strength for methane appears to have been reduced by changes in land use, chronic deposition of nitrogen from the atmosphere and alterations in nitrogen dynamics of agricultural soils (Mosier et al 1991). Ojima et al. (1993) estimated that the consumption of atmospheric methane by soils of temperate forest and grassland eco-systems has been reduced by 30%. Without the temperate soil sink for methane, the atmospheric concentration of methane would be increasing at about 1.5 times the current rate. Since atmospheric methane is currently increasing at a rate of about 30 to 40 million tonnes per year, stabilising global methane concentrations at current levels would require reductions in methane emissions or increased sinks for methane of approximately the same amount. This reduction represents approximately 10% of current anthropogenic emissions. The major agricultural sources of methane are flooded rice paddies, enteric fermentation and animal wastes. Decreasing methane emissions from these sources by 10 to 15% would stabilize atmospheric methane at its present level and is a realistic objective (Duxbury and Mosier, 1993). Anaerobic digestion in the forestomach of ruminants is a major source of methane emissions. The contribution of other livestock such as horses, rabbits, pigs or poultry is much less significant.

The methane in the rumen is produced by methanogenic bacteria and protozoa. The role of protozoa in methane formation is interesting. It has been established that virtually all of the bacteria attached to protozoa are methanogens and that these bacteria are responsible for between 0.25 and 0.37 of the total methane produced. By removing the protozoal population through defaunation, the ruminal bacterial population is modified, VFA production is shifted from acetate and butyrate towards propionate, and methane emissions are decreased. There is also a negative impact on fiber digestion so care must be taken not to unduly disrupt rumen metabolism by this route. The hind-gut has been reported to account for between 0.13 and 0.23 of the total emissions by sheep (Murray *et al.,* 2001). However, it appears that most (0.89) of the methane produced in the hind-gut is absorbed through the gut wall and excreted via the lungs (Murray *et al.,* 2001). In the hind gut, protozoa are absent, and methane is produced by methanogenic bacteria. Methane emissions from the hindgut are lower than from the rumen and it has been speculated that this could be due to hydrogen removal by reductive acetogenesis rather than methanogenesis. Methane contributes 15-20% of total Green House Gases. CH_4 has 21 times more Global Warming Potential than CO_2. By the year 2030 the world is likely to be 1–2 °C warmer than today and The concomitant rise in global mean sea level is 17 to 26 cm due to global warming caused by green house gas effect includes CH_4.

12.1 Factors Influencing Methane Production

Methane yield can thus be related to many and different categories of factors. However, these factors are often interrelated and so are their effects on methane yield in the rumen. This complicates the use of such factors to predict the course of fermentation in the rumen, the extent of organic matter digestion and the productive response of the ruminant. Due to these interrelationships, the contribution of a single feed component or type of carbohydrate to methane yield is not necessarily constant (Smink *et al.,* 2004), but may vary with a change of the dietary characteristics and the fermentation conditions in the rumen (Bannink *et al.,* 2005a). Changing the level of feed intake, the dietary characteristics or the fermentation conditions in the rumen affect the extent of substrate degradation by micro-organisms and the efficiency of microbial synthesis. As a consequence, amounts of microbial matter as well as undegraded feed substrate flowing out of the rumen to the small intestine change. Because of the multiple factors that may have changed simultaneously and have affected rumen fermentation and hence methane yield, the effect of nutritional measures on Volatile Fatty Acid (VFA) and methane production may be difficult to predict and interpret. The observed effect of a nutritional intervention on methane yield is therefore strongly confounded with the concomitant changes brought about in these factors. Some of the principal factors affecting rumen function and methane production discussed below.

Internal factors like Methanogens, protozoa, the composition of the microbial population within the rumen, feed residence time, the dynamics of the passage of particles, fluid and the microbial population, the inflow of saliva and the absorption capacity of the rumen wall, animal species & it's production level affects the rumen fermentation and methane production. External factors - Diet composition, Level of feed intake, the proportion of concentrates in dietary dry matter, the composition and the rate and extent of degradation of individual feed fractions (the types of carbohydrate and protein) in dietary dry matter, Forage processing, Feeding frequency, Environmental factors, the presence of unsaturated long chain fatty acids.

12.1.1 Feed intake

Changes in dry matter intake not only affect the amount of substrate available for microbial degradation, but it also changes fermentation conditions and the size of the microbial population. For example, the fate of ingested starch changes with changes in the amount of dry matter ingested, as increased intake levels will lead to a proportionally higher amount of starch digested in the small intestine rather than fermented in the rumen. Aspects which need to be considered are the storage of starch by micro-organisms with increasing concentrations in the rumen, or an altered passage rate which alters the time available for microbial degradation. Almost all models that predict methane production by ruminants require daily feed intake or a closely related variable as an input.

12.1.2 Intrinsic degradation characteristics

Microbial degradation of substrates in the rumen depends primarily on intrinsic characteristics that determine the susceptibility of the substrate to be attacked, degraded and utilized by micro-organisms. These characteristics differ between types of substrate and between types of feedstuffs that are used as dietary components. Obviously, intrinsic characteristics are important determinants of substrate degradation and utilization by microorganisms, VFA production and the concomitant methane yield. Passage rate of substrate together with their intrinsic degradation characteristics determines the fraction of a substrate that becomes degraded in the rumen or escapes to the small intestine. In present day dairy nutrition, feedstuffs are often selected to increase the quantity of starch and protein escaping rumen fermentation, hence contributing to the nutrient supply of the cow without generating VFA and methane. Hence, a higher passage rate due to a higher feed intake level as well as a less degradable substrate may both increase the escape of substrate and lead to a decrease in methane yield.

12.1.3 Type of substrate fermented and type of diet

Different types of fermented carbohydrate give different profiles of VFA production and hence methane yields. With higher levels of milk production, basal rations of dairy cows are supplemented with concentrates. Independent of the effect of fluid acidity, an analysis of VFA profiles showed about 25 and 15% lower methane yields for fermented sugars and starch, respectively, on concentrate-rich diets compared to forage-rich diets (Bannink *et al.*, 2005a).

12.1.4 Source of carbohydrate and pattern of fermentation

Because proportions of the individual VFAs are influenced by the composition of organic matter (OM) of the diet, mainly by the nature and rate of fermentation of carbohydrates, these dietary characteristics will have large effects on methane production. Diets rich in starch which favour propionate production will decrease the methane/OM ratio in the rumen. Conversely, a roughage-based diet will increase the ratio. As an example, the level of methane losses was6–7% or 2–3% of energy intake when forages were fed at maintenance or when high grain concentrates were fed *adlibitum* respectively (Johnson and Johnson, 1995). It increases when mature dried forages are fed (Sundstol, 1981) or when they are coarsely chopped rather than finely ground or pelleted (Hironaka *et al.*, 1996), and decreases when forages are preserved in ensiled form (Moss, 1993). Because they stimulate the rumen degradation of plant cell walls, alkali-treatments of poor-quality forages have been shown to increase the amount of methane emissions (Moss *et al.*, 1995).

12.1.5 Fermentation rate and fluid acidity

The acidity of rumen fluid (pH) influences rumen fermentation in two ways. Firstly, pH values lower than 6.2 appear to reduce the activity of fibrolytic micro-organisms degrading cell walls (Dijkstra *et al.*, 2005). Therefore, pH determines cell wall degradability and its contribution to microbial growth, and VFA and methane yields. Secondly, pH determines the profile of VFA produced (separate from the type of substrate and the type of diet). An increased rate of substrate fermentation, as a result of an increased feed intake or due to large concentrate meals, leads to increased rates of VFA production, to higher VFA concentrations and a more acidic rumen fluid. As a result, also the profile of VFA shifts towards a propionate oriented fermentation (Dijkstra *et al.*, 2005). This will cause a lower H_2 and lower methane yield. In an analysis of in vivo data on rumen fermentation, a decrease of the pH of the rumen fluid from 6.5 to 5.5 was estimated to lead to about 15% less methane produced from both fermented sugars and starch (Bannink *et al.*, 2005a & 2005b). Not only VFA concentrations determine rumen fluid pH, but also the buffering capacity of saliva flowing into the rumen (Bannink & Dijkstra, 2006). Buffering of rumen contents is enhanced by the inclusion in the diet of ingredients that stimulate rumination (such as straw), by feeding mixed rations, by frequent feeding (particularly of concentrates), and by preparing the rumen wall for strong increases in concentrate intake during the first weeks of lactation. If buffering processes are stimulated at the same time, an increased feed intake rate does not necessarily lead to a strong acidification of rumen fluid.

12.1.6 Residence time in the rumen and level of intake

A reduction in methane production is expected when the residence time of feed in the rumen is reduced since ruminal digestion decreases and methanogenic bacteria are less able to compete in such conditions.

12.1.6.1 Principle of Methane Abatement

- One important consequence of hydrogen utilization by the methanogens is that they maintain a low partial pressure of hydrogen in the rumen.

- If hydrogen accumulates in the rumen, re-oxidation of NADH is inhibited, reduced fermentation end-products such as lactate accumulate, and forage digestion and microbial growth are reduced.

- For digestion to proceed normally to produce acetate, propionate and butyrate as nutrients for animal production, the partial pressure of hydrogen in the rumen needs to be kept low.

- Consequently, reduction or elimination of methanogenesis would require other routes of electron transfer if the animal were to benefit.

- Otherwise hydrogen would act as an inhibitor in the fermentation process and prevent further degradation of organic matter.

- So we need an alternative sink for H_2 Redirect energy into production.

- Management of hydrogen in the rumen is the key to control ruminant methane emissions.

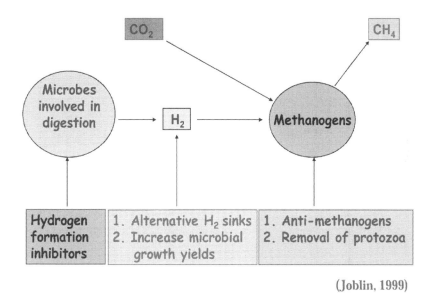

(Joblin, 1999)

Fig. 12.1: Possible intervention sites for lowering ruminant methane

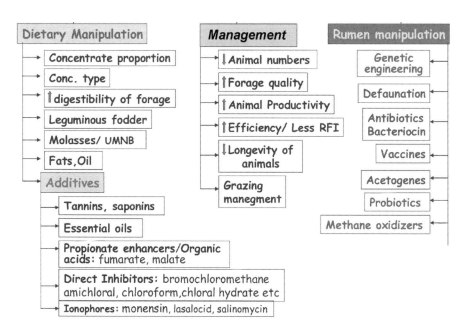

Fig. 12.2: Methane Abatement options in Ruminant

12.1.6.2 Management of Ruminants

- **Animal number:** InAustralia reduction in sheep numbers from 175 million in 1990 to a present population of 116 million has been associated with a reduction in emissions from 1170 to 776 Gg/annum. Further research may be considered in Indian condition also.

- **Animal size:** The metabolisable energy requirements of animals increase with increasing live weight, so use of animals with smaller mature live weight will mean less feed is required to maintain the herd/flock and it can be assumed, less methane emissions will be produced per head per day. Thus, improved productivity of Indigenous breed through better management practices may be suggested.

- **Feed-use efficiency:** Selection of ruminants with a high net feed efficiency offers an opportunity to reduce daily methane emissions without reducing livestock numbers or reducing product output or quality.

- Increase in Animal Productivity

Increasing animal productivity will generally reduce methane emissions per kg of product (milk or meat) because the emissions associated with maintenance are spread over a larger amount of product. However, daily emissions and thus emissions per animal per year are usually increased because the higher productivity is usually associated with higher intake. Methane production is closely related to dry matter (DM) intake. Kirchgesser *et al.* (1995) reported that increasing milk yield from 4000 to 5000 kg/year increases annual methane emissions, but will decrease emissions per kg of milk by 0.16 for a 600 kg cow (Table 12.1).

Table 12.1: Estimates of methane emissions (kg/year and per kg milk in parentheses) from dairy cows as affected by annual milk yield and body weight) (Kirchgessner *et al.,* 1995)

Body weight (kg)	Milk Yield (kg/year)*		
	4000	5000	6000
500	95 (23.8)	100 (20.0)	105 (17.5)
600	103 (25.8)	108 (21.6)	113 (18.3)
700	111 (27.8)	116 (23.2)	121 (20.2)

*310 days of lactation combined with a 55 day dry period

12.1.6.3 Effect of Longevity in Dairy Cows

An example of a 100 cow farm is presented in Figure 12.3, where the average number of lactations varies from 2.5 to 5.0. It is assumed that dairy cow emissions are 118 kg/yr while the rearing of a replacement heifer to calve at 2 years old results in methane emissions of 100 kg. Fig. 12.3 shows that total farm emissions of methane from enteric fermentation decline from 15,800 kg/yr to 13,800 kg/yr as the average number of lactations increases from 2.5 to 5.0.

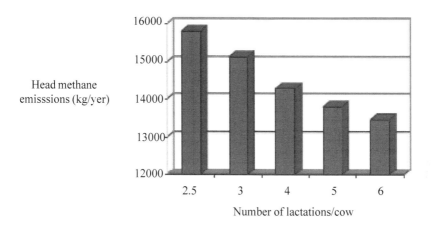

Fig.12.3: Total farm emissions of methane from enteric fermentation as the average number of lactations increases

- The option of decreasing livestock numbers in India is desirable economically also, considering the feed and fodder shortage in the country with respect to the size of the bovine population.

- May not be practically feasible strategy until there is a definite shift in the rural areas from subsistence to commercial dairy farming and the national policy on slaughtering of 'economically unviable' animals. Particularly, cows as religious and social taboos are involved with animal slaughtering in India.

- Longer stay of cow in a herd, the lower the number of replacements required, and thus the lower the total farm methane emissions.

- Any measures which reduce involuntary culling should be encouraged.

12.1.6.4 Effect of Management Practices

- Pasture management, including forage species selection, stocking rate and continuous vs. rotational grazing strategies have all been shown to influence enteric methane emissions.

- Management-intensive grazing (MIG) offers the potential for more efficient utilization of grazed forage crops via controlled rotational grazing and more efficient conversion of forage into meat and milk.

- Proper grazing management practices to improve the quality of pastures by inclusion of legumes in the forage species mixture McCaughey *et al.,* (1999) increases animal productivity and has a significant effect on reducing methane emission from fermentation in the rumen. Enhancing the level of productivity decreases the maintenance subsidy and, thus, decreases the obligatory methane emissions from fermentation of the feed associated with animal maintenance.

- Animal selection for increased production, management for improved reproductive performance, use of growth promoting agents, and application of more refined ration balancing technologies are examples of strategies that will reduce maintenance costs, and thereby methane emissions, per unit of animal product.

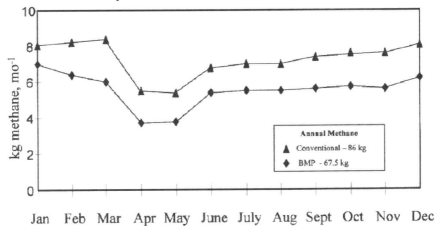

Fig. 12.4: Monthly methane emission projections in beef cows on best management practices (BMP) and conventional forage management systems

12.1.6.5 Effect of Residual Feed Intake

- Metabolizability as well as individual animal differences the methane production in rumen.

- There is a genetic link between methanogens and their hosts such that the presence of methanogenic bacteria in an animal requires a quality of the host that is under phylogenetic rather than dietary constraint (Table 12.2).

Table 12.2: Relationship of feedlot residual feed intake (RFI, kg of DM/d) with methane production (Nkrumah *et al.*, 2006).

Trait	RFI group		
	High	Medium	Low
Methane, L/kg of BW0.75	1.71c	1.68 c	1.28d
Intake energy, kcal/kg of BW0.75	384.77	382.24	387.98
Methane energy, kcal/kg BW0.75	16.08c	15.90cd	12.09d
Methane Energy loss % of GEI	4.28c	4.25c	3.19d (25%)

- Methane production was 28% and 24% less in low-RFI animals compared with high- and medium- RFI animals, respectively (Nkrumah *et al.*, 2006).

- The low RFI group have lower MPR ($P = 0.017$) and reduced methane cost of growth (by 41.2 g of methane/kg of ADG; $P = 0.09$) (Hegarty *et al.*, 2007).

12.1.6.6 Dietary Manipulation (Strategic Feeding)

- **Principles of Reduction in Enteric Methane Production**

  - Reducing H_2 & CO_2 production (Ionophores).

  - Reducing protozoa or Defaunation- prevention of interspecies H_2 transfer

  - Enhancing propionate production (Acrylate, Organic acids)

  - Rechanneling H_2 & CO_2 to acetate (Enhancing acetogenesis).

  - Reducing or killing methanogens (Vaccines).

  - Use of alternate electron sink (OA, Nitrate, Sulfate).

  - Enhancing hydrogen utilizers (Microbes).

12.1.6.7 VFA Yield, Hydrogen Balance and Methane Yield

Together with the production of VFA, and depending on the type of VFA, H_2 is either generated or utilized. With acetate and butyrate production, H_2 is produced, whereas with propionate and valerate production, H_2 is utilized. The goal in addressing livestock methane emissions is not simply stopping methane emissions, but rather, redirecting rumen hydrogen into more beneficial end products. Acetate is the main VFA resulting from rumen fermentation of fibre, starch, sugars and protein (Bannink *et al.*, 2006a), and therefore a net excess of H_2 is produced in the rumen. Although some other sinks of H_2 can be identified (microbial synthesis with NH_3 as the N source and the biohydrogenation of unsaturated fatty acids), the type of VFA produced is the major determinant of the amount of H_2 produced. The addition of compounds such as ionophores, which affect the viability of specific classes of micro-organisms (Chen *et al.*, 1979), may force a shift in the rumen fermentation patterns towards more propionate and a lower H_2 excess (Kohn & Boston, 2000).

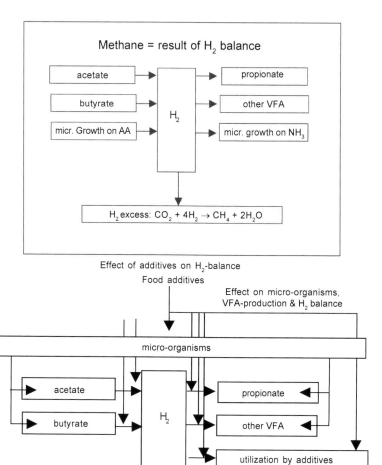

Fig. 12.5: Schematic representation of the effect of VFA profile on the H_2 excess (red arrow) and subsequent methane yield (left), and of the impact of feed additives on this H_2 excess (right). Adapted from Kebreab *et al.* (2006a)

12.1.6.8 Effect of Concentrate in the Diet

Compared to forages, concentrates are usually lower in cell wall components. Due to the presence of non-structural carbohydrates (starch and sugars), concentrates normally ferment faster than forage, giving rise to elevated levels of propionic acid. Veen (2000), suggest that methane production can be lowered by almost 40% (from 272 to 170 g/day) when a forage rich diet is replaced by a concentrate rich diet.

12.1.6.9 Concentrate Proportion in the Diet

Increasing the dietary proportion of concentrates usually reduces methane losses. This effect appeared independent of the genetic merit (Ferris *et al.,* 1999).

- Concentrates contain less structural carbohydrates than forages, and the effect of increasing the proportion of concentrates in the diet causes increase in the proportion of propionate and a decrease in the proportion of acetate (and sometimes butyrate).

- Reduce rumen pH, and as methanogens are pH sensitive, this will also tend to reduce methane emissions.

- Increased use of concentrates also increases animal performance and this will further reduce emissions (Johnson and Johnson, 1995).

- The proportion of concentrate within the diet has been reported to be negatively correlated with methane emissions (Yan *et al.*, 2000).

Limitations

- Required minimum level of physical structure in the diet (to prevent subclinical acidosis)

- The balance energy intake (prevention of overfeeding) in low producing animals (dry and late lactation cows, young stock).

- Feeding large amounts of concentrates is sometimes associated with a higher risk for lameness (Manson & Leaver, 1988).

- High concentrate intake could also result in a high BCS (fat cows) at calving.

- A higher risk for metabolic disorders, and reduced fertility

Table 12.3: The effect of concentrate level on methane emissions and their relationship to animal productivity(Lovett *et al.*, 2005)

Item	Low concentrate (0.87 kg on DM basis)	High concentrate (5.24 kg on DM basis)
Methane (g/d)	346	399
Methane(g/kg of DMI)	19.60	17.83 (9%)
Methane(g/kg of milk)	21.0	17.7 (15%)
Methane (g/kg of FCM)	19.26	16.02
Methane (g/kg of milk protein)	555	509
Methane (g/kg of milk fat)	525	428

Table 12. 4: Methane production by lactating buffaloes under different feeding regimes (Singhal *et al.,* 2006).

Parameters	Berseem + Wheat straw	Berseem+ wheat straw + Concentrate	Wheat straw + Concentrate
Total DM intake (kg)	14.68	9.90	9.28
Milk yield(kg/d/animal)	7.01	7.42	7.61
Methane production			
Total production (g/d)	259.74[a]	162.67[b]	177.03[b]
g /kg DMI	17.76[a]	16.52[ac]	19.13[ab]
g /kg DDMI	22.31[b]	22.08[b]	30.20[a]
g /kg milk yield	38.16[a]	22.12[b]	23.45[b]

12.1.6.10 *Effect of Concentrate Type*

Johnson & Johnson, (1995) was shown that digested cell walls normally lead to higher losses than non cell wall components, and that within non cell wall components soluble sugars are more methanogenic than starch. Moe and Tyrrell (1979) reported that for every gram of cellulose digested, methane emissions are nearly three times that of hemicellulose and five times thatof the soluble residue. However, there has been little work to compare methane production on different concentrates. This could be of interest as there is a large selection of concentrate ingredients available, ranging from cereals (low in fibre, high in starch) to cereal-by-products (high in fibre, low in starch), pulps (high fibre), molasses (high sugar), oilseed meals (high in protein, variable in fibre), etc. Johnson and Johnson (1995) noted that soluble sugars have a higher methanogenic potential than starch. Research is required to establish if concentrates can be formulated to bring about significant reductions in methane production. In feedlot cattle fed with barley or corn based diets in the finishing phase methane losses amounted to 2.8 and 4.0% of GEI for corn and barley, respectively. The low proportion of GEI lost as methane was probably caused by a propionate type of rumen fermentation, also resulted from a larger proportion of starch in corn, escaping digestion in the rumen(Beauchemin & McGinn, 2005).

Conclusion

Feeding more concentrates per cow, especially those with a higher amount of (rumen resistant) starch and less sugars has a very positive effect on the reduction of methane losses.

12.2 Effect of Forage Quality

- Increasing the digestibility of cell walls in forages has also been suggested as a means to lower methane losses. At high intake levels, the proportion of energy lost as methane decreases as the digestibility of the diet decreases (Johnson and Johnson, 1995).

- Increasing the digestibility of pasture for grazing ruminants has been proposed as the most practical means of reducing their methane emissions and improve productivity (Hegarty, 1999a).

- Forage maturity and physical form will further influence fermentation and thus methaneproduction; emissions, % GEI being higher for mature forage vs. immature forage, coarse chopped vs. finely ground or pelleted low quality forage, hay vs. silage.

- Legislation aiming at a reduction of the use of N fertilizer will probably result in forages harvested at a slightly higher maturity which would also not be in favour of reducing methane losses.

Table 12.5: Effect of forage organic matter digestibility on enteric methane emission (Boadi *et al.,* 2004)

Parameter	Forage Quality		
	High	Medium	Low
IVOMD, %	61.5	50.7	38.5
Ad-libitum			
DMI, kg/d	9.7[a]	8.9[a]	6.3[c]
Methane, L/d	281.7[a]	289.8[a]	203.5[b]
Methane, %GEI	6.0	7.1	6.9
Restricted Intake (2% BW)			
DMI, kg/d	6.4	6.1	6.1
Methane, L/d	224.6	193.3	195.6
Methane, %GEI	7.6	7.1	7.1

12.2.1 Effect of forage type

Forage legumes

- Legumes generally have higher intakes and digestibility than grass swards and thus give rise to higher productivity, thus reduce methane emissions.

- Feeding forage legumes like lucerne or red clover also tends to decrease methane losses (g/kg DMI) compared to grass (Ramirez-Restropo & Barry, 2005).

- McCaughey *et al.* (1999) speculated that the reduced emissionscould result from a modified ruminal fermentation pattern combined with higherpassage rates.

- Legumes that contain condensed tannins such as sulla (*Hedysarum coronarium*) also reduce methane loss.

- Leguminous fodder which is rich in N_2 content utilize the available H_2 to form ammonia thus reduce the availability of H_2 for methane production.

Fresh grass

- Methane production in ruminants tends to decrease with the quality of the forage fed. Quality of forages depends predominantly on maturity and less mature forage often has a higher N and lower sugar content.

- Average DCP/DOM ratios in the grass were 0.238, and 0.176 with methane losses (as % of GEI) of 5.8, and 7.3% respectively (Bruinenberg *et al.*, 2002).

Grass silage

- Methane losses in animals fed grass silage are likely to be higher than in animals fed fresh grass.

- Grass silage is usually harvested at a later stage of maturity that results in a lower DOM and a lower N content, lower sugar content and a fraction of lactate as a result of the process of ensiling. Besides, the fat content of grass silage is lower and less unsaturated than that of fresh grass.

Whole Cereal Plant silage (WPS)

Whole Cereal plants having grains could yield starch and consequently reduce methane losses.

Maize silage

- Because of its high starch content, maize silage is expected to result in lower methane losses than grass silage, despite of its low cell wall digestibility.

- Replacing half of the 60% of grass silage in the DM of a control ration with maize silage fed to mid-lactation dairy cattle in the Wageningen respiration chambers reduced methane production (non significantly) from 6.0 to 5.8% of GEI (13.6 and 12.1 g methane/kg milk respectively) (Van Laar and Van Straalen, 2004).

- Harvesting maize silage at a more mature stage reduced rumen degradability of starch may result in a shift of starch digestion from the rumen to the intestine and thereby contribute to a reduction of methane losses.

12.2.2 Effect of dietary oil supplementation

- Unsaponified or otherwise unprotected fats or oils may have a general suppressing effect on rumen fermentation when included at 7% or above in most diets.

- The unsaturated fatty acids are a potential alternative sink for hydrogen because of Bio-hydrogenation of unsaturated fatty acid.

- It also toxic to methanogens.

- Enhances propionic acid production.

- Defaunation or removal of protozoa from the rumen is one method which could reduce methane emissions includes the addition of certain oils/fats (Machmüller and Kreuzer, 1999).

  - In the absence of protozoa, rumen methane output is reduced by 0.13 on average, although this varies with diet. (Hegarty, 1999a).

  - The magnitude of reduction in methane output following dietary supplementation of fats\oils is source dependent, with coconut oil identified as being very effective.

  - More recently, comparison of the efficacy of a range of oils has identified coconut oil as being most effective (Machmuller and Kreuzer 1999). It is the medium chain fatty acids ($C10 - C21$), which cause the greatest reduction in methane production (Dohme et al., 2000) and also the methanogen population (Dong et al., 1997).

  - Rumen protected Fat (Bypass Fat) has very less effect on methane production.

Limitation

- High oil prices may make the inclusion of vegetable oils in dairy diets less competitive.

- Decreases DMI and Fiber digestion (even at 5% inclusion level).

In beef cattle, the addition of sunflower oil (400 g./d or 5% of DMI) decreased methane emissions by 22% with no negative effect on DM intake, but reductions in DM and NDF digestibility were 9% (from 62.0 to 58.2) and 23% (from 44.3 to 34.1), respectively (McGinn et al., 2004).

Table 12.6: Effect of Sunflower oil in methane emissions

Item	Treatment			
	Control	Enzyme (1mL/kg DM)	Monensin (33mg/kg DM)	Sunflower oil (5% of DMI)
DMI, kg/d	7.40	7.55	7.71	6.91
Methaneg/steer	166.2b	164.4b	159.6b	129.0c (22%)
g/kg of DMI	22.64b	22.11b	20.70b,e	18.81c
% GE intake	6.47b	6.32b (3%)	5.91b,e(9%)	5.08c (21%)
% DE intake	10.51bc	11.27b	9.31cd,e	8.76d

The addition of coconut oil (250 g/d) to a 50/50 grass silage to concentrate ratio diet of beef cattle in Ireland either as refined oil or as copra meal decreased methane loss by between 15 and 20% when expressed in L/d, per kg DMI or as % of GEI, without showing negative effects on DMI or digestibilities of DM or NDF (Jordan et al., 2006).

Table 12.7: Effect of refined oil or as copra meal on methane production from beef cattle.

Item	Treatment		
	Control	Refined coconut oil	Copra meal
		(250 g/d)	(250 g/d)
DMI, kg	8.67	8.81	8.66
Methane, L/d	334.4	271.6 (19%)	284.6 (15%)
Methane,kg/DMI	38.8	31.1	33.2
Methane, % of GEI	7.9	6.1 (23%)	6.7 (15%)
Methane, L/kg of ADG	243.7	168.2	192.7
Protozoa, 106/mL	3.2	1.2	1.1

- Canola oil (6% in DM) in the diet of Angus heifers fed diets of barley silage (75%) and concentrates (80% barley grain) reduced methane emissions by 32%, primarily due to a decreased DMI together with a lowered total tract digestibility of DM and fibre (Beauchemin & McGinn, 2006). In short term trials (14 days), adding mixtures of sunflower and fish oil (500 g/ d) to the diet of dairy cows fed pasture based diets, oils had no effect on DMI or milk yield, but reduced methane by 27% (13.5 vs 18.5 g methane kg/ DM). However, in a long term trial (12 weeks) the addition of oil (300 g linseed and fish oil) to the diet of grazing cows had no effect on methane emissions (21.7 vs. 23.0 g methane kg/DM) reported by Woodward et al. (2006).

- Whole Soybean seed feeding having less effect on methane emission than soyoil.

Conclusion: Results of adding fats to diets of cattle on methane emission are variable and influenced bythe type of FA (chain length, degree of unsaturation), type of diet (forage vs. concentrate rich) and length of the experimental period.

12.2.3 Use of molasses/ UMB

- It is the most cost effective measure to reduce methane emission.
- Potential to reduce methane emissions by 25 to 27%.
- In India, methane reduction could be possible from 10 to 15%.
- Increase milk production at the same time.
- Chemical upgrading of poor quality roughage is also possible.

12.2.4 Use of TMR/ Balanced ration

- Total mixed rations (TMR) for dairy cattle are formulated to contain levels of energy, protein and minerals that are required for the desired levels of animal production.
- Use of a nutritionally balanced ration is important to maximize efficiency of nutrient utilization, thereby reducing environmental pollution caused by excess nutrients leaving the animal as waste.
- Environmental pollution from dairy farms can be caused by overfeeding and/or poor synchronization of release of nutrients in the rumen.
- Increasing efficiency of utilization of nutrients and energy is one strategy to increase animal production per unit of feed consumed, thereby reducing environmental impact.

12.2.5 Feed additives

12.2.5.1 Plant Secondary Metabolites

i). Tannin

1. Indirectly through a reduction in fiber digestion and subsequently reduce H_2 production.
2. Directly through an inhibition of the growth of methanogen (Tavendale *et al.*, 2005).

- Methane losses (g methane/kg DMI) were reduced by between 20 and

55% in grazing dairy cows in New Zealand grazing sulla (*Hedysarum coronarium*) or lotus (*Lotus corniculatus*) as compared to animals pastured on ryegrass/white clover mixtures.

- Cichory (*Chichorium intybus*) also a promising forage to reduce methane losses in ruminants (Ramirez-Restropo & Barry, 2005).

- Goats, fed with condensed tannin containing forage Sericea lespedeza (*Lespedeza cuneata*), observed in Oklahoma (USA) a reduction in methane loss of over 30% compared with goats fed with a mixture of crabgrass (*Digitaria ischaemum*) and tall fescue (*Festuca arundinacea*) (Puchala *et al.*, 2005).

- Tannins will inevitably be anti-nutritional when dietary CP concentrations are limiting production because they reduce absorption of Amino acids (Waghorn, 2008).

12.2.5.2 Saponin

- Saponin containing plants and their extracts have been shown to suppress the bacteriolytic activity of rumen ciliate protozoa (Moss *et al.*, 2000). Saponins are considered to have detrimental effects on protozoa through their binding with sterols present on the protozoal surface (Francis *et al.*, 2002).Because of their anti protozoal activity; saponins might have potential to reduce methane (Pen *et al.*, 2006).

- A potential methane-suppressing effect of specific tropical plants could result from direct effects on the methane-forming microbes or from indirect effects on the ruminal protozoa population which are, besides the cellulolytic bacteria, the major hydrogen suppliers to the methanogens (Ushida and Jouany, 1996).

Researchable issue: Investigation of grasses rich in Tannin and saponin.

12.2.5.3 Direct Inhibitors: (Analogues)

- Many halogenated methane analogues such as chloroform, Trichloroacetamide and trichloroethyl adipate (Clapperton, 1977), Bromochloromethane, carbon tetrachloride, chloral hydrate and bromo-ethanesulphonic acid can be very potent methane inhibitors.

- While some of these compounds are volatile and difficult to administer, McCrabb *et al.* (1997) claimed success in inhibiting methane in cattle with bromochloromethane complexed with a-cyclodextrin, which reduced volatility.

- 2-bromoethanesulfonic acid (BES), a bromine analogue of coenzyme involved in methyl group transfer during methanogenesis, is a potent methane inhibitor. BES is a specific inhibitor of methanogens and does not appear to inhibit the growth of other bacteria (Sparling and Daniels, 1987).

- 9,10-anthraquinone- inhibited the reduction of methyl co-enzyme M to methane by uncoupling electron transfer in methanogenic bacteria (Kung *et al*, 1998*)*.

- Halogenated methane analogues have potential as methane inhibitors, provided that problems such as adaptation by rumen microbes, host toxicity and suppression of digestion can be overcome. Most of these inhibitors have been found to suffer reduced efficacy after sustained or repeated administration.

12.2.5.4 Use of Ionophores

- Ionophores are highly lipophilic compounds which are able to shield and delocalize the charge of ions and facilitate their movement across membranes and toxic to many bacteria, protozoa, and fungi.

- Ionophoric antibiotics such as monensin, rumensin, lasalocyde, salinomycin etc. depress methane production by mixed rumen microbes in vitro (Van Nevel, 1992).

- Methanogenesis is not due to a direct effect of the ionophores on methanogenic bacteria but rather results from a shift in bacterial population from gram positive to gram negative organisms with a concurrent shift in the fermentation from acetate to propionate (Chen and Wolin, 1979 and Newbold *et al.,* 1988).

12.2.5.5 Monensin

- Monensin reduces methane production from livestock were attributed to:

  1. Reduction in voluntary feed intake causing reduced fermentation.

  2. Selectively reducing acetate (and therefore H_2) production.

  3. Inhibiting the release of H_2 from formate.

- The methane suppressing effect of monensin is typically not maintained for long periods. Its action has been shown to continue for 35 days in sheep.Cattle studies has shown loss of methane suppressing activity with prolonged (Johnson and Johnson *et al.,* 1995 and McCaughey *et al.,* 1997) or repeated application.

- Short-term *in vivo* trials suggest that the use of monensin can depress methane production by 25 %.

- Monensin supplementation (250 mg /d) in yearling heifers resulted in a 15 % reduction in emissions when heifers were on an alfalfa diet for 28 days. Thereafter, heifers were moved to a smooth bromegrass pasture for a further 16 days. While on pasture, no differences were observed for emissions between control and monensin supplemented heifers (Johnson *et al.*, 1997)

- In a lactating dairy herd, where whole barn emissions declined in the initial month after inclusion of monensin in the lactation ration, with emissions returning to previous levels thereafter.

Table 12.8. Effect of length of time fed ionophore on methane production of steers fed a high grain basal diet (Rumpler *et al.*, 1986)

Ionophoregroup	Days after beginning on ionophore		
	2-3	12-13	22-23
	methane production, l/d		
No ionophore	113a	107a	124a
Monensin	96a	123b	124b
Lasalocid	100a	115a	117a
Avg	103a	115b	120b

12.2.6 Hydrogen acceptors

While reduction of sulfate and nitrate are both thermodynamically more favourable than is the reduction of CO_2 to methane (McAllister *et al.*, 1996), the toxicity of their reduced forms prevent these being practical means of causing a substantial reduction in rumen methane emissions. Stoichiometric calculations show that reducing methane emissions of a sheep by 50% would require ingestion of 0.75 moles of sulphate or nitrate per day. Since sulphate and nitrate are toxic to sheep at approximately 0.1 moles/d and 0.25 moles/d respectively, they cannot safely be fed at levels appropriate to significantly reduce methane emissions (Hegarty 1999b). Leng (1991) also emphasized the critical importance of gradual adaptation of the animal to nitrate and that low-protein diets are the natural background for successful utilization of nitrates as a methane mitigating tool. The rumen ecosystem clearly has to adapt to dietary nitrates and acquire the ability to reduce nitrates rapidly to NH_3. Urea-molasses multinutrient blocks are designed to provide urea, mixtures of minerals, and in some cases slowly degradable protein to animals in rangeland conditions (Sansoucy *et al.*, 1988). Nitrate reducing activities of ruminal fluid of sheep acclimatized to nitrate at 2.5 g/kg BW/day dropped to their initial levels within three weeks after the

KNO$_3$ supplement was withdrawn.Some forages can have nitrate levels as high as 2.6 (corn silage) to 2.9 percent (green-chop sudangrass). Low total dietary N in the basal diet may be an important condition for a successful nitrate application so that enteric methane mitigation is not offset by increased N$_2$O emissions from application of manure in soil or a potential increase in rumen N$_2$O formation.

- Methane is formed as a result of the need to remove hydrogen from the rumen Propionate formation also utilizes hydrogen. Precursors of propionate are added to the diet to reduce methane production by removing some of the hydrogen (acting as a H$_2$ sink) produced during ruminal fermentation.

- The organic acids such as malate, fumarate, citrate, succinate, acrylate etc are propionate precursors.

- It has been demonstrated both in vitro (Newbold *et al.*, 2002) and in vivo (Newbold *et al.*, 2005) that addition of organic acids to the diet reduces methane production, with the response being dose dependent.

- Dicarboxylic organic acidssuch asmalatei) Alter rumen fermentation in a manner similar to ionophores (Martin, 1998).ii) Converted to propionate via fumarate, also stimulated propionate production and inhibited methanogenesis *in vitro*.

- Fumarate- a precursor of propionate added to rumen simulating fermentors, propionate production increased with a decrease in methane production (Lopez *et al.*, 1999).

- Acrylate- an alternative precursor of propionate, depressed methane production in rumen simulating fermentors (Ouda *et al.*, 1999).

Table 12.9: Source of Natural Organic Acids

• DL-Malate	• Fumarate
• Alfalfa	• Quince fruit
• Bermuda grass	• *Artabotrys hexapetulas*
• Barley grain	• *Sarcandra glabra*
• Alfalfa pellet	• Apple pomace & (juice)
• Dehydrated maize + alfalfa mixture	• Can be produced from rice bran by
• mustard leaves,	selective fermentation using fungus.
• *Ficus indica*,	

- In a short-term batch cultures, testing 15 potential precursors of propionate, including pyruvate, lactate, fumarate, acrylate, malate and citrate, it is found that Sodium acrylate and sodium fumarate produced the most consistent effect decreasing methane production by between 8 and 17%.

- Free acids rather than salts were more effective in reducing methane, but also decrease pH with possible negative effects on fibre degradation. In longer term (21 d) in vitro incubations, fumarate addition decreased methane production by 28% while maintaining DM degradation, where as malate was not effective (Newbold *et al.*, 2005).

12.2.7 Essential Oils

- Essential oils are present in many plants and may play a protective role against bacterial, fungal, or insect attack.

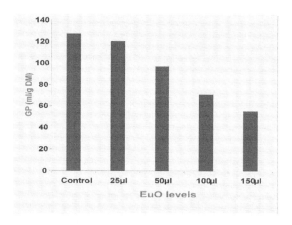

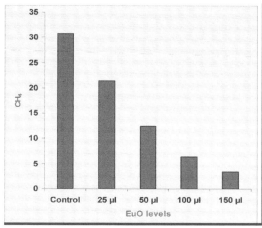

Fig. 12.6: Effect of different levels of EuO (ml/g DM)and methane production (ml/g DOM) *in vitro* for 24 h incubation

- There is an increasing interest in exploiting natural products as feed additives to manipulate enteric fermentation and possibly reduce methane emissions from livestock production.

- Essential oils are a group of plant secondary compounds that hold promise as natural additives for ruminants.

- The antimicrobial activity of essential oils can be attributed to a number of small terpenoids and phenolic compounds, e.g monoterpenes, limonene, thymol, carvacrol. The specific mode of action of essential oil constituents remains poorly characterized or understood.

- The Eucalyptus Oil appears to have a potential to mitigate the methane emission in ruminant *in vitro*.

- Some of the plant extracts have the potential to affect rumen fermentation efficiency, and black seed extract could be a promising methane mitigating agent.

12.2.8 Rumen manipulation

12.2.8.1 Methane Oxidisers

- Methane oxidising bacteria have been isolated from a wide range of environments, including the rumen (Hanson, 1992).

- Studies with 13methane tracers suggest that oxidation of methane to CO_2 is of little quantitative importance in the rumen but may be more important in the gut of pigs.

- Methane oxidising bacteria from the gut of young pigs which decreased methane accumulation when added to rumen fluid in vitro (Valdes *et al.*, 1997).

12.2.8.2 Oxidation of Methane in the Rumen

Oxidation of methane is a normal aerobic reaction in soils (Schnell and King 1995),Small but significant decreases in methane production have also being achieved by addition of an exogenous aerobic methane oxidiser (*Brevibacillus parabrevis*) to short and long-term incubations of rumen fluid (Nelson *et al.*, 2000), while the added organisms did not reduce the methanogen population.

12.2.8.3 Stimulation of Acetogens

Besides methanogenesis, H_2 and CO_2 can be converted to acetate by acetogens, which are also present in the rumen environment. Based on thermodynamics, Kohn & Boston (2000) argued that under normal fermentation conditions in the rumen, methanogenesis is feasible (reduction potential below -0.3 Volt), but for acetogenesis to occur more reducing conditions are required. This means that acetogenesis does not play an important role in the rumen hydrogen balance and the quantities of methane produced. An alternative strategy to reduce ruminal methanogenesis would be to re-channel substrates for methane production into alternative products. Acetogenic bacteria, in the hindgut of mammals and termites, produce acetic acid by the reduction of carbon dioxide with hydrogen and reductive acetogensis acts as importanthydrogen sink in hindgut fermentation (Demeyer and Graeve, 1991).Bacteria carrying out reductive acetogenesis have been isolated from the rumen (Morvan *et al.,* 1994).Acetogens depress methane production when added to rumen fluid in vitro and even if a stable population of acetogens could not be established in the rumen it might be possible to achieve the same metabolic activity using the acetogens as a daily fed feed additive(Lopez *et al.,* 1999). Over 10 acetogenic bacteria have been isolated from the rumen (Joblin, 1999). The affinity of methanogens for hydrogen is 10 to 100 times higher than the affinity of the reductive acetogens, so the acetogens cannot compete with the methanogens in the rumen because the partial pressure of hydrogen is normally too low. Acetogens do not compete well in the rumen compared to methanogens, so experiments are in progress to see if the microbial ecosystem can be manipulated to enhance acetogen activity. Hence after removal of metanogenic bacteria acetogenic bacteria may act. One strategy is to genetically modify acetogens so that they can compete more effectively in the rumen.

12.2.8.4 Defaunation

- Methanogens associated with ciliate protozoa were responsible for between 9 and 25% of the methanogenesis in rumen fluid and the removal of protozoa from the rumen has been associated with decreases in methane production (Newbold *et al.,* 1995 and Ushida *et al.,* 1997).

- Plant secondary metabolites - used as possible defaunating agents. Saponin-containing plants show promise as a possible means of suppressing or eliminating protozoa in the rumen without inhibiting bacterial activity. T he inclusion of fat in ruminant diets depresses protozoal numbers and the use of lipids as a defaunating agent has been suggested (Czerkawski *et al.,* 1995). Lipids have been shown to inhibit methanogenesis even in the absence of rumen protozoa, possibly due to the toxicity of long chain fatty acids to methanogenic bacteria (Dohme *et al.,* 1999).

- Defaunating agents such as manoxol, teric, alkanate 3SL3 and sulphosuccinate can reduce methane emission. They appear to act by disrupting the close symbiotic relationship between methanogenic bacteria and protozoa. The toxicity of many of these defaunating agents restricts their routine use.

12.2.9 Probiotics

- *Aspergillus oryzae* (AO) - reduce methane by 50% which was directly related to a reduction in the protozoal population (45%) (Frumholtz *et al.*, 1989).

- Addition of *Saccharomyces cerevisiae* (SC) to an in vitro system reduced the methane production by 10% initially (Mutsvangwa *et al.*, 1992).

- It would appear that more research is needed to evaluate whether probiotics have any role in methane mitigation strategy, although it seems unlikely that they would be effective with animals grazing pasture.

- **Mode of action of probiotics**
  - Reduce H_2 availability to methanogens.
  - Increased butyrate or propionate production(Martin *et al.*, 1989).
  - Reduced protozoal numbers (Newbold *et al.*, 1998).
  - Promotion of acetogenesis is a sink for hydrogen(Chaucheyras *et al.*, 1995).

12.2.9.1 Immunisation

- It may be possible to immunise ruminants against their own methanogens with associated decreases in methane output (Baker, 1999).

- A vaccine would be a valuable tool in providing a cost-effective and long-acting treatment to reduce methane emission and enhance animal production under grazing.

12.2.9.2 Bacteriocins

- Bacteriocins are antibiotics, generally protein or peptide in nature, produced by bacteria.

- Research is ongoing to see if these compounds can be used to manipulate the rumen ecosystem as a biological control.

- The bacteriocin nisin, which is produced by *Lactococcus lactis*, was used to produce a 36% reduction of methane production *in vitro* and stimulate propionate production (Callaway *et al.*, 1997).

- Many lactic-acid bacteria produce bacteriocins, it may well be that part of the reduced methane production observed at very low pH is due to bacteriocin effects on methanogens rather than a direct pH effect (Russell, 1998). Half of the 50 strains of *Butyrivibrio* produced inhibitory (presumed bacteriocin) activity.

- Further research is required to establish their adaptability and long term effectiveness as a feed additive for methane supressor.

12.2.9.3 Evaluation of Mitigation Strategies in Indian Context

- Among the various options for reducing methane emission discussed above, its mitigation through changes in rumen microflora is not commercially available technologies for dairy animals at present. Thus, the options of improving better animal nutrition which can be applied in the field condition with appropriate technology to mitigate climate change scenario.

12.2.9.4 Strategic Supplementation Using Molasses-Urea Products (MUP)

- This technology is applicable to all the types of dairy and non-dairy animals who are on poor diet. Hence, the cost of technology is worked out for indigenous cows and buffaloes. The assumptions used to work out the cost of methane mitigation from this option are:

(a) The average quantity of the product consumed by an adult animal is 400 g/day and by heifer 300 g/day.

(b) The methane reduction per animal from MUP supplementation is taken as 11%.

(c) Milk yield of animals is assumed to increase by 10%.

(d) The cost of producing one kilogram of the product is taken as Rs 20.00.

12.2.9.5 Dietary Manipulation through Increasing Concentrate Feeding

In India on an average less than 500 grams of concentrate was fed to dairy animals per day.The existing proportion of 7.5% concentrate is not sufficient to cater to the recommended nutritional requirement of 40% concentrate and 60% roughage on dry mater basis for the Indian cattle. For high milk producing dairy animals, the concentrate to roughage ratio is still higher i.e. 50:50.

12.2.9.6 Side-Effects of Attempt to Reduce Methane

12.2.9.6.1 Impact of Dietary Manipulation

Detrimental effects on cow performance specifically mentioned were the risk of (subclinical) rumen acidosis with feeding more concentrates.

12.2.9.6.2 Milk Lactose

Reduction of methane losses by dietary means is not expected to affect milk lactose to a great extent.

12.2.9.6.3 Milk Protein

Nutrition has very little effect on the relative appearance of various protein fractions in the milk.

12.2.9.6.4 Protein Level and Source

Methane production from fermented protein is lower than that from fermented carbohydrates (Bannink *et al.*, 2005a). Increasing the level of dietary crude protein has no consistent or only a small positive effect on milk protein concentration (Beever *et al.*, 2001). Thus methane reduction by increasing protein concentration is expected to give no, or only minor positive, responses in milk protein.

12.2.9.6.5 Energy Level and Source

Milk protein concentration is positively correlated with diet ME concentration except when this energy is provided by lipids (DePeters & Cant, 1992). Increase in ME level is often caused by changed roughage: concentrate ratio. Increasing starch content of the diet to reduce methane may increase milk protein and milk protein production with higher insulin levels in plasma, leading to signals in the body of the cow to produce more milk protein (Jenkins & McGuire, 2006).

12.2.9.6.6 Fat Level and Source

Feeding additional fat often causes a small reduction in milk protein content (Jenkins & McGuire, 2006). The use of unsaturated FA to decrease methane may reduce milk protein content, while lauric and myristic acid will have no or a less pronounced effect on milk protein content.

12.2.9.6.7 Additives and Plant Secondary Metabolites

Yeast supplementation does not affect milk protein concentration (Erasmus *et al.*, 2005). Condensed tannins may lead to a slightly increased milk protein content (Bhatta *et al.*, 2000 and Woodward *et al.*, 2000).

12.2.9.6.8 Milk Fat

The extensive biohydrogenation of FA in the rumen, in which hydrogen is used thereby potentially reduce methane production in the rumen, prevents less proportions of unsaturated FA and more saturated FA in milk.

12.2.9.6.9 Energy Level and Source

Increasing concentrate proportions to reduce methane emissions may reduce milk fatty acid content and increase the proportion of various unsaturated fatty acids. Using more maize silage at the expense of grass silage to reduce methane, however, may increase the proportion of saturated fatty acids in milk.

12.2.9.6.10 Fat Level and Source

Plant oils (e.g., linseed oil, rapeseed oil, sunflower oil) in the diet generally results in a reduction in the proportion of shorter fatty acids (C16 or less) and a usually small increase in unsaturated fatty acid proportions (Chilliard *et al.*, 2000). The use of lauric or myristic acid to reduce methane production has adverse effects on milk fat composition, since addition of these fatty acids to the di*et al*so increased their proportions in milk fat (Dohme *et al.*, 2004). The use of unsaturated fatty acids to reduce methane production, in particular omega-3 C20 or C22 fatty acids, generally shifts the milk fat profile to a more desired profile, whereas the use of lauric or myristic acid does not give a favourable fatty acid profile shift.

12.2.10 Animal health

- In early lactation, feeding more glucogenic nutrients due to more starchy concentrates, to reduce the emission of methane, result in an energy balance and less negative conditions and improved fertility (Van Knegsel *et al.*, 2005).

- Feeding lipogenic nutrients (including fats and oils) had ambiguous results on energy balance and caused an undesired increase in plasma NEFA and BHBA, related to increased fatty liver syndrome.

- Inclusion of rumen protected long chain fatty acids (PUFA) in the diet of dairy cows in early lactation shows positive effects on fertility (Staples & Thatcher, 2001).

12.2.11 Other greenhouse gases and ammonia

- Having more milk per cow not only reduced the emission of methane, but also that of N_2O and NH_3 according to Schils *et al.*, (2005).

- Reduction of methane losses is associated with a reduced cell wall digestion; one may also expect elevated hindgut fermentation with a concomitant increased faecal output of cell walls as well as microbial protein. The resulting shift in N excretion from urine to faeces could even further reduce the release of N_2O.

- Reduction in methane losses is brought about by feeding more concentrates or more starch, associated by an elevated loss of CO_2, because of the use of more fossil energy (Tamminga, 1996).

12.2.11.1 Practical Implementation of Dietary Measures to Reduce Methane Production

Feeding of Balanced Ration: Generally animals in Underfeeding or in overfeeding due to lack of proper scientific knowledge.

12.2.12 Strategies for Minimize Methane Production

- Increase the proportion of maize silage in the diet.

- Harvestmature maize fodder to improve the proportion of starch and by-pass starch in maize silage.

- Increase the proportion of by-pass starch in concentrates.

- The addition of fats and oils

- Provide forage with secondary plant metabolites (e.g. tannins).

- All dietary measures mentioned above can be applied in mid-lactation dairy cows.

- However, a side effect of these measures is an increased energy intake.

- These dietary measures should be applied with care because of the risk of over feeding.

12.2.13 Conclusion

After reviewing the research findings it may be concluded that:

12.2.13.1 On-farm practices will reduce enteric methane and total GHGs emissions by reducing feed costs associated with animal maintenance:

- Increase productivity per cow to reduce methane emission per kg of milk.

- Based on methane emission in relation to milk production a correlation could be established as follows:

- When body weight is constant or same but milk production is higher the productivity of methane is low but if milk production is less the methane production is more.

- When total output levels (e.g. total milk or beef production) remain constant and livestock numbers are reduced total methane production is also reduced.

- Early slaughter age reduces total methane production.

- Best management practices and pasture management also lower methane production.

- Selection of Low RFI animals that achieve similar growth rate and body weight can inhibit total methane production.

12.2.13.2 By Dietary Manipulation Through

- Increasing proportion of concentrate rich in starch,
- High quality leguminous forage,
- Grinding and pelleting of forages,
- Inclusion fats and oils,
- Total mixed ration (TMR) and Urea Molasses Mineral Block(UMMB),
- Organic acids (fumarate, malate)
- Tannin and saponin rich plants,
- Ionophores, like monensin, lasalocid, salinomycin
- Direct inhibition of methanogensis by using halogenated methane analogues (BES, AQ) is not found suitable.

12.2.13.3 Through Indirect Way of Altering Rumen Ecology

- Immunization,
- Elimination of ciliate protozoa,
- Enhanced bacteriocin production,
- Enhanced acetogenesis
- Probiotics or Yeast cultures

Biotechnology can play an important role by manipulating the rumen microbes to enhance the digestibility of poor quality feed stuff and lower or halt the production of methane.

References

Bannink, A. and J. Dijkstra, 2006. Voorspelling van de zuurgraad van pensvloeistof. ASG report 12. ASG, Lelystad.

Bannink, A., Dijkstra, J., Mills, J. A. N., Kebreab, E. and J. France, 2005a. Nutritional strategies to reduce enteric methane formation in dairy cows.In: Emissions from European Agriculture, pp. 367-376 [T. Kuczynski, U. Dämmgen, J. Webb and A. Myczko, editors]. Wageningen Academic Publishers, Wageningen, the Netherlands.

Bannink, A., Dijkstra, J., Mills, J.A.N, Kebreab, E. and J. France, 2005b. A dynamic approach for evaluating farm-specific as well as general policies to mitigate methane emissions by dairy cows.Proc. 4th Int. Symp.non-CO_2 Greenhouse Gases (NCGG-4). Science, Control, Policy & Implementation. A. Van Amstel, Coordinator. Millpress (www.millpress.nl), Utrecht, NL.

Bannink, A., Kogut, J., Dijkstra, J., Kebreab, E., France, J., Tamminga, S. & A.M. Van Vuuren, 2006a. Estimation of the stoichiometry of volatile fatty acid production in the rumen of lactating cows. J. Theoret. Biol., 238, 36-51. on Ruminant Physiology, Ferdinand Enke Verlag, Stuttgart, 1995, pp. 333–348.

Beauchemin, K.A. & S.M. McGinn, 2005. Methane emissions from feedlot cattle fed barley or corn diets. J. Anim. Sci.,83: 653-661.

Beauchemin, K.A. & S.M. McGinn., 2006. Methane emissions from beef cattle: Effects of fumaric acid, essential oils, and canola oil. J. Anim. Sci.,84: 1489-1496.

Beever, D.E., Sutton, J.D., & C.K. Reynolds, 2001.Increasing the protein content of cow's milk.*Austr.* J.Dairy Technol.,56: 138-149.

Bhatta, R., Krisnamoorthy, U. & F. Mohammed, 2000. Effect of feeding tamarind (Tamarindus indica) seed husk as a source of tannin on dry matter intake, digestibility of nutrients and production performance of crossbred dairy cows in mid-lactation.*Anim.* Fd. Sci. Technol.,83: 67-74.

Boadi, D., Benchaar, C., Chiquette, J. & D. Massé, 2004. Mitigation strategies to reduce enteric methane emissions from dairy cows. Update review. *Can.* J. Anim. Sci.,84: 319-335.

Bruinenberg, M.H., Van der Honing, Y., Agnew, R.E., Van Vuuren, A.M. & H. Valk, 2002. Energy metabolism of dairy cows fed on grass. Livest. Prod. Sci.,75: 117-128.

Callaway, T.R., Martin, S.A., Wampler, J.L., Hill, N.S. and Hill, G.M. 1997. Malate content of forage varieties commonly fed to cattle. Journal of Dairy Science.80:1651-1655.

Chaucheyras, F., Fonty, G., Bertin, G., Salmon, Gouet, P.1995. In-vitro H2 utilisation by a ruminal acetogenic bacterium cultivated alone or in association with an Archea methanogen is stimulated by a probiotic strain of Saccharomyces cerevisiae. Appl. Env. Micro. 61: 3466-3467.

Chen, M. & M.J. Wolin, 1979. Effect of monensin & lasacoid-sodium on the growth of methanogenic & rumen saccharolytic bacteria.Appl. Environ. Microbiol.,38: 72-77.

Chilliard, Y, Ferlay, A., Mansbridge, R.M. & M. Doreau, 2000. Ruminant milk fat plasticity: nutritional control of saturated, polyunsaturated, trans and conjugated fatty acids. Ann. Zootech.,49: 181-205.

Clapperton J.L.,1977. The effect of a methane-suppressing compound trichloroethyl adipate on rumen fermentation and the growth of sheep,Anim. Prod. 24 169–181.

Czerkawski J.W., Christie W.W., Breckenridge G., Hunter M.L., 1995. Changes in rumen metabolism of sheep given increasing amounts of linseed oil in their diet, Brit. J. Nutr. 34: 25–44.

Demeyer D.I., De Graeve K., 1991.Differences in stoichiometry between rumen and hindgut fermentation, J. Anim. Physiol Anim. Nutr. 22, 50–61.

DePeters, E.J. & J.P. Cant, 1992. Nutritional factors influencing the nitrogen composition of bovine milk: A review. J. Dairy Sci., 75, 2043–2070.

Dijkstra, J., Kebreab, E., Bannink, A., France, J. and S. Lopez, 2005. Application of the gas production technique in feed evaluation systems for ruminants. Anim. Fd. Sci.Technol., 123-124, 561-578.

Dohme F., Machmuller A., Estermann B.L., Pfister P., Wasserfallen A., Kreuzer M., The role of the rumen ciliate protozoa for methane suppression caused by coconut oil, Lett. Appl. Microbiol. 29 (1999) 187–193.

Dohme, F., Machmuller, A., Sutter, F. & M. Kreuzer, 2004.Digestive and metabolic utilization of lauric, myristic and stearic acid in cows, and associated effects on milk fat quality. Arch. Anim.Nutr., 58: 99-116.

Dohme, F., Machmüller, A., Wasserfallen, A & M. Kreuzer, 2000. Comparative efficiency of various fats rich in medium-chain fatty acids to suppress ruminal methanogenesis as measured with RUSITEC. Can. J. Anim. Sci., 80: 473-482.

Dong.Y., Bae, H.D., McAllister, T.A., Mathison, G.W., Cheng, K.G. 1997. Lipid induced depression of methane production and digestibility in the artificial rumen system (RUSITEC). Can. J.Anim. Sci. 77, 269-278.

Duxbury J.M., Mosier A.R., 1993.,Status and issues concerning agricultural emissions of greenhouse gases, in: Kaiser H.M., Drennen T.W. (Eds.), Agricultural Dimensions of Global Climate Change, St. Lucie Press, Delray Beach, FL, pp. 229–258.

Erasmus, L.J., Robinson, P.H., Ahmadib, A., Hinders, R. & J.E. Garrett, 2005. Influence of prepartum and postpartum supplementation of a yeast culture and monensin, or both, on ruminal fermentation and performance of multiparous dairy cows. Anim. Fd. Sci.Technol.,122: 219-239.

Ferris, C.P., Gordon, F.J., Patterson, D.C., Porter, M.G. & T. Yan, 1999.The effect of genetic merit and concentrate proportion in the diet on nutrient utilization by lactating dairy cows. J. Agric. Sci. (Camb), 132,483-490.

Francis, G., Kerem, Z., Makkar, H.S.P. & K. Becker, 2002. The biological action of saponins in animal systems: a review. Br. J. Nutr., 88: 687-605.

Frumholtz, P. P., Newbold, C. J. and R. J. Wallace, 1989. Influence of Aspergillus oryzae fermentation extract on the fermentation of a basal ration in the rumen simulation technique (Rusitec). J. Agric. Sci. (Cam),113: 169–172.

Hegarty, R. S., Goopy, J. P., Herd, R. M. and B.McCorkell, 2007.Cattle selected for lower residual feed intake have reduced daily methane production. J. Anim. Sci., 85: 1479–1486.

Hegarty, R.S. 1999b. Mechanisms for competitively reducing ruminal methanogenesis.Aust. J. Agric.Res. 50, 1299-1305.

Hironaka R., Mathison G.W., Kerrigan B.K.,Vlach I.,1996. The effect of pelleting of alphalpha hay on methane production and digestibility by steers, Sci. Total Environ. 180: 221–227.

Jenkins, T.C, & M.A. McGuire, 2006. Major advances in nutrition: impact on milk composition. J. Dairy Sci., 89: 1302-1310.

Joblin, K.N. 1996. Options for reducing methane emissions from ruminants in New Zealand and Australia.In Coping with Climate Change. (Eds. W. J. Bouma, G.I. Pearman, M.R. Manning) pp 437-449. CSIRO Publishing, Collingwood, Australia.

Johnson, D.E., Ward, G.M. & G. Bernal, 1997. Biotechnology mitigating the environmental effects of dairying: Greenhouse gas emissions. Page 497-511 In: Milk composition, production and biotechnology, Eds. R.A.S. Welch, D.J.W. Burns, S.R. Davis, A.I. Popay, C.G. Prosser, CAB International, Wallingford, UK.

Johnson, K. A. and D. E. Johnson, 1995.Methane emissions from cattle.J. Anim. Sci.,73: 2483–2492.

Jordan, E., Lovett, D.K., Monahan, F.J., Callan, J., Flynn, B. & F.P. O'Mara, 2006.Effect of refined coconut oil or copra meal on methane output and performance of beef heifers.J. Anim. Sci.,84: 162-170.

Kebreab, E., France, J., McBride, B.W., Bannink, A., Mills J.A.N. & J. Dijkstra, 2006a. Evaluation of models to predict methane emissions from enteric fermentation in North American Cattle.In: Modelling Nutrient Utilization in Farm Animals, Eds, E. Kebreab, J. Dijkstra, J. France, A. Bannink & W.J.J. Gerrits,. CAB International, Wallingford, UK.

Kirchgessner M., Windisch W., Muller H.L.,Nutritional factors for the quantification of methane production, in: Engelhardt W.V., Leonhard-Marek S., Breves G., Giesecke D.(Eds.), Ruminant physiology: Digestion, metabolism, growth and reproduction, Proceedings of the Eighth International Symposium

Kirchgessner, M., Windisch, W. & H.L. Muller. 1994. Methane release from dairy cows and pigs. Pages 399-402 In Proc. 13th Symposium on Energy Metabolism of farm Animals, Ed. J.F. Aguilera, EAAP publ. No. 76 CSIS Publ. Service, Spain.

Kohn, R.A. & R.C. Boston, 2000.The role of thermodynamics in controlling rumen metabolism.Pages 11- 24.In Modelling Nutrient Utilization in Farm Animals.Eds. J.P. McNamara, J. France & D.E. Beever. CAB International, Wallingford, UK.

Kung L. Jr., Hession A.O., Bracht J.P.,1998. Inhibition of sulfate reduction to sulfide by 9,10-anthraquinone in in vitro ruminal fermentations,J. Dairy Sci. 81, 2251–2256.

Leng R.A., Improving Ruminant Production and Reducing Methane Emissions from Ruminants by Strategic Supplementation, United States Environmental Protection Agency, Office of Air and Radiation, Washington (D.C.), 1991,105 p.

Lopez S., McIntosh F.M., Wallace R.J., Newbold C.J., 1999. Effect of adding acetogenic bacteria on methane production by mixed rumen microorganisms, Anim. Feed Sci. Technol. 78: 1–9.

Machmuller, A. and Kreuzer, M. 1999. Methane suppression by coconut oil and associated effects on nutrient and energy balance in sheep.Can. J.Anim. Sci. 79, 65-72.

Manson F.J., & J.D. Leaver, 1988.The influence of concentrate amount on locomotion and clinical lameness in dairy cattle.Anim. Prod.,47: 185-190.

Martin S.A.,1998. Manipulation of ruminal fermentation with organic acids: a review, J. Anim. Sci. 76 : 3123–3132.

Martin, S.A., Nisbet, D.J., Dean R.G. 1989. Influence of a commercial yeast supplement on the in-vitro ruminal fermentation. Nutr.Rep. Int. 40, 395- 403.

McAllister, T.A., Okine, E.K., Mathison, G.W., Cheng,K.N. 1996. Dietary, environmental and microbiological aspects of methane production in ruminants.Can. J. Anim. Sci. 76, 231-243

McCaughey, W. P., Wittenberg, K. and D. Corrigan, 1999.Impact of pasture type on methane production by lactating cows.Can. J. Anim. Sci., 79: 221-226.

McCaughey, W.P., Wittenburg, K.M., Corrigan, D. 1997. Methane production by steers on pasture.Can. J. Anim. Sci. 519-524.

McCrabb G.J., Berger K.T., Magner T., May C., 1997. Hunter R.A., Inhibiting methane production in Brahman cattle by dietary supplementation with a novel compound & the effects on growth,Aust. J. Agric. Res. 48: 323–329.

McGinn, S.M., Beauchemin, K.A., Coates, T., & D. Colombatto, 2004. Methane emissions from beef cattle: Effects of monensin, sunflower oil, enzymes, yeast and fumaric acid. J. Anim. Sci.,82: 3346-3356.

Moe, P.W. & H.F. Tyrrell, 1979.Methane production in dairy cows.J. Dairy Sci.,62: 1583-1586.

Morvan B., Doré J., Rieu-Lesme F., Foucat L., Fonty G., Gouet P., 1994. Establishment of hydrogenutilizing bacteria in the rumen of the newborn lamb, FEMS Microbiol.Lett. 117: 249–256.

Mosier A.R., Schimel D., Valentine D., 1991.Bronson K., Parton W.J., Methane and nitrous oxide fluxes in native, fertilised and cultivated grasslands,Nature 350: 330–332.

Moss A.R., Givens D.I., Garnsworthy P.C.,1995 The effect of supplementing grass silage with barley on digestibility, in sacco degradability, rumen fermentation and methane production in sheep at two levels of intake, Anim. Feed Sci. Technol.55:9–33.

Moss A.R., Methane-Global Warming and Production by Animals, Chalcombe Publications,Canterbury, UK, 1993, 105 p.

Moss, A. R., Jouany, J. P. and C. J. Newbold, 2000. Methane production by ruminants: Its contribution to global warming. Ann. Zootech., 49: 231-253.

Murray, P.J., Gill, E., Balsdon, S.L. & S.C. Jarvis, 2001. A comparison of methane emissions from sheep grazing pastures with differing management intensities. Nutr.Cycl.Agroecosyst.,60: 93-97.

Mutsvangwa T., Edward I.E., Topps J.H., Paterson G.F.M., 1992.The effect of dietary inclusion of yeast culture (Yea-Sacc) on patterns of rumen fermentation, food intake and growth of intensively fed bulls, Anim. Prod. 55-35–40.

Nelson, N., Valdes, C., Hillman, K., McEwan, N.R.,Wallace, R.J., Newbold, C.J. 2000. Effect of a methane oxidizing bacterium isolated from the gut of piglets on methane production in Rusitec. Reprod. Nutr.Devel. 40: 212.

Newbold C.J., Wallace R.J., Watt N.D., Richardson A.J., 1988.The effect of the novel ionophore tetronasin (ICI 139603) on ruminal microorganisms, Appl. Environ.Microbiol. 54: 544–547.

Newbold, C.J., Lopez, S., Nelson, N., Ouda, J.O., Wallace, R. J. and Moss, A.R. 2005. Propionate precursors and other metabolic intermediates as possible alternative electron acceptors to methanogenesis in ruminal fermentation in vitro.Brit. J. Nutr., 94: 27-35.

Newbold, C.I., McIntosh, F.M., Wallace, R.J. 1998. Changes in the microbial population of a rumen simulating fermenter in response to yeast culture.Can. J. Anim. Sci. 78:241-244.

Newbold, C.J., Lassalas, B., Jouany, J.P. 1995. The importance of methanogens associated with ciliate protozoa in ruminal methane production in-vitro.Lett. Appl. Micro. 21: 230-234.

Ojima D.S., Valentine D.W., Mosier A.R., Parton W.J., Schimel D.S., 1993.Effect of land use change on methane oxidation in temperate forest and grassland soils, Chemosphere 26: 675–685.

Ouda J.O., Newbold C.J., Lopez S., Nelson N., Moss A.R., Wallace R.J., Omed H., 1999. The effect of acrylate and fumarate on fermentation and methane production in the rumen simulating fermentor (Rusitec), Proceedings of the British Society of Anim. Sci., 36 p.

Pen, B., Sar, C., Mwenya, B., Kuwaki, K., Morikawa, R. & J. Takahashi, 2006. Effects of Yucca schidigera and Quillaja saponaria extracts on in vitro ruminal fermentation and methane emission. Anim Fd. Sci. Technol.,129: 175–186.

Puchala, R., Min, B.R., Goetsch, A.L. & T. Sahlu, 2005. The effect of a condensed tannin-containing forage on methane emission by goats. J. Anim. Sci.,83: 182-186.

Ramirez, C. A. and T. N. Barry, 2005.Alternative temperate forages containing secondary compounds for improving sustainable productivity in grazing ruminants.Anim. Fd. Sci. Tech., 120: 179- 201.

Rumpler W.V., Johnson D.E., Bates D.B., 1986.The effect of high dietary cation concentrations on methanogenesis by steers fed with or without ionophores, J. Anim. Sci. 62:1737–1741.

Russell, J.B. 1998. The importance of pH in the regulation of ruminal acetate to propionate ratio and methane production in vitro.J. Dairy Sci. 81:3222-3230.

S. 1998. Methane output and lactation response in Holstein cattle with monensin or unsaturated fat added in the diet. J. Anim. Sci. 76:906-914.

Sansoucy, R., Aarts, G. & Leng, R.A.1988. Molasses-urea blocks as a multinutrient supplement for ruminants. In R. Sansoucy, G. Aarts & T.R. Preston (eds). Sugarcane as Feed, vol. 2,pp. 263–278. Proc. of an FAO Expert Consultation held in Santo Domingo, 7-11 July 1986.

Schils, R. L. M., Verhagen, A., Aarts, H. F. M. and. L. B. J. Sebek, 2005. A farm level approach to define successful mitigation strategies for GHG emissions from ruminant livestock systems.Nutr.Cycl.Agroecosyst.,71: 163-175.

Schnell, S., King, G.M. 1995.Stability of methane oxidation capacity to variations in methane and nutrient concentrations.FEMS.Microbiol.Ecol. 4:285-294.

Singhal, K. K., Mohini, M. and N.Senthil, 2006.Newer Strategies for Mitigation of Methane Emission from Ruminants.Lead Paper at Proceedings of the XII Animal Nutrition Conference on Technological Interventions in Animal Nutrition for Rural Prosperity, AAU, Anand, Gujarat, India.

Singhal, K. K., Mohini, M., Jha, A. K. and P. K. Gupta, 2005.Methane emission estimates from enteric fermentation in Indian livestock: Dry matter intake approach. Current Sci.88 (1): 119- 127.

Sparling R., Daniels L., 1987.The specificity of growth inhibition of methanogenic bacteria by bromoethanesulfonate, Can. J. Microbiol. 33: 1132–1136.

Staples, C.R., & W.W. Thatcher, 2001. Nutrient influences on reproduction of dairy cows. Proceedings of Midsouth Ruminant Nutrition Conference, pp. 21–35 (http://www.txanc.org/proceedings.html#2001).

Sundstøl F., 1981. Methods for treatment of low-quality rouhages, in: Kategile J.A., Sundstøl F.(Eds.), Utilization of low-quality roughages in Africa, Agric. University of Norway, As-NLH, Norway, pp. 61–80.

Tavendale, M. H., Meagher, L. P., Pacheco, D., Walker, N., Attwood, G. T. and S. Sivakumaran 2005. Methane production from in vitro rumen incubations with Lotus pedunculatus and Medicago sativa, and effects of extractable condensed tannin fractions on methanogenesis.Anim. Fd. Sci. Tech., 123&124: 403–419.

Ushida K., Jouany J.P.,1996. Methane production associated with rumen-ciliated protozoa and its effect on protozoan activity, Lett. Appl. Microbiol.23: 129-132.

Ushida K., Tokura M., Takenaka A., Itabashi H., 1997. Ciliate protozoa and ruminal methanogenesis, in: Onodera R., Itabashi H., Ushida K., Yano H., Sasaki Y. (Eds.), Rumen Microbes & Digestive Physiology in Ruminants, Japan Sci. Soc.Press, Tokyo, pp. 209–220.

Valdes C., Newbold C.J., Hillman K., Wallace R.J.,1997. Los microorganisms metanotrofos come agentes modifactdores de la fermentacion ruminal, ITEA 18: 157-159.

Van Knegsel, A.T.M., Van den Brand, H., Dijkstra, J., Tamminga, S. & B. Kemp, 2005. Effect of dietary energy source on energy balance, production, metabolic disorders and reproduction in lactating dairy cattle.A review.Reprod. Nutr. Dev.,45: 665-688.

Van Laar, H. & W.M. van Straalen, 2004.Ontwikkeling van een rantsoen voor melkvee dat de methaanproductie reduceert.Schothorst Feed Reseatrch 740.

Van Nevel C.J., Demeyer D.I., Influence of antibiotics and a deaminase inhibitor on volatile fatty acids and methane production from detergent washed hay and soluble starch by rumen microbes in vitro, Anim. Feed Sci. Technol. 37 (1992) 21–31.

Veen, W.A.G., 2000. Veevoedermaatregelen ter vermindering van methaanproductie door herkauwers.Schothorst Feed Research proefverslag. www.robklimaat.nl

Waghorn, G.C., Woodward, S.L., Tavendale, M. & Clark, D.A.2006. Inconsistencies in rumen methane production – effects of forage composition and animal genotype. Int. Congr. Series 1293: 115–118.

Woodward, S. L., Waghorn, G. C. and N. A. Thomson, 2006. Supplementing dairy cows with oils to improve performance and reduce methane– does it work ?Proc. NZ Soc. Anim. Prod.66: 176-181.

Woodward, S.L., Laboyrie, P.J. & E.B.L. Jansen, 2000.Lotus corniculatus and condensed tannins – Effects on milk production by dairy cows.Asian-Austral. J. Anim. Sci.,13: 521-525.

Yan, T., Agnew, R.E., Gordon, F.J. & M.G. Porter, 2000. Prediction of methane energy output in dairy and beef cattle offered grass silage-based diets. Livest. Prod. Sci.,64: 253-263.

13

Impact of Nutrition in Augmenting Production and Reproduction in Small Ruminants

V.P. Maurya, V. Sejian, Gyanendra Singh and Mihir Sarkar

Physiology and Climatology Division, IVRI, Izatnagar, Bareilly
Uttar Pradesh – 243122

The arid and semi-arid regions of India are drought prone and the availability vegetation of these regions is more acute during summer when the quality of the pastures becomes vulnerable. Small ruminants in the semi-arid regions need to adopt special physiological function to maintain thermal equilibrium. The reproductive efficiency of different breeds of sheep inhabiting in the semi-arid regions of India is relatively low (Arora and Garg, 1998).The nature of nutritional resources and management problems differs with climate, soil and vegetation, the physical process which governs the animals reproductive performance are the same; the female must reach puberty, show behavioral estrus and shed one or more ova which have then to be fertilized. Dietary nutrition promotes the programming and expression of the metabolic pathways that enable animals to achieve their genetic potential for reproduction. Nutrition has an important impact on the reproductive performance in sheep, but the magnitude of the effect on reproduction may vary with the season (White *et al.,* 1983). Sheep are more prone to neglect compared to cattle and often suffer from lack of feed and fodder. The animals during the grazing are exposed to combined stress (Sejian *et al.,* 2010 ab, and 2011) .This leads to decrease in body weight and loss of reproductive functions. Concurrently, Rhind and Mc Neilly (1986) showed that ewes with low body fat reserves had fewer large ovarian follicles and therefore the potential for subsequent maturation and ovulation declined.

Sheep, which has a vital economic importance in India, inhabitant in semi-arid region and mostly raised under harsh environmental conditions (high ambient temperature, scarcity of feed and water). In India, there are about 50.8 million (M) sheep, which ranks it fourth in the world, contributing meat (167 M kg), wool (42.5 M kg) and skin (39.0 M kg). This species of livestock is also source of employment for marginal and land less farmers particularly in the arid and semi-arid region of India. The national sheep wealth is about Rs. 24 billion producing an annual income of Rs. 8 billion in the form of meat, wool, skin, milk and manure.

13.1 Nutrition and Early and Later Stage of Pregnancy

Under nutrition during fetal life reduces subsequent litter size (Robinson, 1990[a]). Recent studies have demonstrated an effect of under nutrition on the concentration of oogonia in the ovine ovary as early as day 47 of fetal life (Borwick et al., 1994) and an associated postponement in the arrest of ovarian metabolic activity on day 62 (Borwick et al., 1995). An important component of reproduction in the survivability of lambs is at birth. Thompson et al. (1994) found that birth weights of lamb from embryos cultured in synthetic oviduct fluid medium supplemented with human serum were greater then those from spontaneous ovulating ewes or those derived from embryos cultured in synthetic oviduct fluid supplemented with amino acids and bovine serum albumin. McKelvey et al. (1988) reported that high plane of nutrition during the post mating period can, adversely affect the embryo survival and that this was mediated through the decline in plasma progesterone concentration; low plane of nutrition at this time may affect the embryo survival.

13.2 Nutrition and Puberty

The onset of puberty in sheep is influenced by genetic and environmental factors such as breed and strain differences, the nutrition plane and time of birth. Puberty, the age at first ovulation, occurs approximately at 7 to 9 month of age. Prerequisites of the onset of puberty in the female sheep include the achievement of a critical body size. In sheep (Foster et al., 1988; Dyrmundsson,1987) restricted nutrition during the early life and failure to achieve the necessary live weight before the end of the first breeding season after birth results in delay in onset of puberty at least at the following breeding season when the animal has usually reached the critical live body weight. Any alteration of nutrition before the alters the rate of growth of young animals and which affects the attainment of puberty (Mc Call et. al., 1989). Adequate nutrition is necessary for the proper pubertal development. Under-nutrition at the time of prepubertal development retarded the growth of animals seriously. High plane of nutrition advances the puberty considerably (Moustgaard, 1969). Rearing of ewe lamb on high plane of nutrition will normally promotes both faster growth and development of reproductive organs (Kuliev, 1965). Thus owing to the close association between general body growth and sexual development, raising the level of nutrition will not only results in more ewes lamb attain puberty during the first breeding season, but also ewes lamb growing at faster rate will exhibit their first estrus and are more likely to conceive at lower age and heavier body weight than ewe lamb growing at slower rate. The dietary energy level affects the blood profile of ewes (Maurya et al 1998[a]) which indirectly affects the reproductive potential of ewes. At the ovarian level, low feed intake that delay puberty are accompanied by reduced follicular development which in case of heifer, is associated with smaller dominant follicles (Bergfeld et al., 1994). Blood born metabolites reflect nutritional status and are involved in the control of GnRH release and subsequent ovulation control.

13.3 Nutrition and Incidences of Estrus

It is well established that the follicular development in the many mammals is affected by nutrition. This effects can either be indirect, acting by way of hypothalamic pituitary axis to alter the secretion of gonadotrophin, or can be direct acting on the ovary to mediate the action of gonadotrophin on the follicles. Allison (1977) reported that greater number of follicle >2mm diameter and larger mean follicle diameters in ewes of high live weight compared with ewes of low live weight. Under nutrition affects the incidences of estrus very badly (Maurya et al., 2004) and some times estrus is being completely inhibited (Hafez,1952) or breeding season can be prematurely terminated (Knight et al.,1983). Ovulation rate is one of the most important determinants of reproductive performance in sheep, which depends on the age (Knight, et al., 1975), genotype (Wheeler and Land, 1977) and stage of breeding season (Devis et al., 1976., Maurya et al., 1998[b]) as well as nutrition (Gunn,1983). The small ruminants raised under tropical conditions generally maintained under condition of minimal management in a climatically harsh environment and ewes are having lower ovulation and lambing rate, which can be increased by providing them good nutrition (Naqvi et al., 2002). The body weight of ewes at mating representing the static effect has been shown to influence subsequent litter size (Coop, 1966); effect is mainly the result of different ovulation rate. Cumming (1977) reported that heavier ewes with in the flock were found to have more ovulations than the lighter ones, showing about 2.5-3.0% increase for each 1.0 Kg increase in live weight. The net nutritional status of ewes also affects the ovulation rate. Lindsay (1976) suggested that ovulation rate in ewes depends on the net nutritional status of the ewes. According to him, heavy ewes given poor feeding may still show a good ovulation rate because they have a reasonable endogenous source of energy and protein. On other hand, poor ewes temporarily well-fed will also ovulate well because the contribution of exogenous nutrients. Effect of nutrition on follicle development and ovulation rate could potentially be mediated through changes circulating blood metabolites (glucose, non-etherified fatty acids, amino acids etc), metabolic hormones (insulin, growth hormone etc.), gonadotrophin (LH and FSH) or a combination of several these factors (Robinson, 1996). Table '13.1 describes the effect of feed restriction on reproduction.

Table 13.1: Effect of feed restriction on reproduction.

Attributes	Phase I		Phase II	
	Ad lib.	Restricted	Ad lib.	Ad lib.
No. of animals	8	8	8	8
No. of animals exhibited heat	8[a](100%)	2[b](25%)	8[a](100%)	8[a](100%)
Estrous duration (hrs)	26.08±0.84[a]	15.00±3.00[b]	28.80±0.89[a]	29.00±1.63[a]
Estrous cycle (days)	18.63±0.50 [bc]	31.50±2.02 [a]	17.68±0.43 [c]	20.34±0.79[b]
No. of animals repeated after first insemination	-	-	1 [b](12.5%)	6[a] (75%)

[a, b,c] means with different superscripts in same row differ significantly (P<0.01)

(*Source:* Maurya *et al.*, 2004)

13.4 Nutrition and Embryo Mortality

The first determinants of the rate of reproductive success are the fertilization rate is proper shedding of ovum from ovary. Extremes in the level of feeding are detrimental to embryo survival. The retinoids are the main metabolites of vitamin A and are involved in cell proliferation and differentiation, expression of growth factor, steroidogenesis, which play important role in embryo survival. Recently, attempts have been made to identify the mechanism of involved in the adverse impact of high protein diet on the fertility rate of dairy cow. Elord and Butler (1993) found that high intake of rumen degradable protein (RDP), leading to excess rumen ammonia production, were associated reduction of pH of the uterine environment. This disturbance in the uterus environment leads to the embryonic death. It is well known that progesterone play a crucial role for in maintaining pregnancy in ewe (Denamur and Martinet, 1955) and some workers have use exogenous progesterone to improve embryo survival in sheep. Nutrition in early pregnancy and peripheral progesterone concentration may be inversely related (Williams and Cumming, 1982). Further, Parr *et al.* (1987) demonstrated that sheep fed high energy ration after mating had reduced progesterone levels and showed an increase in embryo mortality. High plane of feeding increase the metabolic clearance rate of (Parr *et al.,* 1993) and the concomitant reduction of progesterone concentration at the time

(Day 11 and Day 12 of post mating in ewes) when the embryo is extremely sensitive to low concentration undoubtedly are involved (Parr,1992). Dietary induced suppression of circulating progesterone during oocyte maturation in superovulated ewes primed with a single CIDR device (0.3 g progesterone) can impart a legacy of developmental retardation which leads to decreased embryo survival (McEvoy, *et al.*, 1995). Severe under nutrition significantly decreases the pregnancy rate of sheep (Abecia *et al.*, 1994, Maurya *et al.*, 2004). Embryo recovery from superovulated ewes on differential feeding is depicted in Table 13.2

Table 13.2: Embryo recovery from superovulated ewes on differential feeding

Treatment	Group 1	Group 2	Group 3
Number of ewes flushed	6	6	7
Tatal number CL	54	51	60
Mean CL/ewe flushed	9.0 ±2.99	8.5± 1.26	8.6± 1.41
Number of ewe yielded egg (%)	6 (100)	6(100)	7(100)
Number of egg recovered	34	34	35
Recovery (%)	63	67	58
Mean egg recovery/ewe flushed	5.6±1.80	5.6± 1.74	5.0±1.48
Mean transferable embryos	4.0± 1.57	4.0± 1.47	4.3± 1.37

Group 1: ewes maintained on grazing, Group2: Grazing+ Concentrate (150g/ewe/day)
Group3: Grazing+ Concentrate (300g/ewe/day)
(Sourse: Naqvi et al., 2002)

13.5 Nutrition and Neonatal Metabolism

Under-nutrition decrease uterine blood flow and accompanying decrease in fetal insulin and IGF-I coupled with increase in growth hormone, adrenocorticotrophin corticosterone impair fetal growth and development. Fetal urea concentration increase during the under-nutrition, reflecting increase in gluconeogenesis by fetus from amino acids (Bell, 1993). In new born lambs, low insulin, high corticosterone and a low selenium and /or iodine status inhibit thermo genesis from brown adipose tissue (Robinson, 1990[b]; Robinson and Symonds, 1995) which leads to a reduced viability at birth. Under nutrition during pregnancy significantly reduces the live birth weight of lambs (Maurya *et at.*, 2004). Advance pregnancy is the most important period of pregnancy with fetal mass increasing in a curvilinear manner such that in the last 8,4,2 weeks of gestation the fetus is gaining equivalent to 85,50 and 25% of the final birth weight (Robinson, 1983). This is the period when pregnancy nutrition can have the greatest effect on total potential lamb production. The degree of under-nutrition can also be observed with the help of plasma glucose, plasma free fatty acids. Table 13.3 describes the effect of feed restriction on lambing rate and birth weight of lambs.

Table 13.3: Effect of feed restriction on lambing rate and birth weight of lambs

Treatment	Phase II	
	Ad lib.	Restricted
No. of animals conceived	8 [a](100%)	8 [a](100%)
No. of animal lambed	8 [a] (100%)	7[a](87.5%)
No of mating /ewe	1.25 [b]±0.16	1.87 [a] ±0.23
No of lamb /ewe	1.00	0.88
Body weight of lamb (kg)	3.59 [a] ±0.35	3.06 [b] ±0.94

[a, b] means with different superscripts in same row differ significantly (P<0.01)

(*Sourse:* Maurya *et al.*, 2004)

13.6 The Body Condition Score and Reproductive Efficiency

The body condition score is Condition scoring the ewe/goat flock can be a useful management tool. Throughout the production cycle, sheep producers must know whether or not their sheep are in condition (too thin, too fat, or just right) for the stage of production: breeding, late pregnancy, and lactation. There is an optimum condition (score) for each ewe in the flock for each stage of the production cycle. Body condition scoring is a simple but useful procedure, which can help producers make management decisions regarding the quality and quantity of feed needed to optimize performance. BCS had a significant influence on conception rate, lambing rate, bith weight and weaning weight of lambs (Sejian *et al.,* 2010). It may be inferred from this study that active management of breeding sheep flock to achieve optimum BCS of 3.0-3.5 may ensure economically viable return from these flocks (Table 13.4).

Table 13. 4: Body condition score and reproductive efficiency

Parameter	Group I	Group II	Group III
	(2.5 BCS)	(3.0-3.5 BCS)	(4.0 BCS)
Ewes in heat (%)	70 (7)	90 (9)	60 (6)
Conception rate (%)	40[b] (4)	90[a] (9)	40[b] (4)
Fetal sac volume (C^3)	167.5±26.1[a]	193.2±18.6[b]	214.2±18.9[c]
Lambing rate (%)	40[b](4)	90[a] (9)	40[b] (4)
Litter size (no.)	1.0±0.00	1.25±0.164	1.0±0.00
Birth weight of lamb (Kg)	2.96±0.21[c]	3.6±0.16[b]	4.4±0.27[a]
Weaning weight of lamb (Kg)	15.67±0.84[b]	18.0±0.76[a]	18.4±0.60[a]

(*Source:* Sejian *et al.,* 2010)

13.7 Nutrition and Male Reproductive Efficiency

The availability of sufficient nutrition is clearly important for the early sexual development of the male lambs and kids. Nutritive deficiency specially low energy intake, will retard sexual development and delay the attainment of puberty

(Foote, 1969). There is a lack of gonadotrophin from the hypophysis though the testis will usually continue to produce testosterone and androgenic function of the testis is retarded more markedly than is spermatogenesis (Mann,1964). Conversely in well fed and rapidly growing lambs it would appear that androgenic function is perfectly advanced (Dyrmundsson, 1987). In rams there is 6 to 7 week delay in the response in spermatozoa's number to diet, reflecting the time it taker for the development of spherical spermatids in the germinal epithelium to fully mature spermatozoa in the distal cauda epididymus . Evidence for the direct effect of the nutrition on testis growth and production of spermatozoa in the absence of alteration in the GnRH gonadotrophin system, led Martin and Wlkden Brown (1995) to develop the concept to both GnRH independent and dependent pathway are involved in nutritional effect on spermatogenesis. Salman (1964) reported that ram fed a high protein diet displayed a more intense sex drive than ram fed a low protein supplement and resulted in the more fertile mating.

13.8 Role of Leptin and Gherlin in Feed Utilization Efficiency

Leptin was discovered in 1994 by Jeffrey M. Friedman and colleagues at the Rockefeller University through the study of mice. Human leptin is a protein of 167 amino acids. It is manufactured primarily in the adipocytes of white adipose tissue, and the level of circulating leptin is directly proportional to the total amount of fat in the body. Leptin is synthesised in many other tissues including the stomach, ovary, placenta and liver and its receptors are also located in a diverse range of tissues (Margetic *et al*, 2002). Leptin is a 16 kDa protein hormone which plays a major role in regulating energy intake and energy expenditure, including appetite and metabolism. It is one of the most important adipose derived hormones. The ''Ob (Lep)'' gene (Ob for obese, Lep for leptin) is located on chromosome 7 in humans. Leptin inhibits food intake by acting in the appetite control centres of the brain as administering leptin into the brain causes a significant reduction in food intake which does not occur after peripheral injection. Leptin acts centrally, telling the brain to stop eating, there is another peripheral signal originating in the stomach which instructs the brain to promote eating. Leptin appears to protect the body against starvation by only allowing energy dense processes to occur when the body is sufficiently ready. Its primary role is complex, is it a satiety signal to prevent overeating or an evolutionarily efficient signal protecting against starvation? Leptin is secreted by white adipose tissue in ruminants, and its secretion is decreased when there is an insufficient supply of nutrients (Ahima and Flier, 2000). Increased levels of leptin are associated with decreased feed intake. Low concentrations of leptin increase appetite and prepare the animal to conserve energy (Ahima, 1996). Ghrelin was discovered in 1999 when it was noted that tissue extracts from the rat stomach activated the GHS receptor (Kojima *et al.,* 1999). It has since been recognized that the stomach is the richest source of ghrelin. Over 90 percent of the body's ghrelin is in the stomach and duodenum. Lower amounts are found in the pancreas, pituitary, kidney, and placenta. A limited region of the arcuate nucleus of the hypothalamus contains small amounts of ghrelin. Ghrelin used to have potent growth hormone releasing and appetite stimulating activities .The name "ghrelin" is based on "ghre", a word root in Proto-Indo-European languages for "grow", in reference to its ability to stimulate GH release. The rat and human ghrelin precursors are both composed of 117 amino acids. Ghrelin is a 28 amino acid peptide that is the natural ligand for the growth hormone secretagogue (GHS) receptor (Kojima, 2010). Based on its structure, it is a member of the motilin family of peptides. When administered peripherally or into the central nervous system, ghrelin stimulates secretion of growth hormone, increases food intake, and produces weight gain (Tajkaya *et al.,* 2000). Ghrelin receptors are expressed in a wide variety of tissues, including the pituitary, stomach, intestine, pancreas, thymus, gonads, thyroid, and heart (Howard *et al* 1996). The diversity of ghrelin receptor locations suggests ghrelin has diverse biological functions. Ghrelin levels increase before meals and decrease after meals. In brief it can be said that Leptin and ghrelin seem to be the big players in regulating appetite, which consequently influences body weight/fat. Both leptin and ghrelin are peripheral signals with central effects. In other words, they're secreted in other parts of the body (peripheral) but affect our brain (central). Leptin decreases hunger and in contrast to leptin, ghrelin increases hunger. Leptin and Ghrelin play a central role in the neurohormonal regulation of food intake and energy homeostasis.

13.9 Conclusion

It has been found out that small ruminants, which have undergone a period of under-nutrition, are capable to regaining body weight loss and reproductive function. It appears that animals fed *ad libitum* could overcome the harsh climate effect on their body weight and reproduction, while animals exposed to thermal and nutritional stress simultaneously have reduced body reserves, body condition score and reproductive efficiency. The losses incurred on body weight and reproduction during coupled stresses (thermal and nutritional) could not be compensated during a post stress period. The coupled stress is very dangerous for the exploiting their reproductive and productive potential.

References

Abecia,J.A., Rhind,S.M.and McMillen,S.R. 1994. Effect of under nutrition on luetial function and the distribution of progesterone in endometrial tissue in ewes. ITEA, Production Animal 90 A 63-71.

Ahima, R. S. D. Prabakaran, C. Mantzoros, D. Qu, . Lowell, E. Maratos-Flier. 1996. Role of leptin in the neuroendocrine response to fasting. Nature (Lond.) 382:250-252.

Ahimas, R. S., and J. S. Flier. 2000. Leptin. Annu. Rev. Physiol. 62:413-437.

Allision,A.J. 1977.effect of nutritionally induced live weight differences on the ovulation rates and the production of ovarian follicles in ewes. Theriogenology 8:19-24.

Arora,C.L and Garg,R.C.1998. In: Sheep Production and Breeding. Reproductive performance of sheep.pp.154-163.

Bell,A.W. 1993. pregnancy and foetal metabolism. In: Forbes,J.M and France,J. (edts.) Aspects of ruminant digestion and metabolism. CAB, International ,Wallingford. pp 405-431.

Bergfeld,E.G.M.Kojima,F.N.Cupp,A.S.and Kinder,J.E. 1994. Ovarian follicular development in prepubertal heifer is influenced by level of dietary energy intake. Biology of Reproduction.51:1051-1057.

Block, S. S., W.R. Butler, R. A. Ehrhardt, A. W. Bell, M.E. Van Amburgh and Y. R. Boisclair. 2001. Decreased concentration of plasma leptin in periparturient dairy cows is caused by negative energy balance. J. Endocrinol.171:339-348.

Borwick,S.C., Rhind,S.M/and McMlillen,S.R.1995. Effect of under nurtition from the time of mating on the ovarian; development in foetal sheep at 62 d of gestation. Journal of Reproduction and Fertility. Abstract Series No. 15,p.52.

Borwick, S.C., Rhind, S.M and McMlillen,S.R. 1994. Ovarian steroidogenesis and development in fetal Scottish Blackface ewes, undernourished in utero from conception . Journal of Reproduction and Fertility Abstract Series No.14,p.14.

Coop, I.E. 1966. Effect of flushing on reproductive performance of ewes. Journal of Agricultural Science, Cambridge 67:305-321.

Cumming,I.A. 1977. Relationship in the sheep of ovulation rate with live weight, breed , season and plane of nutrition. Australian Journal of Experimental Agriculture and Animal Husbandry,17:234-241.

Davis, I.P. and Kenny, P.A and Cummins.I.A. 1976.Effect of time of joining and rate of stocking on the production of Corredale ewes in southern Victoriya II Ovulation rate and embryonic survival. Australian Journal of Experimental Agriculture and Animal Husbandry,16: 11-18..

Denamur, R and Martinet, J. 1955. Efects de, l'ovariectomise chez in brebis pendant in gestation. Cr. Senic Social Biolology140:2105-2107.

Dyrmundsson,O.R.1987. Advancement of puberty in male and female. In: Marai, I.F.M and Woven J.B (eds.) New Techniques in sheep production, Butterworths, London. pp 65-76.

Elrod, C. C. and Butler W.R. 1993. Reduction infertility and alteration of uterine pH in heifers fed access ruminally degradable protein. Journal of Animal Science.71:694-701.

Foote, R.H.1969. In: Reproduction in Domestic Animals. Cole, H.H. and Cupps, P.T (eds.) Academic Press,London.pp.313-333.

Foster, D.L. Ebling, E.B.J. Vannerson, L.A. and Veenvilet, B.A. 1988. Modulation of gonadotrophins secretion during development by nutrition and growth. Proceedings of 11th International Congress on Animal Reproduction and Artificial Insemination.5: 101-108.

Hafez, E.S. E., 1952. Studies on the breeding season and reproduction of the ewe. Journal of Agricultural Science Cambridge. 42: 189-265.

Hall, J.B., Schillo,K.K., Hileman, S .M. and Boling, J.A.1992. Does tyrosine acts as a nutritional signal mediating the effect of increased feed intake on leutenizing hormone patterns in growth restricted lambs. Biology of Reproduction. 46:573-579.

Howard AD, Feighner SD, Cully DF, Arena JP, Liberator PA, Rosenblum CI, Hamelin M, *et al.*.1996. "A receptor in pituitary and hypothalamus that functions in growth hormone release".*Science* 273 (5277): 974–7.

Ingvartsen, K. L. and J. B. Anderson. 2000. Integration of metabolism and intake regulation: A review focusing on periparturient animals. J. Dairy Science 83:1573-1597.

Knight, T.W. Hall, D.R.H and Wilson,I.D. 1983. Effect of teasing and nutrition on the duration of the breeding season in Romney ewes. Proceedings of the New Zealand Society of Animal Production 43:17-19.

Knight, T.W., Oldham, C.M., Smith J.F. and Lindsay, D. R. 1975. Studies in ovine infertility in Agricultural region in western Australia: analysis of reproductive wastage. Australian Journal of Experimental Agriculture and Animal Husbandry 15:183-188.

Kojima M, Hosoda H, Date Y, Nakazato M, Matsuo H, Kangawa K .1999. "Ghrelin is a growth-hormone-releasing acylated peptide from stomach". *Nature* 402 (6762): 656–60.

Kojima, M. 2010. Discovery of Ghrelin and Its Physiological Function Journal of Medical Sciences 3(2): 92-95.

Kuliev,G.K. 1965. Morpho-Biological principles in growth and developments of ovaries of ewes at deferent levels of nutrition. Izvstiya Akedemii Nauk. SSSR. Biologicheskaya 3:51-58.

Lindsay, D.R. 1976. The usefulness to the animal producer of research finding in nutrition on reproduction. Proceedings Australian Society Animal Production 18: pp.217-224.

Mann,T. 1964. The biochemistry of semen and of the male reproductive track. Methan & Co. Ltd. London. pp. 493.

Martin, G.B. and Walkden-Brown, S.W. 1995. Nutritional influence on reproduction in mature male sheep and goat. Journal Reproduction and Fertility. Suppl. 49:437-494.

Maurya, V.P., Naqvi S.M.K. and Mittal. J.P. 1998. Reproductive behaviour of Malpura and BharatMerino ewes lamb under semi-arid condition. Indian Journal of Animal Sciences.68:533-535.

Maurya, V.P., Naqvi S.M.K. and Mittal.J.P. 1998. Influence of dietary energy level on circulating blood constituents of sheep in hot semi-arid region of India. 8th World Conference on Animal Production Seoul.Korea.ON4-10.

Maurya, V.P., Naqvi S.M.K. and Mittal.J.P. 2004. Effect of dietary energy level on physiological responses and reproductive performance of Malpura sheep in the hot semi-arid regions of India. Small Ruminants Research. In press.

Maurya, V.P., Kumar, S., Kumar, D., Gulyani, R ., Joshi, A., Naqvi, S.M.K., Arora, A.L., Singh, V.K. 2009. Effect of Body Condition Score on Reproductive Performance of Chokla Ewes. Indian Journal of Animal Sciences 79:1136-1138.

Maurya, V.P., Sejian, V., Kumar, D and Naqvi, S.M.K (2010). Effect of induced body condition score differences on sexual behavior, scrotal measurements, semen attributes, and endocrine responses in Malpura rams under hot semi-arid environment. Journal of Animal physiology and Animal Nutrition, 94: e308-e317.

McCall, D.G. Clayton. J.B. and Dow, B.W. 1989. Nutrition effect on live weight and reproduction on Cashmere Doe hogget's. Proceedings of the New Zealand Society of Animal Production 49:157-161.

McEvoy, T.J., Robinson, J.J., AitkenFidley,P.A., Plmer, R.M. and Robertson,IS. 1995. Dietary induced suppression of preovulatory progesterone concentrations in superovulated ewes impair the subsequent *in vivo* and *in vitro* development of their ova. Animal reproduction Science.39:89-107.

Mc Kelvey W.A.C., Robinson J.J. , and Iitken, R.P.1988. The use of reciprocal embryo transfer to separate the effect of pre and post mating nutrition of embryo survival in growth of the ovine concetus . Proceedings of the 11th International Congress Animal Reproduction and AI (Dulbin)2.Paper 176.

Moustgaard, J. 1969. In: Reproduction in domestic animals. Cole, H.H. and Cupps, P.T (eds.) Academic Press,London.pp.489-516.

Naqvi, S.M.K., Gulyani,R., Joshi Anil., Das, G.K. and Mittal. J.P. 2002. Effect of dietary regimens on ovarian response and embryo production of sheep in tropics. Small Ruminant Research.46:167-171.

Parr, R.A. 1992. Nutrition progesterone interaction during early pregnancy in sheep. Reproduction Fertility Development.4:297-300.

Parr,R.A.,Davis,I.F. Fairclough,R.J and Miles,M.A.1987. Overfeeding during early pregnancy reduces peripheral progesterone concentration and pregnancy rate in sheep. Journal of Reproduction and fertility.80:317-320.

Parr,R.A., Davis,I.F. Miles,M.A.and Squires,T.J. 1993. Feed intake affects metabolic clearence rate of progesterone in sheep. Research in Veterinary Science.55:306-310.

Rhind, S.M and McNeilly,A.S. 1986. Follicle population, ovulation rate and plasma profiles of LH,FSH and Prolactin in Scottish Blackface ewes in high and low level of body condition. Animal reproduction Science.10:105-115.

Robinson, J.J.1983. Nutrition of the pregnant ewes. In: Haresign, W.(eds.) Sheep Production. Butterworths . London.pp.111-131.

Robinson, J.J. 1990 Nutrition in the reproduction of farm animals. Nutritional Research Reviews.3:253-276.

Robinson, J.J. 1990 The postural animal industries in the 21st century. Proceedings of the New Zealand Society Animal Production 50:345-359.

Robinson, J.J. 1996. Nutrition and reproduction. Animal Reproduction Science.42:25-34.

Robinson, J.J. and Symonds,M.E. 1995. Whole body fuel selection: "Reproduction" Proceeding of Nutritional Society. 54:283-299.

Salamon, S. 1964. The effect of nutritional regimen on the potential semen production of rams. Australian Journal of Agricultural Research. 15:645-646.

Sejian, V., Maurya V.P and Naqvi, S.M.K. 2011. Effect of thermal, nutritional and combined (thermal and nutritional) stresses on growth and reproductive performance of Malpura ewes under semi-arid tropical environment. Journal of Animal Physiology and Animal Nutrition, 95:252-258.

Sejian, V., Maurya V.P and Naqvi, S.M.K. 2010. Adaptability and growth of Malpura ewes subjected to thermal and nutritional stress. Tropical Animal Health and Production, 42:1763-1770.

Sejian, V., Maurya V.P and Naqvi, S.M.K. 2010. Adaptive capability as indicated by endocrine and biochemical responses of Malpura ewes subjected to combined stresses (thermal and nutritional) under semi-arid tropical environment. International Journal of Biometeorology, 54:653-661.

Sejian, V., Maurya, V.P., Naqvi, S.M.K., Kumar, D and Joshi, A. 2010. Effect of induced body condition score differences on physiological response, productive and reproductive performance of Malpura ewes kept in a hot, semi-arid environment. Journal of Animal Physiology and Animal Nutrition, 94(2): 154-161. (DOI: 10.1111/j.1439-0396.2008.00896.x).

Tajkaya K, Ariyasu H, Kanamoto N, et al. 2000. Ghrelin strongly stimulates growth hormone release in humans. J Clin Endocrinol Metab; 85:4908.

Thompson, J.G., Gardner,D.K., Pugh,P., AmcMillan, W.H.and Tervit,H.R. 1994. Lamb birth weight following transfer is affected by the culture system used for pre elongation development of embryos. Journal of Reproduction and Fertility. Abstract Series.13:25.

Wheeler, A.G. and Land R.B. 1977. Oestrus and ovarian activity of Finnish Landrace, Tashmanian Merino and Scottish Blackface ewes. Animal Production 24:363-376.

White, D. H., Bowman, P. J., Morley, F. H. W., Mc Manus, W. R., Filan, S. J., 1983. A simulating model of the breeding ewe flock. Agricultural System 10:149-189.

Williams, A.H. and Cumming, A.I. 1982. Inverse relationship between concentration of progesterone and nutrition in ewes. Journal of Agriculture Science Cambridge 98:517-522.

Zamorano PL; Mahesh VB; De Sevilla LM; Chorich LP; Bhat GK and Brann DW. 1997. Expression and localization of the leptin receptor in endocrine and neuroendocrine tissues of the rat. Neuroendocrinology 65(3):223.

14

Food Safety and Security in Relation to Animal Agriculture and Climate Change

Mahesh Chander[1] and Prakashkumar Rathod

[1]*Principal Scientist & Head, Division of Extension Education*
Indian Veterinary Research Institute, Izatnagar – 243122 (UP) India

The demand for agricultural and livestock products is rising and the challenge to feed 8 billion people in 2020 is considerably enormous. Further, it is also hard to imagine meeting 2050-projected demand by raising twice as many poultry, 78% more small ruminants, 58% more cattle and 37% more pigs, without further damaging natural resources (Rivera and Lopez, 2012). Hence, sustainable development based on balance of ecology, economics, norms and values are to be considered at various levels of the scale: between food and farming systems, regions, nations and continents (Zipp, 2003). The food systems encompass activities related to the production, processing, distribution, preparation and consumption of food; and the outcomes of these activities contributing to food security (Fig. 14.1). The interactions between and within bio-geo-physical and human environments influence both the activities and the outcomes.

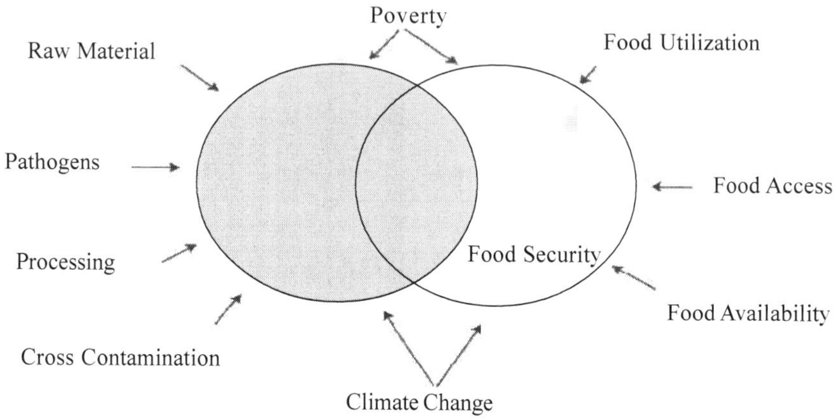

Fig. 14.1: Inter-relationship of food safety and food security (Hanning *et al.*, 2012).

14.1 Climate Change

Climate change can refer to: (i) long-term changes in average weather conditions (ii) all changes in the climate system, including the drivers of change, the changes themselves and their effects or (iii) only human-induced changes in the climate system. However, this indicates that there is no internationally agreed definition of the term "climate change". Climate change has significant potential impacts on the global food system, that include those on the water and energy used in food processing, cold storage, transport and intensive production, and those on food itself, reflecting higher market values for land and water and, possibly, payments to farmers for environmental services (FAO, 2008). The livestock sector in particular is considered to be the major contributor for climate change by emission of green house gases but can also deliver a significant share of the necessary mitigation effort (Gerber, *et al.*, 2013).

14.1.1 Drivers/Factors responsible for climate change

Agriculture and livestock production under various systems are contributing significantly directly or indirectly to the climate change through emission of various green house gases, deforestation and land degradation (Steinfeld *et al.*, 2006). On a global scale, the difference in emission intensities between various production systems is not substantial but industrial systems, however, account for majority of both total production and emissions (Gerber, *et al.*, 2013). Unfortunately, many national policies fail to promote livestock production or consumption in a way that is favourable to small farmers but aim at favouring wealthier producers, focusing on livestock and technical issues rather than on people and poverty reduction (Ahuja *et al.*, 2009). FAO's (2006) report "Livestock's Long Shadow" concluded that directly and indirectly, 18 per cent of the global Greenhouse Gas (GHG) emissions could be linked to animal-based production (Fig. 14.2). Hence, it is urgent to deal with a set of socio-economic drivers like (i) demographic growth, (ii) poor and negligible knowledge, customary practices and institutions in policy-making, and (iii) increasing integration of societies within the market economy (Rivera and Lopez, 2012). The major negative effect of existing livestock production system is in the form of methane emission, a greenhouse gas, which has higher global warming potential than ubiquitous carbon dioxide. There are two main sources in which methane emission takes place: (i) through enteric fermentation of feed in the animal's rumen, and (ii)due to manure management system by which dung-manure is managed. Though, methane emission from enteric fermentation depends upon several factors, quantity and quality of feed fed to animals, and the number of animals are highly responsible factors.

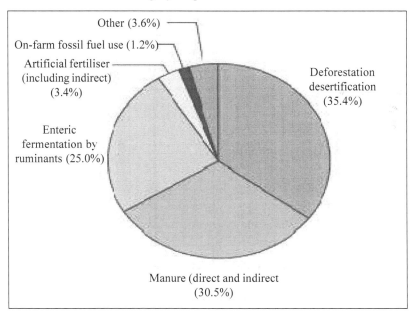

Fig.14.2: Proportion of greenhouse-gas emissions from different parts of livestock production (FAO, 2006)

About 44 per cent of the sector's emissions are in the form of methane (CH_4) followed by N_2O (29 %) and CO_2 (27 %) (Gerber *et al.*, 2013). Among the various species of livestock, cattle contribute about 65 per cent of sector emissions which includes beef cattle and dairy cattle generating almost similar amounts of GHG emissions. Pigs, poultry, buffaloes and small ruminants have much lower emission levels, with each representing between 7 and 10 percent of sector emissions (Gerber *et al.*, 2013). However, in ruminants' production, there is a strong relationship between productivity and emission intensity up to a relatively high level of productivity, emission intensity decreases as yield increases (Gerber *et al.*, 2013). When land use change impacts are included, the GHG contribution from livestock increases further (FAO, 2006). In Sweden and the Netherlands, for instance, it is estimated that consumption of meat and dairy products contributes about 45–50% to the global warming potential of total food consumption (Pathak *et al.*, 2010). Further, a study in India (Pathak *et al.*, 2010), estimates that only 13 per cent of total emissions come from post-farm gate stages. This seems to occur mainly because most people in developing countries consume fresh food mostly produced locally, with much less transport and refrigeration requirements than in rich countries. To date, no full "cradle-to-plate" estimates of global food system greenhouse gas emissions are available (Garnett, 2008). In many of the situations, gradual disruption of local traditional knowledge, abandonment of communal planning and institutions, increase in social differentiation, and overexploitation of the limited resources of rangelands has been very common phenomenon. Further, evident incapacity of modern institutions to adapt to mobile livelihoods, and the obsession of many governments to sedentary small livestock keepers (Rivera and Lopez, 2012) must be overlooked accordingly.

14.2 Food Security

FAO (1996) states that food security exists when all people at all times have physical or economic access to sufficient, safe and nutritious food to meet their dietary needs and food preferences for an active and healthy life. Food security broadly consists of four elements: availability (production, distribution and exchange), access (affordability, allocation, and preference), stability and utilization (nutritional value, social value, food safety). Lack of food worldwide against number of people is shown in Fig. 14.3.

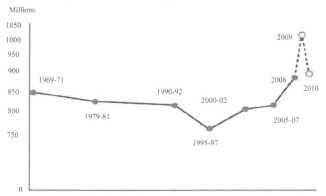

Fig. 14.3: Number of people, worldwide who lack food security 1969-2010 (Hanning *et al.*, 2012)

In recent times, food insecurity has increased due to competing claims for land, water, labour, and capital, leading to more pressure to improve production per unit of land. Rapid urbanization and industrialization in South Asia, for example, has taken away some very productive lands and good quality irrigation water (Fazal, 2000). For the farmers who rely on subsistence agriculture, food security is strongly dependent on local food availability, but for the majority who exchange cash, other commodities or labour for food, the access component is of critical importance, especially in relation to dietary diversity and nutrition. Further, in this context, impact of climate change will be felt in both rural and urban locations where supply chains are disrupted, market prices increase, assets and livelihood opportunities are lost, purchasing power falls, human health is endangered, and affected people are unable to cope (FAO, 2008).

14.3 Food Safety

Food contamination in the form of chemical, physical and microbiological hazards can pose a food safety threat and endanger human health. The food chain can stem from residues of agrochemicals, natural toxins, food-born microorganisms etc. which is the cause of demand for increased food production across the globe. In order to ensure that animal products are safe and marketable, countries need to detect, monitor and control these contaminants and trace the origin of food products. Various methods like food irradiation can reduce risks and also extend the shelf life of foods. The potential impacts of climate change on food safety and security must therefore be viewed within the larger framework of changing earth system dynamics and observable changes in multiple socio-economic and environmental variables (FAO, 2008). In the present scenario, effective and healthy linkages between food production systems and climate are most essential to achieve food safety and security in the changing climatic conditions. The paper highlights a brief discussion on drivers, impact, adaptation and mitigation strategy and role of various stakeholders in countering the climate change issues.

14.4 Impact of Climate Change on Food Safety and Security

Presently, number of undernourished people is among the highest in human history (FAO, 2010) having 925 million people suffering from hunger. When dealing with this issue, the first solution that people and many institutions tend to propose is to produce more food which is not always true. In this context, FAO (2009) has pointed out that, 925 million people are undernourished, not because of global food supply is deficient, but because they cannot afford to buy food or they live in places or societies where it is hard to obtain. Further, there is also a dynamics involved with potential climate change impacts in a holistic food system. The livestock production is likely to be affected both directly like productivity losses (physiological stress) owing to temperature increase, and indirectly by changes in the availability, quality and prices of inputs such as fodder, water etc. (Thornton, 2010). Further, the structure and function of biodiversity will be widely affected for terrestrial, freshwater and marine ecosystems (Walther *et al.*, 2002; Parmesan, 2006).Impacts will also be felt on grassland productivity and species composition and dynamics, resulting in changes in animal diets and possibly reduced nutrient availability for animals (Thornton *et al.*, 2009). It is anticipated that more frequent extreme weather events under climate change will damage infrastructure, with detrimental impacts on food storage and distribution, to which the poor will be most vulnerable (Costello *et al.*, 2009).Further, increasing temperature may cause food quality to deteriorate, unless there is increased investment in cooling and refrigeration equipment or more reliance on rapid processing of perishable foods to extend their shelf-life (FAO, 2008).At the same time, studies (Hesman 2002; Nagarajan *et al.*, 2010) have indicated that there are reports indicating that the nutritional value of food, especially cereals, may also be affected byclimate change and climate change will alsoaffect the ability of individuals to use food effectively by altering the conditions for food safety and food-borne diseases (Schmidhuber and Tubiello, 2007).

The impacts of mean temperature increase will be experienced differently, depending on location (Leff *et al.*, 2004). For example, moderate warming (increases of 1 to 3 °C in mean temperature) is expected to benefit crop and pasture yields in temperate regions, while in tropical and seasonally dry regions, it is likely to have negative impacts, particularly for cereal crops. For climate variables such as rainfall, soil moisture, temperature and radiation, crops have thresholds beyond which growth and yield are compromised (Porter and Semenov, 2005).A change in the geographic distribution of foods resulting from changing rainfall and temperatures could therefore have an impact on the availability of food. Conflicts over water resources will have implications for both food production and people's access to food in conflict zones (Gleick, 1993). Climate change will cause new patterns of pests and diseases to emerge, affecting plants, animals and humans, and posing new risks for food security, food safety and human health (FAO, 2008) which further, increase incidence of diseases could affect both the food chain and people's physiological capacity to obtain necessary nutrients from the foods consumed (FAO, 2008). Since majority of the crops have annual cycles, and yields fluctuate with climate variability, particularly rainfall and temperature which may cause a threat to food stability and couldbring about both chronic and transitory food insecurity (FAO, 2008). Viewing food security from a livelihoods perspective makes it possible to assess the different components of food security holistically at the household level. Although, Morton (2007) has remarked that, the impact of climate change on small farming systems will be difficult to model or predict, it may affect both the physical and the economic availability of certain preferred food items affecting the ability to buy food and may influence their price also (Edame *et al.*, 2011). High prices may make certain foods unaffordable and can have an impact on individuals' nutrition and health.

14.5 Adaptation Strategies to Climate Change

Adaptation to climate change involves deliberate adjustments in natural or human systems and behaviour to reduce the risks to people's lives and livelihoods. Acclimatization is a powerful and effective adaptation strategy. In simple terms, it means getting used to climate change and learning to live comfortably with it. All living organisms, including humans, adapt and develop response to changes in climate and habitat. The main climate change related hazards affecting small-scale livestock farming systems that call for adaptation strategies are: increased temperature, changes in seasonal rainfall patterns and more erratic rainfall, higher prevalence of weather extreme events and higher atmospheric concentrations of CO_2 (Rivera and Lopez, 2012). Adaptation to climate variability is a never-ending process, because vulnerabilities and impacts are permanently evolving, which means that some forms of adaptation that prove appropriate now, may not prove so in the future (Rivera and Lopez, 2012). Hence, the major challenge is therefore, to enable accelerated adaptation without threatening sensitivelivelihood systems as they strive to cope with stress. The key question for both food security and the agricultural economy is whether the food system can keep pace with growing demand in the face of climate and other drivers (Hazell and Wood, 2008). Hence, protecting local food supplies, assets and livelihoods against the effects of increasing weather variability and increased frequency and intensity of extreme events is very essential. There is a need to manage risks specific to particular ecosystems, have need based research and dissemination suitable to changing climate and further introduce cultivation of tree crops for food, fodder and energy and enhance cash incomes. Protecting ecosystems is the most priority area which is possible by use of degraded or marginal lands for productive planted forests or other cellulose, overcoming abiotic stresses in crops through crop breeding is proven to be an effective means of increasing food production (Evenson and Gollin, 2003), and arguably mitigating climate change effects (Burney *et al.*, 2010). Watershed protection, prevention of land degradation, protection and preservation of biodiversity are also important adaptation strategies to be considered. Effective and efficient utilization of water must be seriously addressed , since the proportion of people living in water-stressed regions is expected to rise to 64 per cent by 2025 in comparison with 38 per cent in 2002 (Rosegrant *et al.*, 2002). Since, livestock sector is a major user of fresh water, it is estimated that 1 kg of edible beef requires 12,000 litres of water in grazing systems, while up to 53,200 litres in intensive systems (Steinfeld *et al.*, 2010). More efficient management of irrigation water, improved management of cultivated land, improved livestock management and use of new, more energy-efficient technologies can also be effective adaptation measures.

Today's farming system has broad-scale analyses that identify regions and crops that are sensitive to progressive climate change (Parry *et al.*, 2007; Lobell *et al.*, 2008), but there is sparse scientific knowledge as to how current farming systems can adapt, and which current farming systems and agricultural practices will enable adaptation. However, the knowledge, institutions and customary practices, highly adapted to the local conditions and developed throughout centuries of co-evolution with changing environments, can be of great value to adapt the whole livestock sector to the current situation of increased climate variability (Rivera and Lopez, 2012).

14.6 Mitigating Strategies Against GHG Emissions

IPCC (2001) defines mitigation as an anthropogenic intervention to reduce the sources or enhance the sinks of greenhouse gases. Mitigation of climate change involves actions to reduce greenhouse gas emissions and sequester or store carbon in the short term, and development choices that will lead to low emissions in the long term. Adoption of best practices for mitigation in livestock sector can have multiple payoffs for food security, including contributing to the stability of global food markets and providing new employment opportunities in the commercial agriculture sector, as well enhancing the sustainability of vulnerable livelihood systems. Therefore, although there are opportunities for reducing the carbon footprint of food at all stages of the food chain, the focus of mitigation efforts in the food system should be on introducing agricultural production practices that reduce emissions or increase carbon sequestration. The positive contributions of livestock to the environment include (i) saving of chemical fertilizers due to dung used as manure and prevention of greenhouse emission (ii) saving of natural resource, land, due to use of dung as fuel (iii) saving of land due to recycling of agricultural by-products as animal feed: and (iv) prevention of greenhouse gases emission especially of carbon dioxide, due to animal draught power used in agriculture (Dikshit and Birthal, 2013). Emissions of CO_2 can be reduced by controlled deforestation, wildfires, adoption of alternatives to the burning of crop residues after harvest etc. Further, reducing emissions of methane and nitrous oxide is possible through improved nutrition for ruminant livestock, more efficient management of livestock waste and efficient applications of nitrogen fertilizers on the fields. Carbon sequestration involves increasing the carbon storage in terrestrial systems, above or below ground. Carbon sequestration is possible by improved management of soil organic matter, with conservation agriculture involving permanent organic soil cover, minimum mechanical soil disturbance and crop rotation and improved management of pastures and grazing practices. Many grasslands increase biomass production in response to frequent grazing, which, when managed appropriately, could increase the input of organic matter to grassland soils (Calvosa et al., 2009). However, Schader et al.(2013) has indicated that global environmental impacts can be mitigated if livestock production is grassland-based. At the global level, however, meaningful carbon sequestration through reforestation and afforestation would require that more new trees be planted each year than were lost to deforestation in the previous year. Selection of faster growing breeds has better efficiency in converting energy from feed into production and losses through waste products can be reduced. The composition of feed has some bearing on enteric fermentation and the emission of CH_4 from the rumen or hindgut (Dourmad et al., 2008). Since many developing countries are striving to increase production from ruminant animals, improvements in production efficiency are urgently needed to meet goals while avoiding increased methane emissions (FAO, 2008). Dietary factors such as composition of diet, type of forage, processing, feeding frequency, nature of concentrate, type of starch (slowly versus rapidly degraded), level of intake influences methane emission from the rumen by altering surface area for microbial activity, ruminal pH, flow rate of digesta and fermentation pattern (Malik et al., 2012). Soliva et al. (2008) from their study concluded that various plant materials, including accessions of Acacia angustissima, Sesbaniasesban and Cajanuscajan, were promising to approach the goal of mitigating the methane emission from small ruminants. Although manure is considered a waste product, it contains important amounts of nitrogen, phosphates and potassium that provide valuable soil nutrients when applied to farmers' fields. Hence, animal waste management through different mechanisms such as the use of covered storage facilities is very important since the level of GHG emissions from manure depends on the temperature and duration of storage. The widespread and often poorly controlled use of animal waste as fertilizer can also lead to substantial emissions of nitrous oxide from agricultural soils (FAO, 2008).Various biological strategies to reduce ruminal methanogenesis such as reductive acetogenesis and immunization play an essential role in mitigation(Immig et al., 1996; Leahy et al., 2010).Very recently, mitigation through market mechanisms has been a changing trend across the globe. In the past few years, the global carbon trading has increased by 83 per cent (Paul et al., 2009). The strategy based on the notion of 'sustainable intensification', means improving yields without damaging ecosystems (Garnett, 2011). It includes measures such as crop and animal breeding, feed optimization and dietary additives, and pest and disease management. This productivity-oriented approach is promoted on the principle of 'land sparing',whereby intensive production takes place on the smallest possible area in order to maximize exploitation of available land for conservation or forestry (Garnett, 2010). The mitigation of GHG emissions through behavioural modification is probably the least employed and least investigated group of mitigation strategies. The fundamental purpose of the behavioural modification is to foster ever-growing 'climate-smart diets' focussing on reduction in amount of meat consumption and consuming organically-produced food (Rivera and Lopez, 2012). On average, 25 kcal fossil energy is used per kcal of meat produced, compared with 2.2 kcal for plant-based products (Pimentel and Pimentel, 2003). However, McMichael *et al.* (2007) has suggested that meat consumption should decrease in industrialized countries and increase in developing countries as an effective mitigation strategy. Unmitigated climate change will, in the long term, exceed the capacity of natural and human systems to adapt. Given the magnitude of the challenge to reduce GHG concentrations in the atmosphere, it is imperative to receive the contribution of all sectors with significant mitigation potential.

14.6.1 Reorienting farming systems

The agriculture and livestock production system has to reorient to meet the changing climatic conditions across the globe.

14.6.1.1 Conservation Agriculture

Conservation agriculture is an approach to manage agro-ecosystem for improved and sustained productivity, increased profits and food security while preserving and enhancing the resource base and the environment. It is characterized by three linked principles *viz.* Continuous minimum mechanical soil disturbance, permanent organic soil cover and diversification of crop species grown in sequence or associations (FAO, 2012). Mechanized soil tillage allows higher working depths and speeds and involves the use of such implements as tractor-drawn ploughs, disk harrows and rotary cultivators. This initially increases fertility because it mineralizes soil nutrients and makes it easier for plants to absorb them through their roots. In the long term, however, repeated ploughing and mechanical cultivation breaks down the soil structure and leads to reduced soil organic matter and loss of soil nutrients. This structural degradation of soils results in compaction and the formation of crusts, leading to soil erosion. Farming systems that successfully integrate crop and livestock enterprises stand to gain many benefits that can have a direct impact on whole farm production. Ruminant animals are especially desirable due to their ability to convert forages, browse and crop residues high in cellulose to useful food and fibre products. Such animals provide for: system diversification; recycling of nutrients; soil enhancing rotation crops; power and transportation; and act as biological "savings accounts" for farmers during periods of stress.

14.6.1.2 Climate-Smart Agriculture

Climate-smart agriculture seeks to increase productivity in an environmentally and socially sustainable way, strengthen farmers' resilience to climate change, and reduce agriculture's contribution to climate change by reducing greenhouse gas emissions and increasing carbon storage on farmland. The climate-smart agriculture includes proven practical techniques in many areas, especially in water management but also innovative practices such as better weather forecasting, early warning systems, and risk insurance. It is about getting existing technologies into the hands of farmers and developing new technologies such as drought or flood tolerant crops to meet the demands of a changing climate. Achieving Climate-Smart agriculture needs an integrated approach, tackling productivity and food security, risk and resilience, and low carbon growth together, but integration and institutional coordination remains a challenge in many countries (World Bank, 2011).

14.6.1.3 Sustainable Agriculture

There is a need for "rapid and significant shift from industrial monocultures and factory farming towards mosaics of sustainable production systems that are based on the integration of location-specific organic resource inputs; natural biological processes to enhance soil fertility; improved water-use efficiency; increased crop and livestock diversity that is well adapted to local conditions and integrated livestock and crop farming systems" (IAASTD, 2009). Producing more crops from less land is the single most significant means of jointly achievingmitigation and food production in agriculture, assuming that the resulting "spared land" sequesters more carbon or emits fewer GHGs than farm land (Robertson et al., 2000). This"land sparing" effect of intensification is uneven in practice and requires policies and price incentives to strengthen its impacts (Angelsen and Kaimowitz, 2001). More efficient use of inputs, more sustainable alternatives and breeding for efficiency will be required to reduce the carbon intensity (emissions per unit yield) of products, as well as reduce land areas and inputs that damage environmental health (Tillman et al., 2002).

14.6.1.4 Precision Livestock Farming

Spilke and Fahr (2003) stated that Precision Farming aims for an ecologically and economically sustainable production with secured quality, as well as a high degree of consumer and animal protection with specific emphasis on technologies for individual animal monitoring. Precision farming is based on information technology, which enables the producer to collect information and data for better decision making. This concept is considered by some as the future of agriculture and allied sectors. This concept is also called as spatially prescriptive farming; computer aided farming; farming by satellite; high-tech sustainable agriculture;

soil-specific crop management; site-specific farming; and precision farming. The main objectives of precision farming are maximizing individual animal potential, early detection of disease, and minimizing the use of medication through preventive health measures. Although, very vast research has been conducted on precision agriculture, but livestock sector has negligible studies on precision farming. International studies have also shown slow adoption rates of precision farming (Batte and Arnholt, 2003) due to small farm size, farmer age, education level, computer illiteracy etc. The advantages posed by the technology are often not immediately apparent and they require more management expertise along with an investment of time and money to realize (Bell, 2002).

14.6.1.5 Organic Livestock Farming

Organic animal husbandry has been defined as a system of livestock production that promotes the use of organic and biodegradable inputs from the ecosystem deliberately avoiding the use of synthetic inputs such as drugs, feed additives and genetically engineered breeding inputs, while ensuring the welfare of animals (Chander et al., 2011). There are four principles of organic farming viz; principle of ecology, principle of health, principle of fairness, and principle of care, which organic systems must always take into consideration. In order to achieve the animal welfare, environmental protection, resource-use sustainability and other objectives, certain key principles are adhered to under organic livestock production systems. Subrahamanyeswari and Chander (2008) have depicted that majority of the livestock production practices were in line with what the organic standards recommend, and thus, are compatible to organic systems. Since, majority of the animal husbandry practices followed by the farmers were favourable to or closer to the recommended organic livestock production standards which is a clear indication of becoming organically certified by effective interventions. In this context, public and private organizations have a major role to make livestock farmers more compatible with the organic standards and also act as an effective solution for the changing climatic conditions.

14.6.1.6 Role of Science and Society

It is almost unrealistic to mitigate the methane emission using customary approaches, and hence, different stakeholders have to realize their role and responsibilities in effective adaptation and mitigation strategies. Hence, a multi-stakeholder analysis about knowledge and their contribution from local to policy levels must be effectively oriented. The present study has attempted to address different stakeholders like policy makers, researchers/scientists, extension/ outreach departments, farming community etc.

14.6.1.7 Government and Policy Makers

The government can promote financial instruments such as incentives, subsidies, loans, microfinance schemes and regulations for improving natural resource management and livestock production systems. Effective care must be taken for proper risk management mechanisms and preparedness to cope with the impacts of more frequent and extreme climatic events. Significant additional research is needed to further assess the costs and benefits of mitigation practices, to help policy-makers understand which policy options are better placed to incentivize their uptake. Improvements in production efficiency can have strong ramifications for land-use change, because they can lower the amount of inputs required, including land for grazing and feed production, to produce any given level of output. The labelling and certification programmes can help to incentivize mitigation by informing consumers (including livestock producers) about the emission attributes of products at different stages along livestock supply chains. Research into mechanisms to create influential knowledge generally requires active collaborations between researchers and particular decision-makers, with trusted intermediaries or "boundary spanners" often playing a crucial integrative role (Agrawala *et al.*, 2001, Cash *et al.*, 2003).

14.6.1.8 Research and Development Organizations

A major challenge for the research community is to understand not only the impacts, but also the interactions among components of the farming system (Tubiello *et al.*, 2007). The research and development wing has to build the evidence base for mitigation technologies/practices and play an important role in refining existing technologies/practices to increase their applicability and affordability. Efficient and affordable adaptation practices may include.

(i)   provision of shade and water to reduce heat stress from increased temperature.

(ii) rearing of less number of more productive animals lead to more efficient production and lower GHG emissions (Batima, 2007).

(iii) changes in livestock/herd composition (selection of large animals rather than small).

(iv) improved management of water resources etc.

Further, research institutes have to develop need based models like SOL-m (Schader *et al.*, 2013) which can analyze the impacts of different production scenarios on land use, food availability, material flow (i.e. N, P, energy, GHG) and other environmental impacts to meet the challenges and propose strategies accordingly.

14.6.1.9 Extension/Outreach Departments

The research and extension institutes must realize the importance of community involvement in the identification of solutions to ensure the long-term sustainability of interventions. Further, dissemination of information on crucial issues of climate change and its effects must be ensured on timely basis.Giving farmers and other agricultural stakeholders a more effective voice in the design of climate information products and services can effectively bridge this gap. Greater investment is also needed in the capacity of rural communities to access, interpret and act on climate-related information. Although, tools for linking knowledge with action are increasingly tested and applied by interdisciplinary, multi-organizational research-for-development teams (Kristjanson *et al.*, 2009), approaches like communication, training, demonstration farms and networks must facilitate linkages among sector stakeholders. The extension attempts have to raise awareness about livestock's role in tackling climate change to influence and promote mitigation policy development for the sector. Various methods viz. participative mapping of impact pathways (Douthwaite *et al.*, 2007), negotiation tools informed by research (van Noordwijk *et al.*, 2001), social network analysis, innovation histories, cross-country analyses and game-theory modelling (Spielman *et al.*, 2009) must be emphasized.

14.6.1.10 Farming Community

Agriculture by effective utilization of land and its resources, promoting agro-biodiversity is an option for adaptation and mitigation through improving resilience to changing environmental conditions and stresses. Low feed quality is a major obstacle to optimal livestock production and is a challenge for many farmers to feed high quality feed and fodder. But, balanced diets can fetch efficient growth and production. Further, optimum use of fertilizer and water and land use can be a major step from the farming community. FAO (2008) has indicated that vulnerable people can strengthen resilience by protecting existing livelihood systems, diversify their sources of food and income, change their livelihood strategies or migrate if there is no other option. Overriding human-influenced factors like overgrazing, deforestation etc. can cause serious issues to all the stakeholders in the coming future. Hence, improved grazing management, such as optimized stock numbers, improved pasturing and rotational grazing can result in substantial increases in carbon pools. A major study, *Water for food, water for life*, released in 2007 by Earth scan and the International Water Management Institute (IWMI), reveals that one in three people today face water shortages (CA, 2007) which indicated that there is a need for effective utilization of water also. In short, climate change demands rethinking and effective linkages between various stakeholders for addressing the issue of food security in the era of changing climate.

14.7 Conclusion

The challenge of food security is to assure that all people have access to quality and quantity food assuring the safety from chemical, physical or biological aspect. In this context, research and extension in livestock sector, food security and climate change must continue to improve understanding of uncertainty, to allow more confident decision-making and allocation of limited resources towards new climatic futures. The strategies to be developed must be more dependent on the stakeholders, which include both consumers and citizens. The challenge is, therefore, to produce possible combinations of traditional animal production systems with extra quality features; combinations of livestock farming with resource management; organic farming including nature conservation; and experience-based animal production for medical, educational and religious functions. Hence, such a combination of systems can effectively give rise to food security and safety in the context of climate change.

References

Angelsen, A. and Kaimowitz, D. 2001. Agricultural technologies and tropical deforestation (eds). CABI Publishing, Oxon, UK.

Agrawala, S., Broad K. and Guston D.G. 2001. Integrating climate forecasts and societal decision making: Challenges to an emergent boundary organization. Science, Technology and Human Values, 26(4): 454–477.

Ahuja, V., Gustafson, D., Otte, J. and Pica-Ciamarra, U. 2009. Supporting livestock sector development for poverty reduction. Pro-Poor LivestockPolicy Initiative Research Report, FAO, Rome.

Batte, M. T. and Arnholt, M. W. 2003. Precision farming adoption and use in Ohio: Case studies of six leading-edge adopters. Computers and Electronics in Agriculture, 38 (2):125-139.

Batima, P.2007. Climate change vulnerability and adaptation in the livestock sector of Mongolia. Assessments of impacts and adaptations to climate change. International START Secretariat, Washington DC, US.

Bell, C. J. 2002. Internet delivery of short courses for farmers: A case study of a course on Precision Agriculture, Publication No. 02/085, Project No. GOC-1A, A report for the Rural Industries Research and Development Corporation, accessed 30 June, 2008 from http://www.rirdc.gov.au/reports/HCC/02-085.pdf

Burney, J. A., Davis, S. J. and Lobel, D. B. 2010.Green house gas mitigation by agricultural intensification. Proceedings of National Academy of Science, USA,107(26): 12052- 12057.

CA. (2007).Water for food, water for life: A comprehensive assessment of water management inAgriculture. London, Earthscan and Colombo, IWMI.

Calvosa, C.,Chuluunbaatar, D. and Fara, K. 2009.Livestock and climate change. Livestock Thematic Paper of International Fund for Agricultural Development, Rome.

Cash, D.W., Clark, W. C., Alcock, F., Dickson, N.M.,Eckley, N.,Guston, D.H.,Jager J. and Mitchell, R.B. 2003.Knowledge systems for sustainable development. Proceedings of National Academy of Sciences, 100(14): 8086-91.

Chander, M., Subrahmanyeswari, B., Mukherjee, R. and Kumar, S. 2011. Organic livestock production: An emergingopportunity with new challenges for producersin tropical countries.Rev. Sci. Tech. Off. Int. Epiz., 30(3): 969-983.

Costello, A., Abbas, M., Allen, A., Ball, S., Bell, S., Bellamy, R.,Friel, S.,Groce, N., Johnson, A.,Kett, M., Lee, M., Levy, C., Maslin, M., McCoy, D., McGuire, B., Montgomery, H.,Napier, D.,Pagel, C., Patel, J.,Antonio, J., de Oliveira, P.,Redclift, N., Rees, H.,Rogger, D., Scott, J., Stephenson, J.,Twigg, J., Wolff J. and Patterson, C. 2009. Managing the health effects of climate change.*The Lancet*, 373(9676): 1693-1733.

Dikshit, A.K. and Birthal, P.S. 2013.Positive environmental externalities of livestock in mixed farming systems of India.Agric. Econ. Res. Review, 26(1): 21-30.

Dourmad, J., Rigolot, C.and Hayo van der Werf. 2008. Emission of greenhouse gas: Developing management and animal farming systems to assist mitigation. Livestock and Global Change conference proceeding. May 2008, Tunisia.

Douthwaite, B., Alvarez, B.S., Cook, S., Davies, R., Howell, G.P., Mackay, R. and Rubiano, J. 2007. The impact pathways approach: A practical application of program theory in research-fordevelopment.Canadian J. Program Evaluation, 22(2): 127-159.

Edame, G.E., Ekpenyong, A. B., Fonta, W. M. and Duru, E.J.C. 2011. Climate change, food security and agricultural productivity in Africa: Issues and policy directions. Int. J. Humanities and Social Sci. Special Issue,1 (21):205-223.

Evenson, R.E. and Gollin, D.2003. Assessing the impact of the green revolution, 1960 to 2000.Science, 300:758–762

FAO.2008.Climate change and food security: A framework document.Food and Agriculture Organization of the United Nations, Rome.

FAO.2012. http://www.fao.org/ag/ca/1a.html

FAO.1996. Rome Declaration and World Food Summit Plan of Action, Rome. Available at: www.fao.org/docrep/003/X8346E/x8346e02.htm#P1_10.

Fazal, S. 2000. Urban expansion and loss of agricultural land: A GIS based study of Saharanpur City, India. Environment and Urbanization, 12(2): 133-149.

FAO.2006. Livestock's long shadow: Environmental issues and options.Food and Agriculture Organisation, Rome.

FAO.2010. The State of Food Insecurity in the World. Food and Agriculture Organisation, Rome.

FAO.2009. Grasslands: Enabling their potential to contribute to greenhouse gas mitigation. Submission by the UN Food and Agriculture Organization to the Intergovernmental Panel on Climate Change. FAO, Rome.

Garnett, T.2010. Where are the best opportunities for reducing greenhouse gas emissions in the food system (including the food chain)?.Food Policy (In press).

Garnett, T. 2008. Food climate research network, centre for environmental strategy, cooking up a storm: Food, greenhouse gas emissions, and our changing climate (University of Surrey, UK).

Garnett, T. 2011. Where are the best opportunities for reducing greenhouse gas emissions in the food system (including the food chain)?.Food Policy, 36(1): S23–S32.

Gerber, P.J., Steinfeld, H., Henderson, B., Mottet, A., Opio, C., Dijkman, J., Falcucci, A. andTempio, G. 2013.Tackling climate change through livestock – A global assessment of emissions and mitigation opportunities.Food and Agriculture Organization of the United Nations (FAO), Rome.

Gleick, P.H. 1993. Water in crisis: A guide to the world's fresh water resources. New York, Oxford University Press.

Hanning, I. B., O'Bryan, C. A., Crandall, P. G. andRicke, S. C. 2012.Food safety and food security.Nature Education Knowledge. 3(10): 09

Hazell, P. and Wood, S.2008. Drivers of change in global agriculture. Philosophical Transactions of the Royal Society B, 363(1491): 495-515.

Hesman, T. 2002. Carbon dioxide spells indigestion for food chains. Science News, 157: 200–202.

IAASTD.2009. Agriculture at a crossroads. The Global Report.International Assessment of Agricultural Knowledge, Science and Technology for Development, Island Press.

Immig, I., Demeyer, D., Fielder, D., Van Nevel, C. and Mbanzamihigo.1996. Attempts to induce reductive acetogenesis into a sheep rumen. Arch Tiererahr,49(4): 363–70.

IPCC.2001. Climate change 2001: Impacts, adaptation and vulnerability. McCarthy, J.J., Canziani, O.F., Leary, N.A., Dokken, D.J., and White, K.S., (eds.), Cambridge: Cambridge University Press.

Kristjanson, P., Reid,R., Dickson,N., Clark,W.C., Romne,D., Puskur,R., MacMillan, S. and Grace, D.2009.Linking international agricultural research knowledge with action for sustainabledevelopment. Proceedings of the National Academy of Sciences 9(13): 5047-5052.

Leahy, S. C., Kelly, W. J., Altermann, E., Ronimus, R. S., Yeoman, C. J., Pacheco, D. M., Li, D., Kong, Z., McTavish, S., Sang, C., Lambie, S. C.,Janssen, P. H., Dey, D. and Attwood, G. T. 2010. The genome sequence of the rumen methanogen methanobrevibacterruminantium reveals new possibilities for controlling ruminant methane emissions. PLoS ONE e8926 5(1): 1-17.

Leff, B., Ramankutty, N. and Foley, J. 2004. Geographic distribution of major crops across the world. Article No. GB1009 in Global Biogeochemical Cycles, 18(1).

Lobell, D.B., Burke, M.B., Tebaldi, C.,Mastrandrea, M.D., Falcon, W.P. and Naylor, R.L. 2008.Prioritizing Climate Change Adaptation Needs for Food Security in 2030. Science, 319(5863): 580-581.

Malik, P. K., Singhal, K. K., Deshpande, S. B. and Siddique, R. A.2012. Mitigation strategies for enteric methane emission with special emphasis on biological approaches: A review. Indian J. Anim. Sci., 82(8): 794–804.

McMichael, A.J., Powles, J.W., Butler, C.D. and Uauy, R. 2007. Food, livestock production, energy, climate change, and health. The Lancet, 370 (9594) : 1253–1263.

Morton J.F. 2007. The impact of climate change on smallholder and subsistence agriculture.Proceedings of National Academy of Science, USA.104(50): 19680- 19685.

Nagarajan, S., Jagadish, S.V.K., Hari Prasad, A.S., Thomar, A.K., Anand, A., Pal M. and Aggarwal, P.K. 2010. Local climate affects growth, yield and grain quality of aromatic and non-aromatic rice in northwestern India. Agriculture, Ecosystems & Environment,38(3-4): 274-281

Parmesan, C. 2006. Ecological and evolutionary responses to recent climate change. Annual Review of Ecology, Evolution and Systematics, 37: 637-69.

Parry, M.L., Canziani, O.F.,Palutikof, J.P., van der Linden, P.J. and Hanson, C.E. (eds.) 2007. Climate change, 2007: Impacts, adaptation and vulnerability. Contribution of Working Group II to the Fourth Assessment Report of the Intergovernmental Panel on Climate Change. Cambridge University Press, Cambridge.

Pathak, H., Jain, N., Bhatia, A., Patel, J., Aggarwal, P.K.2010.Carbon footprints of Indian food items. Agriculture, Ecosystems and Environment, 139: 66-73.

Paul, H., Ernsting, A., Semino, S., Gura, S., Lorch, A.2009. Agriculture and climate change: Real problems, false solutions. Report published for the 15th Conference of the Parties of UNFCCC in Copenhagen, December. Available at: www.econexus.info/sites/econexus/files/Agriculture_climate_change_copenhagen_2009.pdf.

Pimentel, D. and Pimentel, M.2003.Sustainability of meat-based and plant-based diets and the environment. American J Clin.Nutr.,78(suppl):660S–3S.

Porter, J.R. and Semenov, M.A. 2005. Crop responses to climatic variation. Philosophical Transactions of the Royal Society B: Biological Sciences, 360: 2021-2035.

Robertson, G.P., Paul, E.A. and Harwood, R. R. 2000. Greenhouse gases in intensive agriculture: Contributions of individual gases tothe radiative forcing of the atmosphere. *Science* 289(5486): 1922.

Rivera, F. M. G. and Lopez-i-Gelats, F.2012.The role of Small-Scale Livestock Farming in Climate Change and Food Security.

Rosegrant, M. W., Cai, X., Cline, S.A. and Nakagawa, N. 2002. The role of rainfed in the future of global food production. EPTD Discussion Paper N0. 90. Environment and Production Technology Division (EPTD), International Food Policy Research Institute (IFPRI), Washington, D.C.

Schader, C., Muller, A. and Scialabba, N. E. 2013. Sustainability and organic livestock modelling (SOL-m)-Impacts of a global up-scaling of low-input and organic livestock production.Preliminary Results of Food and Agriculture Organization of the United Nations (FAO), Rome.

Schmidhuber, J. and Tubiello, F. N. 2007.Global food security under climate change.Proceedings of the National Academy of Science, 104 (50):19703-19708.

Soliva, C. R., Zeleke, A. B., Clement, C., Hess, H. D., Fievez, V. and Kreuzer, M. 2008.In- vitro screening of various tropical foliages, seeds, fruits and medicinal plants for low methane and high ammonia generating potentials in the rumen. Anim. Feed Sci. and Tech., 147: 53–71.

Spielman, D.J., Ekboir, J. and Davis, K. 2009. The art and science of innovation systems inquiry: Applications toSub-Saharan African agriculture. Tech. in Society, 30 (2009): 1–7

Spilke, J. and Fahr,R. 2003.Decision support under the conditions of automatic milking systems using mixed linear models as part of a precision dairy farming concept.pp. 780-785 in EFITA Conference, Debrecen, Hungary.

Steinfeld, H., Gerber, P., Wassenaar, T., Castel, V., Rosales, M. and Haan, C. D.2006. Livestock's long shadow: Environmental issues and options. FAO, Rome.

Steinfeld, H., Mooney, H.A., Schneider, F., Neville, L.E., (Eds.) 2010.Livestock in a changing landscape.drivers, consequencesand responses (Volume 1). Island Press, Washington.

Subrahamanyeswari, B. and Mahesh, C.2008. Compatibility of animal husbandry practices of registered organic farmers with organic animal husbandry standards (OAHS): an assessment in Uttarkhand. Indian J. Anim. Sci.,78 (3): 322-327.

Tillman, D., Cassman, K.G., Matson, P.A., Naylor, R. and Polasky, S.2002. Agricultural Sustainability and Intensive Production Practices. Nature, 418: 671–677.

Thornton, P.K. 2010. Livestock production: Recent trends, future prospects. Philosophical Transactions of the Royal Society B, 365: 2853-2867.

Thornton, P.K., Jones, P.G.,Alagarswamy, A. and Andresen, J.2009. Spatial variation of crop yield responses to climate change in East Africa. Global Environmental Change, 19(1): 54-65.

Tubiello, F.N., Amthor, J.A., Boote, K., Donatelli, M., Easterling, W.E., Fisher, G., Gifford, R., Howden, M., Reilly, J. and Rosenzweig, C.2007. Crop response to elevated CO2 and world food supply. European J. Agronomy, 26: 215-228.

vanNoordwijk, M., Tomich, T. and Verbist, B.2001. Negotiation support models for integratednatural resource management in tropical forest margins. Conservation Ecology, 5(2): 21.

Walther, G.R., Post, E., Convery, P., Mezel, A., Parmesan, C., Beebee, T.J.C., Fromentin, J.M., Hoegh-Guldberg, O. and Bairlein, F. 2002. Ecological responses to recent climate change. Nature, 416: 389-95.

World Bank.2011.Climate-smart agriculture: Increased productivity and food security, enhanced resilience and reduced carbon emissions for sustainable development. Report of Agriculture and Rural Development, World Bank.

Zipp, J. V. 2003.Future of livestock production in Latin America and cross-continental developments. Arch. Latinoam. Prod. Anim., 11(1): 50-56.

15

Conservation of Native Breeds Under Projected Climate Change Scenario

K.C. Raghavan and M. Manoj

Centre for Advanced Studies in Animal Genetics and Breeding
College of Veterinary and Animal Sciences, Kerala Veterinary and Animal
Sciences University, Mannuthy, Thrissur, Kerala – 680651

In India, being predominantly an agricultural country, livestock is emerging as a driving force in the growth of agricultural sector. Contribution of livestock to agriculture GDP has been rising steadily during the last three decades from 13.88% in 1980-81 to 27.28% in 2010-11 (BAHS 2012). About 70% population of the Country is engaged in agriculture and more than 50% of people below poverty line are associated with livestock production. India is blessed with a rich genetic resource of bovines, with an estimated number of 304.76 million (Livestock Census, 2007), which is 19% of the world. India continues to be the highest milk producing country in the world with an estimated production of 121.8 million tonnes during 2010-11 of which more than 53% is contributed by buffaloes (105.34 millions). Similarly, India ranks first in buffalo, second in cattle and goat, third in sheep, fourth in duck, fifth in chicken and sixth in camel population of the world. India is the third largest egg producer in the world (BAHS 2012). India posses a large treasure of domestic animal biodiversity with 37 breeds of cattle, 13 breeds of buffalo, 23 breeds of goat, 39 breeds of sheep, 6 breeds of horse/pony, 8 breeds of camel, 2 breeds of pig and 15 breeds of chicken registered under National Bureau of Animal Genetic Resources, Karnal.

The economics of any livestock enterprise is influenced by the production, reproduction and health status of animals which is largely influenced by the adaptability of the animals to the prevailing climatic conditions. Lifetime performance and longevity of breeding stock are highly desirable characteristic that immensely influences the overall profitability of an animal. The conservation of highly adapted native breeds is gaining importance in the prevailing climate change situation. This article attempts to give a brief account on the domestic animal biodiversity and climate change under the following heads:

- Livestock and climate change, its effects, adaptation and mitigation strategies around the world.

- Conservation of animal genetic resources, threats to animal genetic resources, needs and methods of conservation.

- Conservations of Indian cattle under changing climate scenario with a special reference to various conservation programmes.

- Measures taken by Kerala Veterinary and Animal Sciences University in conserving domestic animal biodiversity of Kerala.

15.1 Livestock and Climate Change

Evidence from the Intergovernmental Panel on Climate Change (IPCC, 2007) is now awesomely convincing that climate change is real and worsening truth, and that the poorest people will be the most vulnerable and worst affected. The International Fund for Agricultural Development (IFAD, 2009) acknowledges climate change as one of the factors affecting rural poverty and as one of the challenges to be addressed. While climate change is a global phenomenon, its negative impacts are more severely felt by poor people in developing countries who rely heavily on agriculture and livestock for their livelihoods, the most climate sensitive economic sectors. The IPCC predicts that by 2100 the increase in global average surface temperature may be between 1.8°C and 4.0°C. With increases of 1.5°C to 2.5°C, approximately 20 to 30 percent of plant and animal species are expected to be at risk of extinction (FAO, 2007) with severe impact on food security in developing countries. Responses to climate change include (i) adaptation, to reduce the vulnerability of people and ecosystems to climatic changes, and (ii) mitigation, to reduce the magnitude of climate change impact in the long term. Adaptation includes all activities that help people and ecosystems to reduce their vulnerability to the adverse impacts of climate change and minimize the costs of natural disasters. There is no one-size-fits-all solution for adaptation; measures need to be tailored to specific contexts, such as ecological and socio-economic patterns, and to geographical location and traditional practices. Mitigation activities are designed to reduce the sources and enhance the sinks of green house gases in order to limit the negative effects of climate change. However, neither adaptation nor mitigation alone can offset all climate change impacts, the focus should be made on both. At present, very few development strategies promoting sustainable agriculture and livestock related practices have clearly included measures to support local communities in adapting to or mitigating the effects of climate change. Activities aimed at increasing the resilience of rural communities will be needed to raise their capacity to adapt and to respond to new hazards. At the same time, while small scale agricultural producers and livestock keepers, especially poor farmers, are relatively small contributors to greenhouse gas (GHG) emissions, they have a key role to play in promoting and sustaining a low-carbon ruralpath through proper agricultural technology and management systems. The possible effects of climate change on food production are not limited to crops and agricultural production. Climate change will have far-reaching consequences for dairy, meat and wool production, mainly arising from its impact on grassland and range land productivity. Heat distress in animals will reduce the rate of feed intake and result in poor growth performance (Rowlinson, 2008). Lack of water and increased frequency of drought in certain countries will lead to a loss of resources. Consequently, as exemplified by many African countries, existing food insecurity and conflict over scarce resources will be exacerbated.

15.2 Climate Change and Livestock

In pastoral and agro-pastoral systems, livestock is a key asset for poor people, fulfilling multiple economic, social and risk management functions. The impact of climate change is expected to heighten the vulnerability of livestock systems and reinforce existing factors that are affecting livestock production systems, such as rapid population and economic growth, rising demand for food and livestock products, conflict over scarce resources etc.

15.2.1 Direct effects

The direct effects of climate change will include higher temperatures and changing rainfall patterns, which could translate into the increased spread of existing vector-borne diseases and macro-parasites, accompanied by the emergence and circulation of new diseases. In some areas, climate change could also generate new transmission models. These effects will be evident in both developed and developing countries, but the pressure will be greatest on developing countries because of their lack of resources, knowledge, veterinary and extension services and research technology development. The effects of rising temperatures vary, depending on when and where they occur. A rise in temperature during the winter months can reduce the cold stress experienced by livestock remaining outside. Warmer weather reduces the amount of energy required to feed the animals and keep them in heated facilities.

15.2.2 Temperature humidity index (THI)

THI is the combination of temperature and humidity that is a measure of the degree of discomfort experienced by an individual in warm weather. It was originally called the discomfort index. The index is essentially an effective temperature based on air temperature and humidity; it equals 15 plus 0.4 times the sum of simultaneous readings of the dry and wet-bulb temperatures. Thus, if the dry-bulb temperature is 90° F (32° C) and the wet-bulb temperature is 50° F (10° C), the discomfort index is 15 + 0.4 (140), or 71. Most animals are quite comfortable when the index is below 70 and very uncomfortable when the index is above 80 to 85. The effects of different Temperature–humidity index (THI) values on livestock are presented in Table 15.1

Table 15.1: Effect of Temperature-Humidity Index (THI) on livestock

THI	Livestock response	Effect on livestock
< 75°F (23°C)	No stress	No heat related problems
75°- 78°F (23°-25°C)	Livestock alert	Reduced weight gain, lower milk production, increased respiration rate
79°- 83°F (26°-28°C)	Livestock danger	Reduced weight gain, lower milk production, potential mortality if animals further stressed
84°F (28°C)	Livestock emergency	Mortality possible if animals further stressed by activity, lack of water, lack of external cooling

(Livestock stress index, American Meteorological Society, 2012)

Heat stress could be reason of the significant increase of production cost in the dairy industry. Armstrong (1994) noticed that the relative daily cows' production is constant when temperatures are low and medium, while after passing a threshold, starts to decrease. The rate of decline increases with rising temperatures. Exposition of dairy cattle to high ambient temperatures (Ta), high relative humidity (RH) and solar radiation for extended periods decrease the ability of the lactating dairy cow to disperse heat. At the same time, lactating dairy cows create a large quantity of metabolic heat. So the accumulated and produced heat joined with decreased cooling capability induced by environmental conditions, causes heat stress in the animals. Finally, heat stress induces increase of body temperature. Johnson (1980) observed that when the body temperature is significantly elevated, feed intake, metabolism, body weight and milk yields decrease to help alleviate the heat imbalance. Johnson *et al.* (1962) determined that with the termination of the hot season, in high-producing cow the productivity does not completely return to normal since the energy deficit cannot be fully compensated. The permanent drop in the current lactation is proportional to thelength of the heat stress. The temperature-humidity index (THI) could be used to determine the influence of heat stress on productivity of dairy cows.

Milk production is affected by heat stress when THI values are higher than 72, which corresponds to 22°C at 100 % humidity, 25 °C at 50 % humidity, or 28 °C at 20 % humidity (Du Preez *et al.*, 1990a). Johnson (1980) reported that, when THI reaches 72, milk production as well as feed intake begins to decrease. Beside changes in milk yield, heat stress could also cause changes in milk composition, somatic cell counts (SCC) and mastitis frequencies (Rodriguez *et al.*, 1985, Du Preez *et al.*, 1990b).Highly significant (P<0.01) decrease of dailymilk yield due to enhanced THI was observed in all cows regardless the parity class (Gantner *et al.*, 2011). Regarding the effect of THI value above critical on daily fat and protein content, highly significant (P<0.01) decrease was observed in all cows and in all analysed regions.

15.2.2.1 Limitations of THI

The THI does not include important climatic variables such as solar load and wind speed. Likewise, it does not include management factors (the effect of shade) or animal factors (genotype differences). Gaughan *et al.* (2008) developed and tested a heat load index (HLI) for feedlot cattle. A new HLI incorporating black globe (BG) temperature (°C), relative humidity (RH, decimal form), and WS was initially developed by using the panting score (PS) of 2,490 Angus steers. A related measure, the accumulated heat load(AHL) model, also was developed after the development of the HLI. The AHL is a measure of the animal's heat load balance and is determined by the duration of exposure above the threshold HLI. The THI and THI-hours (hours above a THI threshold) were compared with the HLI and AHL. They reported that the HLI and the AHL were successful in predicting panting score (PS) responses of different cattle genotypes during periods of high heat load.

15.3 Indirect Effects

These are mainly changes in feed resources, the buffering abilities of ecosystems, intensified desertification processes, increased scarcity of water resources and decreased grain production. Other indirect effects will be linked to the expected shortage of feed arising from the increasingly competitive demands of food, feed and fuel production, and land use systems. Thornton *et al.* (2008) examined some of the direct and indirect impacts of climate change on livestock and reported that water, feeds, biodiversity and livestock and human health are the four factors likely to have maximum impacts on livestock production systems.

15.3.1 Water

Water scarcity is increasing at an accelerated pace and affects between 1 and 2 billion people. Climate change will have a substantial effect on global water availability in the future. Not only will this affect livestock drinking water sources, but it will also have a bearing on livestock feed production systems and pasture yield.

15.3.2 Feed

As temperature and CO_2 levels change, optimal growth ranges for different species also change; species alter their competition dynamics, and the composition of mixed grasslands changes. Rising temperatures increase lignifications of plant tissues and thus reduce the digestibility and the rates of degradation of plant species. The resultant reduction in livestock production may have an effect on the food security and incomes of smallholders. Interactions between primary productivity and quality of grasslands will require modifications in the management of grazing systems to attain production objectives.

15.3.3 Biodiversity

In most of the places there will be acceleration in the loss of genetic and cultural diversity in agriculture as a result of globalization. This loss will also be evident in crops and domestic animals. As mentioned earlier, a rise of 2.5° C in global temperature would lead to high risk of extinction of 20 to 30 percent of all plant and animal species. Local and rare breeds could be lost as a result of the impact of climate change and disease epidemics. Biodiversity loss has global health implications and many of the anticipated health risks driven by climate change will be attributable to loss of genetic diversity.

15.3.4 Livestock and health

Vector-borne diseases could be affected by: (i) the expansion of vector populations into cooler areas (in higher altitude areas: malaria and livestock tick-borne diseases) or into more temperate zones (such as bluetongue disease in northern Europe); and (ii) changes in rainfall pattern during wetter years, which could also lead to expanding vector populations and large-scale outbreaks of disease (e.g. Rift Valley fever virus in East Africa). Temperature and humidity variations could have a significant effect on helminth infestations. Trypanotolerance, an adaptive trait which has developed over the course of millennia in sub-humid zones of West Africa, could be lost, thus leading to a greater risk of disease in future. Changes in crop and livestock practices could produce effects on the distribution and impact of malaria in many systems, and schistosomiasis and lymphatic filariasis in irrigated systems. Heat-related mortality and morbidity also could increase.

15.4 Adaptation and Mitigation Strategies

Livestock can play an important role in both mitigation and adaptation. Mitigation measures could include technical and management options in order to reduce GHG emissions from livestock, accompanied by the integration of livestock into broad environmental services. Livestock has the potential to support the adaptation efforts of the poor. In general, livestock is more resistant to climate change than crops because of its mobility and access to feed. However, it is important to remember that the capacity of local communities to adapt to climate change and mitigate its impacts will also depend on their socio-economic and environmental conditions, and on the resources they have available.

15.4.1 Adaptation strategies

Livestock producers have traditionally adapted to various environmental and climatic changes by building on their in-depth knowledge of the environment in which they live. However, the expanding human population, urbanization, environmental degradation and increased consumption of animal source foods have rendered some of those coping mechanisms ineffective (Sidahmed, 2008). In addition, changes brought about by global warming are likely to happen at such a speed that they will exceed the capacity of spontaneous adaptation of both human communities and animal species. The following have been identified by several experts (FAO, 2008; Thornton *et al.*, 2008 and Sidahmed, 2008) as ways to increase adaptation in the livestock sector:

15.4.1.1 Production Adjustments

Changes in livestock practices could include: (i) diversification, intensification and/or integration of pasture management, livestock and crop production;(ii) changing land use and irrigation; (iii)altering the timing of operations; (iv)conservation of nature and ecosystems; (v)modifying stock routings and distances; (vi)introducing mixed livestock farming systems, such as stall-fed systems and pasture grazing.

15.4.1.2 Breeding Strategies

Many local breeds are already adapted to harsh living conditions. However, many developing countries are usually characterized by a lack of technology in livestock breeding and agricultural programmes that might otherwise help to speed adaptation. Adaptation strategies address not only the tolerance of livestock to heat, but also their ability to survive, grow and reproduce in conditions of poor nutrition, parasites and diseases (Hoffmann, 2008). Such measures could include: (i) identifying and strengthening local breeds that have adapted to local climatic stress and feed sources and (ii) improving local genetics through cross-breeding with heat and disease tolerant breeds.

15.4.1.3 Market Responses

The agriculture market could be enhanced by the promotion of interregional trade and credit schemes.

15.4.1.4 Institutional and Policy Changes

Removing or introducing subsidies, insurance systems, income diversification practices and establishing livestock early warning systems and other forecasting and crisis-preparedness systems could benefit adaptation efforts.

15.4.1.5 Science and Technology Development

Working towards a better understanding of the impacts of climate change on livestock, developing new breeds and genetic types, improving animal health and enhancing water and soilmanagement would support adaptation measures in the long term.

15.4.1.6 Capacity Building for Livestock Keepers

There is a need to improve the capacity of livestock producers and herders to understand and deal with climate change by increasing their awareness of global changes. In addition, training in agro-ecological technologies and practices for the production and conservation of fodder improves the supply of animal feed and reduces malnutrition and mortality in herds.

15.4.1.7 Livestock Management Systems

Efficient and affordable adaptation practices need to be developed for the rural poor who are unable to afford expensive adaptation technologies. These could include (i) provision of shade and water to reduce heat stress from increased temperature. Given current high energy prices, providing natural shade instead of high cost air conditioning is more suitable for rural poor producers; (ii) reduction of livestock numbers – a lower number of more productive animals leads to more efficient production and lower GHG emissions from livestock production (Batima, 2006); (iii)changes in livestock/herd composition(selection of large animals rather than small); (iv) improved management of water resources through the introduction of simple techniques for localized irrigation (e.g. drip and sprinkler irrigation), accompanied by infrastructure to harvest and store rainwater, such as tanks connected to the roofs of houses and small surface and underground dams.

15.4.2 Mitigation of GHGs emissions

The main greenhouse gases are: carbon dioxide (CO_2), methane (CH_4) and nitrous oxide (N_2O). Less prevalent but very powerful greenhouse gases are hydro fluorocarbons(HFCs), perfluorocarbons (PFCs) and sulphurhexafluoride

(SF$_6$).Given the magnitude of the challenge to reduce GHG concentrations in the atmosphere, it is imperative to receive the contribution of all sectors with significant mitigation potential. Agriculture is recognized as a sector with such potential, and farmers, herders, ranchers and other land users could and should be part of the solution. Therefore, it is important to identify mitigation measures that are easy to implement and cost effective in order to strengthen the capacity of local actors to adapt to climate change. The livestock production system contributes to global climate change directly through the production of GHG emissions, and indirectly through the destruction of biodiversity, the degradation of land and water and air pollution. According to "Livestock's Long Shadow" (FAO, 2007), livestock is responsible for 18% of global warming. Livestock contributes 9% of all GHG emissions measured in CO_2 equivalents,

65% of human-induced nitrous oxide (which has 296 times the global warming potential of CO_2) and 20% of methane (which has 23 times the global warming potential of CO_2).There are three main sources of GHG emissions in the livestock production system: the enteric fermentation of animals, manure (waste products) and production of feed and forage in the field (Dourmad *et al.*, 2008). In direct sources of GHGs from livestock systems are mainly attributable to changes in land use and deforestation to create pasture land. For example, in the Amazon rain forest, 70 per cent of deforestation has taken place to create grazing land for livestock. In general, smallholder livestock systems have a smaller ecological footprint than large scale industrialized livestock operations. The ecological footprint is a means of evaluating human demand on the earth's overall ecosystem. Ecological footprints measure human demand for energy and air, against the planet's capacity to regenerate resources and provide services. The measurement is based on the area of biologically productive land and sea needed to regenerate the resources consumed by the human population and to absorb and render harmless the corresponding waste. As per FAO (2008) possible mitigation options include:

15.4.2.1 Selection of Faster Hrowing Breeds

The efficiency of conversion of energy from feed into products has to be improved and the losses through waste products have to be reduced by selection of better breeds. Increasing feed efficiency and improving the digestibility of feed are potential ways to reduce GHG emissions. It can also maximize production and gross efficiency by lowering the number of animals. All livestock practices – such as genetics, nutrition, reproduction, health and dietary supplements and proper feeding& grazing management - that could result in improved feed efficiency need to be taken into account.

15.4.2.2 Improved Feeding Management

The composition of feed has some bearing on enteric fermentation and the emission of CH_4 from the rumen or hindgut (Dourmad, *et al.*, 2008). The volume of feed intake is related to the volume of waste product and the higher the proportion of concentrate in the diet, the lower the emissions of CH_4.

15.4.2.3 Better Waste Management

Improving the management of animal waste products through different mechanisms, such as the use of covered storage facilities is also important. The level of GHG emissions from manure (CH_4, N_2O and CH_4 from liquid manure) depends on the temperature and duration of storage. Long-term storage at high temperatures results in higher GHG emissions. In the case of ruminants, pasture grazing is an efficient way to reduce CH_4 emission from manure because no storage is necessary. It is possible not only to mitigate GHG emissions but also to create an opportunity for renewable energy and generation of carbon credits through soil conservation.

15.4.2.4 Grazing Management

One of the major GHG emission contributions from livestock production is from forage or feed crop production and related land use. Proper pasture management through rotational grazing would be the most cost-effective way to mitigate GHG emissions from feed crop production. Animal grazing on pasture also helps reduce emissions attributable to animal manure storage. Introducing grass species and legumes into grazing lands can enhance carbon storage in soils.

15.4.2.5 Lowering Livestock Production and Consumption

Lowering the consumption of meat and milk in areas with a high standard of living is a short-term response to GHG mitigation.

15.4.2.6 Livestock Grazing and soil Carbon Sequestration

Experts have estimated that the soil contains carbon more than twice the quantity in the atmosphere and demonstrated that enhancing carbon sequestration by soils could make a potentially useful contribution to climate change mitigation.

15.4.2.7 Threat to Genetic Diversity and its Reasons

The different indigenous breeds of farm animals and poultry are essentially the result of evolutionary processes, they have adapted to harsh climatic conditions with low management inputs in terms of feeds/fodder and health care, capable to convert low quality feeds and fodder more efficiently into animal products (milk, meat, eggs, wool *etc.*) and better adapted to withstand tropical diseases. They are integral part of agriculture. However, they are now getting diluted and facing degeneration. It is the livestock keepers who conserve the animal genetic resources all over the world. Scientists define a particular group as breed and recognize specific breeds. Others are all nondescript because they are not described by the scientists and approved by the concerned agency. Majority of livestock keepers are poor and or illiterate have their own animal varieties apart from the defined breeds. According to farmers who raise animals under severe harsh environment, it is not the production potential alone but feed and system efficiency are also of greater relevance. In addition, many breeds are appreciated for characteristics that have little to do with productivity, such as ritual significance, social role and aesthetic aspects. Traditional breeds are the collective property, products of indigenous knowledge and cultural expression.

15.4.2.8 Economic and Market Driven Threats

Economic and market driven threats are among the major threats to the preservation of biodiversity. The demand for milk, meat and eggs is now much more than forother livestock products and services as such as hair, wool, animal traction and transport. In the competition camels, donkeys, horses, buffalo, elephants, llama's, yak, wool sheep, etc are loosing. Gradually nomads and farmers replace traditional species and breeds by species and breeds that have a greater productivity and therefore higher economic value in the shortterm.

15.4.2.9 Climate change and Other Factors

Climate change and the higher incidences of natural calamities are also causing endangerment of some breeds available in the susceptible area. Increase in livestock population is also causing deterioration due to inadequate availability feeds, health care measures etc.

15.5 Needs of Conservation of Animal Genetic Resources

The native breeds need to be conserved to meet the following objectives:

15.5.1 Genetic insurance

There is growing awareness to conserve the endangered breeds for future use. The native breeds have been developed over thousands of years, have better adaptability to harsh climate, diseases, heat tolerance and utilization of locally available poor quality roughages. They have unique genes or gene combinations which are associated with adaptability and producing capacity under adverse climate and inadequate feed resources. Thus loss of such breeds means the loss of DNA sequences responsible for specific adaptive traits which may be required to introduce in highly productive germplasm evolved through new techniques of biotechnology. It is not known that what might be needed in future. Thus as a genetic security the conservation of native breeds for their use in future is required.

15.5.1.1 Scientific Study

The conservation of native breeds provides useful research material in genetics, biochemistry, physiology, morphology and anatomy. It is also useful to understand the process of evolution, domestication and the effects of natural and artificial selection.

15.5.1.2 Economic Potential in a Particular Environment

The crossbred animals fail to exploit their genetic potential under Indian conditions of low input and harsh climate, are more susceptible to tropical diseases, and their production level goes down beyond F_1's. Moreover to produce F_1's also requires the conservation of indigenous breeds for future use.

15.5.1.3 Environmental Considerations

Every organism has its own role to play within an ecosystem and the loss of germplasm has adverse effect on the ecosystem which is hazardous to the existence of mankind. The domesticated breeds are integral part of our ecosystem.

15.5.1.4 Cultural and Ethical Requirements

If a breed is not economical viable, it should be preserved for cultural and public interest of historical importancesince the animal genetic resources are the part of the natural heritage, culture and ecosystem. Moreover, to take care of the existence of all forms of life is required on ethical and moral grounds.

15.5.1.5 Energy Source

India has a number of good draught breeds of cattle used to perform a number of agricultural operations before intensification of mechanized agriculture. However, the non-renewable source of energy may be exhausted sooner or later. In such situation, the animal draught power will be required and hence these breeds should not be allowed to extinct. Principles of conservation: The stock for conservation should be maintained with optimum population size above the level of risk. It should be pure, diverse and having special traits. Maintain and conserve the locally adapted breeds and that too in the same location. Maintain live animals providing the similar feeding, management and environmental conditions under which they had been traditionally kept. The genetic merit and diversity of the population should be maintained using appropriate breeding programme.

15.5.1.6 Conservation of Indian Cattle Under Changing Climate Scenario

India has about 37 well defined breeds of cattle so far registered under National Bureau of Animal Genetic Resources, Karnal which are found in different parts of the country. The cattle breeds are classified into three types i.e. dairy, dual purpose and draft/drought animals. These Indian breeds have been evolved over the centuries in the hot humid climate of tropics and are known for their endurance to heat, resistance to tropical diseases, capacity to thrive on coarse fibrous fodder residues and excellent draft power. The highest number of drought breeds is the testimony of their capacity for draft power and their role in the agriculture production scenario of the country in the past. However, with the advancement of mechanization of agricultural processes, the number of the animals of majority of indigenous breeds especially the draft animals has declined over the years. Further, the number of descript animals is very low and majority of cattle population (80%) in India is non-descript.

The introduction of systematic crossbreeding program of our native cattle with exotic breeds in sixties has further led to the dilution of prized germplasm of all the indigenous breeds under farm and field conditions. However, considering the importance of indigenous cattle, some of the breed improvement programs have been initiated to conserve and multiply these breeds. The associated herd progeny testing program has been going on at NDRI Karnal for the genetic improvement of Sahiwal cattle. All India Coordinated Research Project on Hariana and Ongole cattle has been undertaken at different centres by Project Directorate on Cattle, Meerut. Lately, three more breeds namely Sahiwal, Gir and Kankrej have been included in the indigenous breed improvement program (IBP) and different centres have been identified as germplasm units and data recording units for different breeds under the program(Gandhi, 2013). India has witnessed a white revolution due to largely by crossbred cows. Estimated milk production in the country was 121.8 million tonnes during 2010-11(BAHS, 2012). The annual growth rate in milk production of our country over the years was around 3-4%. According to an estimate the milk production of our country would be 170 million tons during 2020. India has 199.08 million cattle, out of which the number of breedable females is 72.95 million (including 10.72 million crossbreds). There are 26.22 million crossbred females, out of which 16.16 million are breedable and 10.72 millions are in milk. The respective figures for indigenous cattle are 89.24, 56.76 and 30.69 million, respectively. About 54.5 million adult buffaloes produce 59.2 million tonnes of milk (52.6%), while 72.95 million cows produce 47.9 million tonnes of milk (42.6% of 112.5 million tonnes). Out of 42.6% of total milk produced by cows, 20% is produced by indigenous cows (30.69 million) and 22.6% by crossbred cows (10.72 million). It is obvious that about 26% of crossbred cattle are producing nearly 53% of total cow milk. The impact of crossbreeding is more pronounced under field conditions in most of the states of the country. This was attributed to higher milk production of crossbred cattle as compared to local cattle and is evident from milk yield/animal/day (6.87 versus 2.14 kg) during the year 2009-10. The performance of crossbreds in terms of milk yield is about 3-4 times more than that of local cows in majority of the states. Lower production of our cattle is mainly attributed to the poor genetic makeup with respect to production traits in majority of animals, which are non-descript and inadequate availability of feed and fodder. However, with respect to adaptability they are superior to any other exotic breeds. Unfortunately the milk production oriented cross breeding programme led to the loss of majority of these adapted breeds of animals.

15.6 Climate Change and Sustainability of Indian Cattle

As mentioned earlier, *Livestock's Long Shadow*, the widely-cited 2002 report by FAO estimated that 7,516 million metric tonnes per annum of CO_2 or 18 per cent of annual worldwide greenhouse gas (GHG) emissions are contributed by cattle, buffalo, sheep, goats, camels, horses, pigs and poultry. Methane emission from indigenous cattle (42%) has been found to be higher than exotic cattle (8.8%), higher population of the indigenous animal seems to be one of the reasons. Different livestock species like cattle, buffalo, sheep and goat contribute 54.72, 30.37, 5.38 and 9.70%, respectively, of total emission from enteric fermentation (Table 15.2).

Table 15.2: Indian livestock population and methane emission

Species	Population (million)	Methane emission/ animal/day(gm)	Contribution (%)
Cattle	199	76.7	54.7
Buffalo	105.3	97.0	30.3
Goat	140.5	10.1	9.7
Sheep	71.5	11.6	5.3

(*Source:* INCCA, 2010)

The level of GHG is expected to increase in India as the number of livestock will increase in future to support the ever increasing human population in India. The higher level of green house emissions will increase the global temperature. As mentioned earlier, the predicted increase in global temperature by the year 2100 will be 1.8-3.9°C (IPCC, 2007). According to United Nations Environment Plan Department (UNEP) and World Meteorology Organization (WMO), due to greenhouse effect 0.5-2.0 °C average increase in air temperature of earth can happen and it may lead to changing pattern of rain and snow. Rainfall increase would lead to higher evaporation to the extent of 2-13% attributing to strengthening of Asian monsoon system. All these changes are expected to influence land use pattern in our country and indirectly affecting the animal production sectors. The potential influence of climate change to livestock sector in India is:

-The anticipated 2.3 to 4.8 °C rise in temperature over the whole country along with increased precipitation resulting from climate change, which is likely to aggravate heat stress on farm animals, adversely affect their productive and reproductive performance.

- Considering the vulnerability of India to rise in sea level, the impact of increased intensity of extreme events in livestock sector would be more devastating for the low income rural masses.

- The predicted negative impact of climate change on Indian agriculture will also adversely influence the livestock production by aggravating the feed and fodder shortage.

The climate change has produced detrimental effect on Indian cattle production system directly and indirectly contributing to economical losses. About 6-9 months (60-75% of days) in a year are either stressful or not congenial for optimum milk production in most parts of the country. Further, the THI ranges 75-85% during noon hours at more than 85% places in India during summer months (April – June). The various studies have shown that the rate of decline in milk production is more pronounced in crossbred cattle (Shinde *et al.*, 1990; Kulkarni *et al.*, 1998; Mandal *et al.*, 2002) as compared to indigenous cattle (Lal *et al.*, 1987; Mandal *et al.*, 2002).

The crossbred cattle perform better at 5-25°C temperature, while indigenous cattle feel comfortable even at a temperature of 38°C. Indian breeds from Rajasthan and Gujarat have the innate potential to tolerate desert conditions and temperature up to 50°C. The major challenge for the high producing crossbred cattle is the heat stress under Indian conditions. As per one of the estimates, the milk yield of crossbreds decrease by 100 litres/cow/lactation, while for indigenous cattle the decline was five folds lesser, merely 20 litres/cow/lactation. The reason is that our indigenous cattle have been evolved over the generations of natural selection under the stressful environment of tropics under low-input system. The dry matter intake is comparatively lower in Indian cattle as compared to crossbred cattle (2kg versus 8 kg) as the average size of local cattle is small and hence indigenous cattle produce less GHG as compared to crossbred cattle and cattle from developed countries.

The adverse effect of excessive heat load on production and reproduction is due to combination of environmental (ambient temperature, relative humidity, wind speed, solar radiation, nutrition) and animal factors (genetic makeup) that result in drastic change in physiology and behaviour. These changes include reduced feed intake, decreased general activity, lower growth and conception rates, increased heart and respiratory rates, panting activity, increased peripheral blood flow and sweating. Collectively, all these changes put more strain on the animal that ultimately results in reduced health status. Hence, exploring distinct adaptive alleles in cattle breeds would be useful in identifying animals with improved thermo-tolerance ability to overcome the influence of heat stress.

15.6.1 Adaptation and mitigation strategies

- Higher use of organic inputs into the prevailing farming system may attribute to the reduction in the GHGs.

- Better management of the irrigation system for the crops like rice and sugarcane would reduce methane emission.

- Reducing the use of indiscriminate fertilizers may also be one of the mitigation strategies.

- Crop cultivation by diversification, changing the planting dates and cultivating various crops varieties would reduce the impact of climate change to some extent.

- Developing genetically engineered crops capable to sustain under extreme climatic change.

- Manure management using covered storage facilities of the manure and livestock waste to lower the methane emission into the environment.

- Mitigation through shelter management, providing comfortable micro-environment and lowering stress by dietary regime manipulations.

- Grading-up of local non-descript cattle with well defined breeds which are adapted to the local climate and reducing the number of unproductive animals over the years in phased manner.

- Drought resistant breeds such as Rathi, Sahiwal, Tharparkar, etc should be propagated in different pockets of the country having harsh climate.

- Molecular genetics approach can be used to identify genes more resistant to thermal stress. Some of the genes like slick gene coding for short, sleek hair, ATP1A1 gene, uncoupling protein (UCP) and heat shock protein (HSP) genes have been identified as the heat tolerance genes in cattle and buffalo by different workers (Madhusudan, 2007; Liu et al., 2010 and Loredana et al., 2011). Identification of polymorphism in such genes can offer a scope for selection of animals resistant to heat stress leading to the development of heat tolerant breeds.

Venkatachalapathy and Iype (2010) studied the adaptive response of Vechur cattle by exposing them to sunlight for three hours (between 10.00 A.M. and 3.00 P.M) and recording the physiological parameters such as body temperature, respiration and pulse rate before and after the exposure. They estimated values of Heat Tolerance Index and reported that the animals did not show any symptoms of heat stress even after exposure to sunlight. There had been only slight increase in body temperature after exposure to sun. But pulse and respiration showed increase. The heat stress occurs from 80°F for dairy cows in US with the signs of rapid shallow breathing. Beyond 90°F there is generally open mouth breathing with panting. In Kerala panting is observed in crossbred cows in summer even within sheds. Summer temperature rises to beyond 100° F in Thrissur area where the studies were made. But the fact there was no signs of stress and not much change in body temperature indicate the heat tolerance of native Vechur cattle. These studies indicated the high heat tolerance of Vechur cattle and this trait is one to be preserved for future. Brody (1956) observed that the higher heat tolerance of the Indian cattle seemed to be due to lower heat production, greater surface area per unit weight, shorter hair, and other body-temperature regulating mechanisms not visually apparent. The larger the animal and the higher the productive level, the lower was the comfort zone temperature. At 105° F environment the near-lethal temperature of 108° F was reached in the Holsteins, 106° F in the Jerseys, and 105° F in the Indian cows of the same weight as the Jerseys. At environmental temperature of 105° to 107°F the European cows were near collapse.

In Kerala 96% of livestock population is cattle producing 98.50% of total milk. More than 82% of the cattle are crossbreds except very few indigenous animals like Vechur. Vechur cattle once thought to be extinct due to extensive crossbreeding are the first native cattle of Kerala to be saved, multiplied and approved as a distinct breed and this is the smallest breed in the world.

15.6.2 Breeding programs for genetic improvement in tropics

The major factors that determine the success or failure of breeding programme are the relevance of the breeding programme for genetic improvement in the prevailing scenario, objectives/goals of the programme and breeding strategies (Barker, 1992). The relevance of the breeding programme is of paramount importance. Bringing about genetic improvement beyond the nutritional or disease-limited environment in which the population lives can be of disadvantage and often counterproductive (McDowell, 1985). Another important factor is the availability of sufficient infrastructure to support the programme like AI and veterinary services, procurement of milk and supply of other inputs at the doorsteps of the farmers. The breeding objectives must be relevant to the prevailing production system of an area. In tropical countries like India, the selection goal must identify animals that perform well under hot humid conditions and can cope up with the seasonally available fodders/crop residues and have the resistance for diseases prevalent in such environment. The goal of any breed improvement programme should be to change the genotypes of that animal population to best fit the total environment of its production system including climate, management, social and economic factors (Barker, 1994). Above all, breeding policy should not allow the 'heritage' breeds of cattle to extinct. It has been suggested that single purpose breeding goal should not be practiced, rather multi-purpose breeding objectives should be considered for the overall 'adaptive value' of an individual in the population. The determinants for recommending the breeding policy are description of production system, definition of breeding objectives (trait or combination of traits) and knowledge of breed resources (genetic variation in production and adaptive traits). Unfortunately, the production systems vary considerably from area to area and even from farmer to farmer. Data on performance traits, adaptive traits and unique traits like resistance to a particular disease/parasite and tolerance to environmental stresses are absent or inadequate for most of the breeds. Under the prevailing circumstances, a short term interim breeding policy is proposed (Mangurkar, 1998):

- Conservation and improvement of well defined indigenous dairy and dual purpose breeds of cattle and buffaloes in their respective breeding tracts.

- Pure breeding in draft breeds with emphasis on both draft and milk production.

- Grading up of non-descript animals with locally adapted dairy/dual purpose breeds, if the farmers opt for.

- Crossbreeding of non-descript cattle in urban and semi-urban areas, and as per farmers choice and resource availability in other areas; the selection of exotic breed(HF or Jersey) depending upon the availability of resources and market demand for milk.

- Inter-se mating amongst crossbreds to maintain the exotic blood level between 50–75%.

The genetic improvement programs on cattle are mainly based on crossbreeding of nondescript cattle with exotic breeds and selective breeding in well defined breeds of indigenous cattle.

15.6.3 Performance and genetic improvement of indian cattle breeds

The breeding policy for improving the indigenous cattle is selective breeding in well defined breeds and grading up of non-descript animals with the indigenous breeds prevalent in the area or adjoining area. Various reports on production performance of indigenous breeds indicated that the milk production of Indian cattle breeds ranged from low to medium and there is an ample scope to improve upon these breeds through selective breeding. The nondescript cattle have to be up-graded with the native available breeds in that area or adjoining area.

15.6.4 Brief history of Kerala livestock sector

The Kerala state came into existence in 1956 by the merger of the Princely States of Travancore, Cochin and the Malabar province of the British ruled Madras State and the Kasargod Taluk of the South Kanara District on Linguistic basis, had a rich biodiversity as a whole. Kerala with its tropical climate and evergreen nature of rain forests is rich in its flora and fauna. 'SilentValley' in the Western Ghats is declared as a 'Biological hotspot' by the United Nations. The Kerala community was traditionally a farming community and they had their own sustainable agriculture (mainly paddy-based) and animal husbandry systems, which were the backbone of rural economy. Kerala had rich domestic animal diversity with respect cattle, goats, pigs, buffaloes, ducks and fowl. Cattle breeds in Kerala were small in size to adapt to the hot humid climate. Among these, Vechur was the most important one. Efforts to improve the productivity of cows started in the 1940s in the Travancore State. The milch breeds of the Indian subcontinent, mainly Red Sindhi were used for this. In 1956, the Department of Animal Husbandry came into existence with a view to provide better animal health care and to improve the livestock and poultry production in the entire state of Kerala. As far as the animal wealth of this state was concerned, all the animals were considered non-descript. No efforts were made to describe them or assess their merits. All the local animals were considered as a hindrance to the economic development of the state. More importance was given to maximum production than optimum production. For improving the livestock, exotic germplasm were introduced into the state. With regard to the cattle of Kerala, in 1961 the Livestock Improvement Act came into existence. The Act banned all local germplasm from getting multiplied. Local Bulls were subjected to forceful castration. The cultivated preference among the farmers also paved way for the drastic reduction in the local germplasm. By the 80's almost more than 80% of the cattle were crossbreds. The reductions in the ethnic germplasm were drastic in the South and Middle areas of Kerala where cross breeding facilities were more. Only some animals were left in the Northern Kerala as the regional Co-operative milk union was not yet established in the Malabar area. In the south and central regions, only few animals were left that too due to geographical isolation or inaccessibility. Similar measures were taken for improving the goats, pigs, buffaloes and poultry. As a result of these, the local livestock breeds face the verge of extinction.

15.6.4.1 Conservation Efforts of Various Indigenous Animals

Cattle

The Silent Valley movement started in 1973 gained momentum in the year 1980. The importance of preserving the biodiversity became a talk in every part of the state. Finally in 1984 Silent Valley was declared a National Park. The 'Silent Valley' campaign gave fuel for conserving the genetic diversity in domestic animals also. The conservation efforts were initiated by then Kerala Agricultural University and are being continued by Kerala Veterinary and Animal Sciences University. Currently, the Vechur Cattle Conservation Centre of the University is working primarily as a conservation unit with the mandate of conserving Vechur Cattle with the importance of keeping our heritage, a valuable tool for future studies on adaptability to climatic stress and disease resistance and as a small cow for the farmers who need milk only for their home consumption. A few Kasargod dwarf animals are also maintained at the Vechur cattle farm as a part of conservation of indigenous cattle breeds. A brief description of the Vechur and other local cattle germplasm of cattle is given below.

Vechur cattle

The Vechur cattle, the smallest of Indian cattle breeds, derived its name from the Vechur village of Kottayam district of Kerala which was the home tract of these animals. The geographical distribution of these animals was spread over Kottayam, alappuzha and eranakulam districts. These cows are maintained as house cows. The fodder available in and around the house, kitchen wastes, rice gruel, bran and small quantity of oil cakes are the main feed for these animals. The grasses available as the undergrowth of coconut gardens form the major roughage source for these animals. The milk production is around 2-3 litres and is used mainly for home consumption. The common belief of medicinal properties for the milk makes it an essential component in ayurvedic treatments. The easy digestibility conferred on Vechur milk is due to the small size of fat globules. Some owners are selling small quantities of the milk at premium price. The easy digestibility and higher fat content of the milk fetch higher price. Even though the animal is not maintained for the economic returns from the milk, the women folk of the house can maintain these docile cows. These aspects of the production system make it sustainable. The data available at Centre for Advanced Studies in Animal Genetics and Breeding reveals that most of the owners of Vechur cattle are middle class farmers and environmentalists. In males the hump and dewlaps are very prominent with short thick neck. Small hump, medium sized dewlap and long thin neck are seen in females. Small naval flap and triangular rump are also the characters of Vechur animals. The ears are medium in size without any prominent hairs and horizontally oriented with an average length of 16.64 ± 0.15 cm. Long and narrow head with a long forehead and prominent blackish eyes are characteristic to this breed. The average face length is 34.59 ± 6.24 cm and face width is around 70 cm. The horns of Vechur cattle are curved or straight with brown or black colour. The orientation of the horn is characteristic to the breed with curving forward, then downward and backward with pointing tips. In some cases the horns are very small and stumpy. The coat colour of Vechur cows is solid without ant patches or strips. In general, red and its shades are the main colour of these animals. Black and its shades and white are also observed;other colours are rarely seen. The udder is symmetrical with cylindrical teats and round teat tips. Prominent milk veins are another character of the Vechur cows.

Kasargod cattle

These are the small local cattle present mainly in the Kasargod, the northern most district of Kerala. These are low producers of milk with solid body colour. Major purpose of these animals is as organic manure producer. Kasargod animals are larger than Vechur but are smaller than most of other Indian breeds including Highrange dwarf and Vatakara cattle. These animals are also maintained in low input system. Most of them find their own fodder by grazing. The gobbora system seen in these areas is putting tree leaves as litter material in the floor of the cattle shed, which will collect the dung and urine. Fresh layer of leaves are added frequently, which are removed after one or two months as manure. The milk production from these animals is less and is used for household purpose only. Selling of milk from these animals is almost absent. The Kasargod animals are with prominent large hump, medium sized dewlap and horizontally placed ears. Black coloured short and stumpy horns with lateral pointing tips are mainly seen in Kasargod animals. Predominant coat colour is black and its shades; a few white and other coloured animals are also noticed.

Vatakara cattle

These animals are seen in and around Vatakara area of Calicut district, some areas in Malappuram district and parts of Kannur district. These small animals are mainly used for homestead milk production. These animals are fed small quantity of concentrates, mainly as oil cakes. The peak yield of some of these local animals is better than those of other cattle and is around 3-5 kg. These are solid coloured animals without patches or spots. Predominant color is black or shades of black followed by red and its shades and grey. Small or medium sized humps and dewlaps are commonly seen in these animals. Black or brown coloured horns, straight and curved with upward or forward pointing tips are common.

High range dwarf cattle

The habitat of High range dwarf cattle is Idukki district of Kerala. The colour of Highrange dwarf cattle is very distinct with brown or shades of brown colour. They have straight and comparatively long black coloured horns. Peak yield recorded was 3.5 liters in some animals.

Goats

Goat rearing in Kerala is one of the most popular animal production enterprises among low income groups. There are mainly two breeds of goats i.e.Malabari and Attapadi Black and majority of the population are either desi or crosses of desi with these two breeds. The native breeds have evolved through the years, after being exposed to stressful environmental conditions such as great heat or shortages of feed or water and are rich in survival and fitness traits, which have disappeared from high performance breeds. The main challenge before animal scientists is to increase the production performance of these breeds by systematic breeding and utilizing the locally available feed and fodder and exploit this gene pool for elucidating the genetic mechanism involved in higher disease resistance and hardy nature of these breeds.

Malabari

The main breed used is Malabari which has evolved centuries ago in northern parts of the Kerala coast. This breed derives its name from its native habitat of "Malabar" area of northern Kerala and evolved few centuries ago by crossing Jamnapari, Surti and Arab goats with local goats (Kaura, 1952). The breed is widely distributed in the districts of Kasargod, Trichur, Kannur, Kozhikkode and Malappuram and is mainly concentrated in and around Tellicherry and therefore is sometimes called the Tellicherry breed. These goats are medium-sized animals. Coat colour varies widely from completely white to completely black. Males and a small percentage of females are bearded. Both sexes have small, slightly twisted horns, directed outward and upward. Ears are medium-sized, directed outward and downward. Tail is small and thin. Udder is small and round, with medium-sized teats. Malabari is a dual purpose breed and well known for high prolificacy, good growth rate, milk yield and adaptability to tropical humid conditions of Kerala.

Conservation of Malabari breed

For conserving a breed it is necessary to evaluate the breed in its home tract under native management system through extensive surveys following appropriate sampling methods. In order to conserve and improve the production potential of Indian goats in terms of growth, milk production and reproductive traits a

Coordinated Research Project was established with its headquarters at Central Institute of Research in Goats (CIRG), Makhdoom, Agra, U.P. There are different centres under this and the focus in each centre is on different breeds. The breeds included are Jamnapari, Barabri, Bikaner, Sirohi, Black Bengal, Ganjam, Surthi, Osmanabadi, Sangamneri, Beetal and Malabari. This evaluation process was successfully carried out in the AICRP on Goat improvement Scheme (Malabari unit) by the Centre for Advanced Studies in Animal Genetics and Breeding, College of Veterinary and Animal Sciences, Mannuthy. AICRP on Goat Improvement Scheme (Malabari unit) has six field centres in northern Kerala, i.e., Thalassery and Thaliparambu in Kannur District, Badagara and Perambra at Kozhikkode and Kottakkal and Thanur at Malappuram and all efforts are taken to conserve the native breed Malabari in their home tract.

Attappady Black goat

Attappady Black goats,found exclusively in the Attappady area inthe Palakkad district of Kerala and rearedmostly by the tribes in Attappady, are not yet recognized as a breed despite their unique characteristics. The home tract of the Attappady Black goats lies in the north-east of the Palakkad district of Kerala, constituting three villages -Agali, Pudur and Sholayur. The total population of Attappady Blacks in the area was estimated to be 9351, which represented only 40% of the total goat population in the area. These goats are reared mainly by the tribes of Attappady and maintained on an extensive grazing system. They are identified by their solid black body colour, bronze coloured eyesand long strong legs. Other goats in the areaare crosses of the Attappady Black with other breeds, mainly the Malabari, as well as with a small number of Saanen goats and other exotic breeds. The Attappady Black is a meat type breed with a milk yield of less than 200 ml daily. Distinguishing the Attappady Black from other groups is not at all difficult because of the distinctive features of the Attappady Black, Malabari and other exotic breeds. There is an Attappady Black goat breeding farm run by the Government of the State in Attappady, the native tract of these animals.

Pigs

Pigs are important meat producing livestock capable of with standing diverse management practices and agro ecological conditions. Among the domestic animals, pigs are the most prolific breeders and fast growers. The biological advantages such as short generation interval, better reproductive efficiency and its ability to thrive on agroindustrial byproducts and kitchen waste made them to play an important role to make up the animal protein deficiency, at a very low cost. Traditionally the indigenous pigs produce bristles and meat. Before the advent of nylon, the pig bristles had a great demand and the meat was a by-product. With the fall in bristle market, pork has become an important commodity. In India, pig and pork industry is in the hands of traditional pig keepers, belonging to the lowest socio-economic stratum with no means to undertake intensive pig farming of pure bred stock and improved methods of breeding, feeding and meat handling. Pig farming will provide employment opportunities to seasonally employed rural farmers and supplementary income to improve their living standards. Thus pigs contribute substantially to the economy of the socially backward and tribal population of this country. The pig population in the country increased from 4.40 millions in 1951 to 10.10 millions in 1983 and now stands at 13.92 millions as per 2003 census. In contrast to other livestock itregistered an impressive growth rate of over 3%. This constitutes around 1.30% of the total world's population. Kerala has a total pig population of about 76,000 of which mostly are crossbreds of Desi, Large White Yorkshire, Landrace and Duroc. The most important desi pig breed available in Kerala is Angamali pig.

Importance of Conservation of Ankamali Pigs

The native pigs are the product of breed evolution andgermplasm improvement through natural and artificial selection over the centuries. They are resistant tovarious diseases and can withstand wide range of environmental extremities. There ispotential to identify genes for disease tolerance and other adaptive traits such as heat tolerance in wild/indigenous varieties and these genes can be transferred to otherwise vulnerable stock by marker associated introgression. About 84% of the total pig population in the country is indigenous and not characterized properly. As a result of industrialization and globalization, the indigenous animals are getting replaced with crossbred animals. For preserving the biodiversity, the indigenous animals need to be conserved either *ex situ* or *in situ*.

Kuttanad buffaloes

In Kerala the contribution of buffaloes in milk production is very less. The buffaloes in the state were mainly used for ploughing the paddy fields. The origin and habitat of Kuttanad buffaloes can be described as Kuttanad area which is situated in Kottayam and Alappuzha districts of Kerala. The present day population of kuttanad buffaloes is around five hundred. They are good swimmers and can travel kilometers in water. These animals can feed on grasses which are seen above 5-6 feet water. The identification feature of Kuttanad buffaloes is mainly by its grey coat colour and the two chevrons. In Kerala Surthi buffaloes and Murrah buffaloes were used to upgrade the local buffaloes. It is not known whether the buffaloes that existed in Kerala before the introduction of milk genes from Murrah and Surti were reverine or swamp type.

15.7 Conclusion

To sum up, it is beyond doubt that crossbreeding has contributed significantly towards the enhanced milk production and played a vital role in white revolution as well as making our country number one in milk production. However, the increased reproductive disorders, susceptibility to the tropical diseases and poor adaptability of crossbreds to tropical hot and humid climate, demand the long term conservation of adaptive breeds with marginal production performance. The cross breeding should be limited to some resourceful areas and too without losing our pride indigenous domestic animal biodiversity. The changing climate also necessitates the conservation of locally adapted breeds of various domestic animals. The characterization of Indian breeds for thermo-tolerance needs to be initiated to explore their innate potential for performing under adverse climatic conditions.

References

Acharya, R. M. 2011. Breeding cattle under different production systems and in different agroecologies in India. Indian Dairyman 63(8): 52-57.

Anilkumar, K. and Raghunandanan, K. V. 2003. The dwarf cattle and buffalo of Kerala. CASAGB, Kerala Agricultural University, Thrissur.

Armstrong, D.V. 1994. Heat stress interaction with shade and cooling. J. Dairy Sci. 7, 2044-2050.

BAHS, 2012. Basic Animal Husbandry Statistics, Ministry of agriculture, Depatment of Animal Husbandry, Dairying & Fisheries, KrishiBhavan, New Delhi.

Barker, J.S.F. 1992. Practical issues for conservation and improvement of priority breeds : General considerations. FAO Animal Production and Health Paper 104, FAO, Rome.

Barker, J.S.F. 1994.Animal breeding for tolerance to adverse environments. Proceedings of 7th AAAP Animal Science Congress, Bali, Indonesia. Pp 29 – 33.

Batima, P., 2006. Climate change vulnerability and adaptation in the livestock sector of Mongolia. Assessments ofimpacts and adaptations to climate change. International START Secretariat, Washington DC, US.

Bindu, K.A., 2006. Study of genetic diversity in Malabari goats (Capra hircus) utilising biochemical and immunological markers. PhD thesis submitted to Kerala Agricultural University.

Bindu, M. 2004. Growth and survivability of GH/ Msp I genotypes in Malabari goats. M.V. Sc. Thesis submitted to Kerala Agricultural University.

Botstein, D., White, R.L., Skolnick, M. and Davis, R.W. 1980. Construction of genetic linkage map in man using restriction fragment length polymorphisms. Am. J. Hum. Genet. 32: 314-331.

Brody, S. 1956. Climatic physiology of cattle. Journal of Dairy science, 39(6): 715-725.

Coppoeters, W., Van de Weghe, A., Peelman, L., Depicker, A., Van Zeveren, A., and Bouquet.Y. 1993. Characterization of porcine polymorphic microsatellites loci. Anim. Genet. 24(3): 163-70.

Dourmad, J., Rigolot, C. and Hayo van der Werf, 2008. Emission of Greenhouse Gas: Developing management andanimal farming systems to assist mitigation. Livestock and Global Change conference proceeding. May 2008, Tunisia.

Du Preez, J.H., Giesecke, W.H. and Hattingh, P.J. 1990a. Heat stress in dairy cattle and other livestock under Southern African conditions. I. Temperature-humidity index mean values during the four main seasons. Onderstepoort J. Vet. Res. 57: 77-86.

Du Preez, J.H., Hatting, P.J., Giesecke, W.H. and Eisenberg, B.E. 1990b. Heat stress in dairy cattle and other livestock under Southern African conditions. III. Monthly temperature-humidity index mean values and their significance in the performance of dairy cattle. OnderstepoortJ. Vet. Res. 57: 243-248.

FAO statistics.1980, 2002, 2007, 2008 and 2009. Food and Agriculture Organization, UN

Gandhi, R.S., 2013. Genetic improvement of Indian cattle under changing climate scenario, Compendium of lectures, National training programme on Molecular genetic data generation, analysis and utilization in animal breeding. NDRI, Karnal pp 1-9.

Gantner, V., Mijic, P., Kuterovac, K.,Solic, D. and Gantner, R. 2011.Temperature-humidity index values and their significance on the daily production of dairy cattle.Daily production of dairy cattle, Mljekarstvo 61 (1), 56-63

Gaughan, J. B., Mader,T. L., Holt, S. M. and Lisle,A. 2008.A new heat load index for feedlot cattle. J. Anim. Sci. 86: 226-234

Giuffra, E., Kijas, J.M.H., Amarger, V., Carlborg, O., Jeon J.T. and Andersson, L. 2000. The origin of the domestic pig: independent domestication and subsequent introgression. Genet.154: 1785-1791

Hoffmann, I. 2008. Livestock Genetic Diversity and Climate Change Adaptation. Livestock and Global Change conference proceeding. May 2008, Tunisia.

http://www.climate.org/2002/programs/washington_summit_temperature_riseshtml

IFAD. 2009, Livestock and climate change.www.ifad.org/lrkm/index.htm.

INCCA (Indian networks for climate change and assessment), 2010.Ministry of Enviroment and forestry.GOI. (http://moef.nic.in/modules/others/?f=event)

IPCC. 2007. Climate Change 2007: The Physical Science Basis. Paris: Intergovernmental Panel on Climate Change. (http://www.ipcc.ch/ipccreports/ar4-wg1.htm).

Jimsy, J. 2007. Genetic and phenotypic variations of geographically differenr goat populations of Kerala. M.V. Sc. thesis submitted to Kerala Agricultural University.

Johnson, H.D. 1980. Environmental management of cattle to minimize the stress of climate changes.Int. J.Biometeor. 24(Suppl. 7, Part 2), 65-78.

Johnson, H.D., Ragsdale, A.C., Berry, I.L. andShanklin, M.D. 1962.Effect of various temperature humidity combinations on milk production of Holstein cattle. Res. Bull. Missouri Agric. Exp. Station, 791.

Kaura, R. L. 1952. Indian breeds of Livestock. Prem Publishers, Lucknow 165p.

Kijas, J.M.H. and Andersson, L. 2001. A phylogenetic study of the origin of the domestic pig estimated from near complete mtDNA genome. J. Mol. Evolution.5.

Kulkarni, A. A.,Pingle, S. S., Atakare, V. G. and Deshmukh, A. B. 1998.Effect of climatic factors on milk production in crossbred cows.Indian Veterinary Journal 75: 846-847.

Lal, S. N., Verma, D. N. and Hasain, K. Q. 1987.Effect of air temperature and humidity on the feed consumption, cario respiratory response and milk production in Hariana cows.Indian Veterinary Journal, 64: 115-121.

Liu, Y. X., Zhou, X., Li, D. Q., Cui, Q. W. and Wang, G. L. 2010.Association of ATP1A1 gene polymorphism with heat tolerance traits in dairy cattle.Genet. Mol. Res. 9(2): 891-896.

Livestock Census, 2007.Depatment of Animal Husbandry, Dairying & Fisheries,Ministry of agriculture, Govt. of India,KrishiBhavan, New Delhi.

Loredana, B., Patrizia, M., Valentina, P., Nicola, L. and Alessandro, N. 2011. Cellular thermo-tolerance is associated with heat shock protein 70.1 genetic polymorphisms in Holstein lactating cows. Cell Stress and Chaperones.DOI 10.1007/s12192-011-0257-7

Madhusudan, C. S. 2007. Molecular characterization of HSP 70 gene in buffalo, M.Sc. Thesis, IVRI, Izatnagar.

Mandal, D. K., Rao, A. V. M. S., Singh, K. and Singh, S. P. 2002a. Effects of macroclimatic factors on milk production in Frieswal herd. Indian Journal of Dairy Science, 55: 166-170.

Mandal, D. K., Rao, A. V. M. S., Singh, K. and Singh, S. P. 2002b. Comfortable macroclimatic conditions for optimum milk production in Sahiwal cows. Journal of Applied Zoological Researches, 13: 228-230.

Mangurkar, B. R. 1998. Breeding policy for sustained milk production.Proceedings of XXIX Dairy Industry Conference, NDRI Karnal. pp 18-21.

Mathew, S., Rai, A.V. and Govindaiah, M.G. 1994. Body weights atbirth and three months of Malabari goatsand its crosses with Alpine and Saanenbreeds. Current Research, 23: 84-86

McDowell, R.E. 1985. Crossbreeding in gtropical areas with emphasis on milk, health and fitness.Journal of Dairy Science, 68: 2418-2435.

Nei, M. 1972. Genetic distance between population. Am. Naturalist. 106: 283-29 52: 302-308

Ott, J. 1992. Strategies for characterizing higly polymorphic markers in human gene mapping.Amer. J. Hum. Genet., 51: 283-290

Rodriguez, L.W., Mekonnen, G., Wilcox, C.J., Martin, F.G. and Krienk, W.A. 1985. Effects of relative humidity, maximum and minimum temperature, pregnancy and stage of lactation on milk composition and yield. J. DairySci., 68: 973-978.

Rowlinson, P., 2008. Adapting Livestock Production Systems to Climate Change – Temperate Zones. Livestock andGlobal Change conference proceeding. May 2008, Tunisia.

Shinde, S., Taneja, V. K. and Singh, A. 1990. Association of climatic variables and production and reproduction traits in crossbreds. Indian J. Anim. Sci., 60: 81-85.

Sidahmed, A. 2008. Livestock and Climate Change: Coping and Risk Management Strategies for a Sustainable Future. InLivestock and Global Climate Change conference proceeding, May 2008, Tunisia.

Sodhi, M., Mukesh, M., Mishra, B.P., Kishore, A., Prakash, B., Kataria, R.S. and Joshi, B.K. 2012. Indian Zebu cattle: a natural resource for A2 milk. Pamphlet published by National Bureau of Animal Genetic Resources, Karnal.

Thomas, N., Joseph, S., Alex, R., Raghavan, K.C., Radhika, G., Anto., L. and Mohan, S. G. 2011. Genetic variation in resistance to caprine foot rot by Dichelobacternodosus in goats of Kerala, India.Biotechnology in Animal Husbandry, 27(2): 235-240.

Thornton, P., Herrero M., Freeman A., Mwai O., Rege E., Jones P., and McDermott J., 2008. Vulnerability, Climate Change and Livestock – Research Opportunities and Challenges for Poverty Alleviation. ILRI, Kenya.

Tomar, S.S., 2004. Text book of animal breeding, Kalyani publishers, New Delhi.

Usha, A.P. and Venkatachalapathy, R.T. 2010. Conservation and genetic diversity of native pigs of Kerala.Proceedings of National conference on native livestock breeds and their sustainable uses, Kottayam.

Venkatachalapathy, R.T. and Iype. S. 2010. Adaptability studies on Vechur cattle of Kerala. Proceedings of National conference on native livestock breeds and their sustainable uses, Kottayam.

16

Climate Resilient Livestock and Poultry Production

N.Maragatham, R.Karthikeyan, D. Rajakumar and R. Mathivanan

Tamil Nadu Agricultural University, TNAU, Coimbatore, Tamil Nadu

Livestock farming is an integral part of crop farming and contributes substantially to household nutritional security and poverty alleviation through increased household income. Indian agriculture is an economic symbiosis of crop and livestock production with cattle as the foundation. Dairy animals produce milk by converting the crop residues and by products from crops which otherwise would be wasted. Dairy sector contributes by way of cash income, drought power and manure. Livestock provides for human needs by way of food, fibre, fuel, fertilizer, skin and traction. India has the largest livestock populations in the world. It has 57 per cent of the world's buffalo population and 16 per cent of the cattle population. It ranks first in respect of cattle and buffalo population, third in sheep and second in goat population in the world. 70 per cent of the livestock are owned by 67 per cent of small and marginal farmers. 76 per cent of the milk is produced by weaker sections of society. Out of total livestock in the country, 37.28 per cent are cattle, 21.29 per cent are buffaloes, 12.71 per cent are sheep, 2640 per cent are goats and only 2.01 per cent are pigs. All other animals are less than 0.37 per cent of the total livestock. The species-wise breakup of livestock population in India (Table 16.1)

Table 16.1: Species-wise breakup of livestock population in India

S. No.	Species	Livestock population (in 000)		
		2007	2012	
		Number	Number	Growth %
1.	Crossbred cattle	33,060	39,732	20.18
2.	Indigenous cattle	1,66,015	1,51,172	- 8.94
Total cattle		**1,99,075**	**1,90,904**	**- 4.10**
3.	Buffaloes	1,05,343	1,08,702	3.19
4.	Sheep	71,558	65,069	- 9.07
5.	Goats	1,40,537	1,35,173	- 3.82
6.	Pigs	11,134	10,294	- 7.54
Total livestock		**5,29,698**	**5,12,057**	**- 3.33**
7.	Poultry	6,48.829	7,29,209	12.39

(*Source:* 18th and 19th Livestock Census, 2007 and 2012)

Tamil Nadu ranks 2nd in respect of poultry, 4th in sheep, 7th in goats, 13th in cattle and 14th in buffalo population in the country. Tamil Nadu contributes 4.44% of Indian livestock population. Similarly 16.09% of Indian poultry is in Tamil Nadu. Livestock and poultry population in Tamil Nadu (in millions) is shown in Table 16.2.

Table 16.2: Livestock and poultry population in Tamil Nadu (in millions)

Species	2007	2012	Growth %
Crossbred cattle	3.81	6.35	66.66
Indigenous cattle	7.38	2.46	- 66.66
Total cattle	**11.19**	**8.81**	**- 21.22**
Buffaloes	2.01	0.78	-61.15
Sheep	7.99	4.78	- 40.10
Goat	9.28	8.14	- 12.20
Pigs	0.28	0.18	- 35.29
Total livestock	**30.75**	**22.72**	**- 26.12**
Backyard poultry	29.47	13.91	- 52.79
Farm poultry	101.77	103.42	1.62
Total Poultry	**131.25**	**117.34**	**- 10.59**

(*Source*: Department of Animal Husbandry and Veterinary Services, Chennai - 600 006)

Indian livestock sector provides sustainability and stability to the national economy and food security. During the last decade, the annual growth rate of livestock production has maintained a steady growth of 4.8 to 6.6 % and even poultry attained 8-12% growth rate.

16.1 Breeds of Tamil Nadu

Tamil Nadu is native of the five indigenous breeds cattle and one buffalo. Eight breeds of sheep and 4 goat breeds are also found in Tamil Nadu. Apart from the indigenous breeds, Jersey and Holstein Friesian cross breeds of cattle are the commonly reared in Tamil Nadu. Among goats, Tellichery and Jamunapari also find a position among the farmers. Nellore and Mandya are the sheep breeds commonly found in district adjoining to Andra Pradesh and Karnataka state respectively.

16.1.1 Cattle breeds

Kankeyam, Umbalachery, Burghur, Pulikulam, Alambadi and Toda buffalo are the native breeds of Tamil Nadu. These breeds constitute 5.07 % of the total bovine population in Tamil Nadu. These breeds are highly adapted to local climate and require low management practices. These animals are resistant to diseases, tolerant to high temperature, adapted to tropical climate and have ability to survive even with poor quality feed.

Kangayam

i.   Native of Kangayam, D h a r a p u r a m , Perundurai, Erode, Bhavani and part of Gobichettipalayam taluk of Erode and Coimbatore district.

ii. **Colour:** Coat is red at birth, but changes to great at about 6 months of age. Bulls are grey with dark colour in hump, fore and hind quarters. Bullocks are grey. Cows are grey or white or grey. However, animals with red, black, fawn and broken colours are also observed. Horns, muzzle, eyelids, tail switch and hooves are black.

iii. They are short with stout legs and strong hooves.

iv. The horns are spread apart, nearly straight with a slight curve backward.

v.   The eyes are dark and prominent with black rings around them.

vi. The dewlap is thin. The sheath is well tucked up to the body.

vii. The average milk yield 600 to 700 kg in a lactation.

Bargur

i. Native of Bargur hills in Bhavani taluk of Erode district.

ii. Bargur cattle are of brown colour with white markings. Some white or dark brown animal are also seen.

iii. Animals are well built, compact and medium in size.

iv. Muzzle is moderate and black in colour.

v. Horns are of light brown colour, moderate length, closer at the roots inkling backward, outward and upward with a forward curve and sharp at the tip.

Umblacherry

i. Native of Thanjavur, Thiruvarur and Nagappattinam districts of Tamil Nadu.

ii. Umblacherry calves are generally red or brown at birth with all the characteristic white marking on the face, on limbs and tail.

iii. The colour changes to grey at about 6 months of age. In adult females, the predominant coat colour is grey with white marking on the face and legs.

iv. All the legs below hocks have white marks either socks or stockings.

v. Horns are very small, curving outward and inward an sometimes spreading laterally.

vi. The practice of dehorning bullocks is peculiar in Umblacherry cattle. Horn buds are removed at 6 months of age by singing with red hot iron.

vii. Ears are pruned and hot iron branding is done.

Pulikulam / Jellicut Breed

i. Pulikulam is seen in Madurai and Theni districts. They are also raised in Cumbum valley and the Periyar river, where there are grazing grounds of vast extent.

ii. These animals depend extensively on forest grazing. This is a quick trotting (5-6 miles per hour) breed.

iii. They are comparatively small, but active and capable of much endurance. Selected bulls are utilized as Jellicut purposes.

Alambadi

i. This breed is restricted to Salem and Coimbatore district of Tamil Nadu and part of Bangalore district in Karnataka and presently very few animals are available.

ii. Alambadi bulls are dark grey, almost black and cows grey or white. They have the typical backward curving horns of Mysore type cattle.

iii. They are active, useful drought animals but not fast trotter.

Toda

i. Toda breed of buffaloes is named after an ancient tribe, Toda of Nilgiris of south India.

ii. Coat colour of the calf is generally fawn at birth. In adult the predominate coat colours are fawn and ash-grey.

iii.   These buffaloes are quite distinct from other breeds and are indigenous to Nilgiri hills.

iv.   The animals have long body, deep and broad chest, and short and strong legs.

v.   The head is heavy with horns set well apart, curving inward outward and forward.

vi.   Thick hair coat is found all over the body. They are gregarious in nature.

16.1.2 Sheep breeds

1.   **Madras Red:** Chennai and Kancheepuram districts of Tamil Nadu. Body colour predominantly brown, the intensity varying from light tan to dark brown. Some animals may have white markings on forehead, inside the thigh and lower abdomen. Medium sized drooping ears. Tail short and thin. Rams have strong, corrugated and twisted horns. Ewes polled. Body covered with short hairs. Adult male: 35 kg, female: 23 kg.

2.   **Kilakarsal or Kilakarisal:** Ramnad, Madurai and Tanjore districts of Tamil Nadu Brown/ dark tan in colour with black spots on head belly and legs. Medium sized ears. Males have thick twisted horns. Most animals have wattle.

3.   **Mecheri:** Salem, Namakkal, Erode, Tirupur and Coimbatore districts of Tamil Nadu. Medium sized light brown in colour. Both sexes are polled. Body covered by very short hairs. Adult male: 35 kg, female: 22 kg.

4.   **Ramnad White:** Ramnad and Sivagangai districts of Tamil Nadu. Medium sized predominantly white. Ears medium sized and directed outward and downward. Males have twisted horns. Ewes polled short and thin tail. Adult male: 31 kg, female: 22 kg.

5.   **Trichy black:** Trichy district of Tamil Nadu. Small animals. Body is completely black. Males horned, ewes polled fleece extremely coarse, hairy and open. Ears and tail small. Adult male: 25 kg, female: 18 kg.

6.   **Vembur:** Tirunelveli district of Tamil Nadu. Tall animals, coat colour is dark tan with black spots on head, belly and legs. Medium sized drooping ears. Males horned. Ewes polled. Body covered with short hairs. Adult male: 34 kg, female: 27 kg.

7.   **Coimbatore:** Coimbatore and Erode districts of Tamil Nadu. Medium sized animals white with black or brown spots. 30% of males polled. Fleece white hairy and open. Adult male: 24 kg, female: 20 kg.

8. **Chevaadu:** National Bureau of Animal Genetic Resources (NBAGR) had recently recognised this animal as one of the breed of sheep. Native of Tirunelveli district. Animal are small and medium in size. Present in light brown and dark brown or tan coat colour. Adult body weight varies from 18-39 kg.

Madras Red

Kilakarsal

Mecheri

Ramnad White

Trichy Black

Vembur

Coimbatore   Chevaadu

16.1.3 Goat breeds

Kanni Aadu: Native of Thirunelveli area, Tamil Nadu. Coat colour: Predominantly black or black with white or brown spots and hence called as palkanni and chenkanni respectively. Body and legs: Tall and stout. Ears: Medium. Horns: Bucks horned and does polled. Live weight: Average, buck 36 and doe 29 kg. Kidding: Twice in 18 months, single or twins; age at first kidding 15 months. Meat: Good. Milk: Enough for kids.

Kodi Aadu: Native of part of Tuticorin and adjoining areas of Ramanathapuram, Tanjore and Pudukottai districts. They are tall, long animals with slender body. White in colour with splashes of black or red colour. Based on the colour they are classified as KarumPorai (Blakish) and ChemPorai (Reddish brown). Both sexes are horned. Suitable for coastal areas. Used as leader in sheep flock while grazing. Adult body weight: Buck: 39.5kg and Doe 32.2 kg.

Salem Black: Native of Salem, Dharmapuri and Vellore district of Tamil Nadu. Salem Black goats are tall animal with a lean body and a completely black coat. The head is medium in length with a medium to broad forehead. The ears are leaf like and pendulus. Both male and female are horned. The tail is thin, medium in length and curled upwards. The adult body weight is 38.5 (male) and 29.5(female) kg.

Molai Aadu: Native of Erode district. The animals are polled and hence it is called as molaiadu.

Tamil Nadu produced 6.831 million tonnes of milk out of 121.8 million tonnes of Indian milk production in 2010-11. Among this 5.247 million tonnes from cross breed, 0.774 million tonnes from local and 0.810 from buffalo. Tamil Nadu contributes 18.27 percent of egg, 8.78 percent of meat and 5.61 percent of milk production in India and stands 2nd in egg, 5th in meat, 8th in milk production in the country (19th Livestock census, GOI, 2012)

Kanni Adu

Kodi Adu

Salem Black

Molai Adu

16.2 Impact of Climate on Livestock

Climate change has complicated impacts on animals affecting distribution, growth, incidence of diseases, availability of prey, productivity and even extinction of species in extreme cases due to habitat loss. Climate change has complex impacts on domestic animal production system affecting feed supply, challenging thermoregulatory mechanism resulting thermal stress, emerging new and old diseases due to change in epidemiology of diseases and causing many other indirect impacts. Global warming has two way effects on animal production system. In one hand, it directly affects the health, reproduction, nutrition etc. of the animals resulting in poor performance, inferior product quality, outbreak of novel diseases, etc. while on the other hand there are indirect effects on animal production due to change in soil fertility, decrease in preferred vegetation, rangeland degradation, desertification and decrease in production of feed stuffs. Most of the impacts of climate change are attributable to increased ambient temperature.

16.2.1 Heat stress

The environmental conditions that induce heat stress can be calculated using the temperature humidity index (THI). Heat stress begins to occur in dairy cattle when the THI is > 72.*Bosindicus* (Zebu) cattle are more thermotolerant than *Bostaurus* cattle due to possession of thermotolerant gene by zebu cattle (Table 16.3).

Table 16.3: Effect of Heat Stress on Dairy Cattle

THI	Stress Level	Comments
< 72	None	
72–79	Mild	Dairy cows will adjust by seeking shade, increasing respiration rate anddilation of the blood vessels. The effect on milk production will be minimal.
80–89	Moderate	Both saliva production and respiration rate will increase. Feed intake maybe depressed and water consumption will increase. There will be an increase in body temperature. Milk production and reproduction will be decreased.
90–98	Severe	Cows will become very uncomfortable due to high body temperature,rapid respiration (panting) and excessive saliva production. Milk production and reproduction will be markedly decreased.
> 98	Danger	Potential cow deaths can occur.

The severity of heat stress experience by an animal depends on actual temperature and humidity, length of the heat stress period, degree of night cooling, ventilation and air flow, size of the cow, level of milk production, housing type, overcrowding, water availability, breed, coat color, hair coat depth etc.The animal can able to dissipate heat through conduction, convection, radiation and evaporationto maintain a normal body temperature. The cow can primarily control or regulate only the evaporative cooling mechanism. Heat stress may be minimized by number of options including ration adjustments like feeding quality fodder, balanced ration, feeding anti-stress medicines and feeding methods like feeding on cooler part of the day *etc.* The quality water must be available throughout the day. Housing management like providing fans, sprinklers, foggers and proper ventilation might reduce the heat stress substantially.

16.2.2 Impact on milk and meat production

One of the direct impacts of climate change on livestock is the milk and meat production and can negatively affect the production performances like milk yield, growth rate in meat animals. The milk yield attribute effect in most predominant in high genetic merit animals like exotic milch animals. At all India level, an estimated annual loss due to direct thermal stress on livestock is about

1.8 million tonnes of milk which is nearly 2% of total milk production in the country.

16.2.3 Impact on reproduction

Heat stress due to high temperature accompanied with excess humidity during summer causes infertility and has adverse effect on reproductive performances of animals. These included decreased length and intensity of the estrus period, decreased conception rate, decreased growth, size and development of ovarian follicles, higher embryonic death, decreased fetal growth and calf size. In male, it reduced quantity and quality of sperm production and results in infertility.

16.2.4 Impact on feed and fodder availability

Climate change affects livestock production by altering the quantity and quality of feed and fodder available for the animals. Reduction in dry matter intake (DMI) indirectly helps to maintain core body temperature by reducing generation of heat during ruminal fermentation and nutrient metabolism. Decrease in dry matter intake is more prominent in animals fed with roughage based diet than in animals fed with concentrate based diet. Also, decrease in feed intake is more prominent in *Bostaurus* (Exotic) cattle than in *Bosindicus* (Indigenous) cattle.

16.2.5 Impact on livestock health

Reduced disease resistance of the animals, enhanced multiplication of microorganisms and altered vector population causes increased incidence of certain diseases like mastitis during summer when ambient temperature is high. Altered metabolic status of the animals may also cause reduction in immunity making animals more susceptible to diseases.

16.2.6 Impact of climate on poultry

Climate can be defined as the sum of environmental factors which influence the normal physiological functions. The climate directly surrounding to the birds is called the micro-climate. The micro-climate is of importance for the birds rather than macro-climate. The following environmental factors are known to affect the performance of poultry,

- Temperature
- Relative humidity
- Air composition
- Air velocity
- Light

16.3 Temperature

Poultry are homeothermic and the average body temperature is between 41°C and 42.2°C. The thermoregulatory mechanism is not fully developed in young chicks so as to they need artificial heat source in the form of brooding. The comfort zone is one where birds are able to keep their body temperature constant with minimum effort. The highest and lowest critical temperature depend very much on age, bodyweight, housing system, feeding level, relative humidity, air velocity and health status of the bird. When temperatures are not within the comfort zone, birds tend to have several mechanisms which enable them to keep their body temperature constant without having to produce extra heat. This may be done by altering its activities like feed intake, water intake, dust bathing in case of deep litter, moulting in elevated temperature etc.

The most efficient temperatures for adult layers are between 20 – 24°C. The recommended microclimate for young chicks is 32-34°C on the day one and it has to reduce for 4°C for every week until it reaches its 20-24°C. Behaviour of the birds in the house is the best tool to assess the comfortable temperature in the house.

16.4 Relative humidity

Humidity is controlled by the intense heating or cooling of house air in response to the temperature outside the house. When outside temperatures are low, relative humidity in the house is low, which often results in dry dust circulating in the air within the house. If the relative humidity is too high, this may result in wet litter. The ideal relative humidity for poultry is 60-80%.

16.5 Air composition

The concentration of O_2 usually varies between 19-21 percent while its concentration below 6% is lethal for birds as well as humans. CO_2 at concentrations between 0.1-0.3%. The air quality thus requires to be monitored frequently in order to ensure safety of the birds and the workers employed in the poultry farm. Since air quality directly reflects the sanitary and hygienic status of the poultry house its assessment from time to time can be taken as an indicator for scheduling manure removal operations and also for assessing the ventilation require. Ammonia emissions were highest from poultry houses as compared to those of cattle and swine. Ammonia in combination with dust is the most significant respiratory hazard to the occupational health of poultry workers. At a concentration of 15 ppm, ammonia is uncomfortable for the workers and above 50 ppm it causes injury. While 30 ppm concentration of the gas in the poultry house affects the general health of the birds reducing egg production and at 0.01% it produces higher incidence of breast blisters and increased water consumption. H_2S at concentrations above 0.05% causes death of chicken. With a pungent odour it causes irritation of eyes/nose, headache and dizziness in humans at concentrations between 0.01- 0.05% and also causes death at 0.1% H_2S when combines with humidity in the air forms corrosive sulfuric acid and damages metal cages thus reducing their durability.

16.6 Air velocity

Young are more sensitive than older, heavier one. Considering the recommended temperatures and relative humidity, the air velocity at bird level is allowed to vary between 0.1 and 0.2 m/second. The air movement patterns depend on the ventilation within a house, the house width, the slope of the roof and the way the house is organized.

16.7 Light

Lighting is an essential factor in the success of the commercial production of layers and broilers. General light intensity required is 20 lux and above for brooding stage and 5 to 8 lux for growing and laying stage at the birds head level. The recommended photoperiod is 24 hours for layer chick, 12 hours for growers and 16 hours for layers. Broilers require 23 hours photoperiod. Heat stress reduced the egg production, egg quality, meat quality, feed efficiency, growth rate, immunity, fertility, hatchability and predisposes the disease outbreak which leads mortality.

16.8 Mitigation Strategies

- Selection of suitable animal and breed to the particular agro-climatic condition can reduce adverse impact on livestock and poultry due to climate change.

- Better housing management effectively reduce the adverse climatic effect on animal performances.

- Manipulating feed and its feeding methods significantly decrease both heat and cold stress.

- Providing potable drinking water to the livestock and poultry effective control water related problems.

16.8.1 Contribution of livestock and poultry to climate change

India emerged as the largest contributor to the methane emission because of its large population even though the emission rate per animal was much lower than in developed countries. Ruminants account for a large share of total livestock green gas emissions because they are less efficient in converting forage in to products than monogastric animals and poultry. Amount of feed consumed and its digestibility are two important factors determine the total methane emission. Feed-crop production and management of pastures give rise to emissions associated with the production and application of chemical fertilizer and pesticides and with the loss of soil organic matter. Further emissions occur because of the use of fossil fuels in the transport of animal feed. Further emissions occur directly from the animals as they grow and produce especially ruminant animals emit methane as a by-product of the microbial fermentation through which they digest fibrous feeds. Emissions of methane and nitrous oxide occur during the storage and use of animal manure. Processing and transport of animal products give rise to further emissions, mostly related to use of fossil fuel and infrastructure development. On a commodity-basis, beef and cattle milk are responsible for the most emissions, respectively, contributing 41 percent and 20 percent of the sector's overall GHG outputs (This figure excludes emissions from cow manure and cattle used as drought). They are followed by pig meat (9 per cent of emissions), buffalo milk and meat (8 per cent), chicken meat and eggs (8 per cent), and small ruminant milk and meat (6 per cent). The remaining emissions are sourced to other poultry species and non-edible products. Emission intensities (i.e. emissions per unit of product) vary from commodity to commodity. They are highest for beef (almost 300 kg CO_2-eq per kilogram of protein produced), followed by meat and milk from small ruminants (165 and 112kg CO_2-eq kg respectively). Cow milk, chicken products and pork have lover global average emission intensities (below 100 CO_2-eq/kg.) (At the sub-global level, within each commodity type there is very high variability in emission intensities, as a result of the different practices and inputs to production used around the world. Enteric emissions and feed production (including manure deposition on pasture) dominate emissions from ruminant production. In pig supply chains, the bulk of emissions are related to the feed supply and manure storage in processing, while feed supply represents the bulk of emissions in poultry production, followed by energy consumption. Greenhouse gas emissions by the livestock sector could be cut by as much as 30 percent through the wider use of existing best practices and technologies, according to a new study released by the UN Food and Agriculture Organization (FAO), 2013.

16.8.2 Challenges associated with changing climate on livestock production system

Livestock production system is expected to be exposed to many challenges due to climate change in India. They are:

16.8.2.1 Challenges Associated with the Direct Effects of a Changing Climate and its Alleviation

Direct effect of climate change through raised temperature, humidity and solar radiation may alter the physiology of livestock, reducing production and reproductive efficiency of both male and female and altered morbidity and mortality rates. Heat stress suppresses appetite and feed intake, however, animals' water requirements is increased. In general, the high-output breeds especially crossbreds, which provide the sizable amount of Indian production, are more vulnerable to heat stress as compared to indigenous one. Also, as people are lured by immediate money making methods, indiscriminate cross breeding is adding to the concern, however, this approach is not sustainable. Options for alleviating heat stress include adjusting animals' diets to minimize diet-induced thermogenesis (low fibre and low protein) or by increasing nutrient concentration in the feed to compensate for lower intake; taking measures to protect the animals from excessive heat load shading/improving ventilation by using fans) or enhance heat loss from their bodies (Sprinklers/misters); or genetic selection for heat tolerance or bringing in types of animals that already have good heat tolerance . All these options require some degree of initial investment, some require access to relatively advanced technologies, and all except simple shading require ongoing input of water and/or power. The practicality of implementing cooling measures depends on the type of production system. They can most easily be applied in systems where the animals are confined and where the necessary inputs can be afforded and easily accessed. In extensive grazing systems, it is difficult to do more than provide some shade for the animals and possibly places for them to wallow. Livestock producers, in areas where relative humidity is high (north-eastern part of India), face additional problems as there is less potential for the use of methods based on evaporative cooling. Small-scale producers who have adopted high-output breeds, but struggle to obtain the inputs needed to prevent the animals from becoming overheated, may find that their problems are exacerbated by climate change.

16.8.2.2 Challenges Associated with Livestock Feeding and Nutrition and its Alleviation

Livestock production and its economic efficiency depend on quantity and quality of feed and water that animals need to survive, produce and reproduce. About 10% of cropland is used for producing animal feed and other agriculture land provides crop residues used for feeding livestock. The future of livestock production systems depends on the continued productivity of these various feed-producing areas – all of which are potentially affected by climate change. The influence of the climate on the distribution of plant variety and type is complex.

The effects of climatic interaction with soil characteristics and its direct effect on plants influences the distribution of the various other biological components of the agro-ecosystem – pests, diseases, herbivorous animals, pollinators, soil microorganisms, *etc.* – all of which in turn influence plant communities. All these processes have the potential to influence directly or indirectly the growth of the plants on which livestock feed. Pressure on feed resources and other constraints to traditional livestock-keeping livelihoods have promoted the spread of agro-pastoralism (i.e. livelihoods that involve some crop production in addition to livestock keeping) at the expense of pastoralism. In production systems where animals are fed on concentrates, rising grain prices (may be driven by climate change) increase the pressure to use animals that efficiently convert grains into meat, eggs or milk. Thus, within such systems climate change may lead to greater use of poultry and pigs at the expense of ruminants, and greater focus on the breeds that are the best converters of concentrate feed under high external input conditions. Increases in the price of grain may also contribute to the further concentration of production in the hands of large-scale producers.

16.8.2.3 Challenges associated with the Effects of Diseases and Parasites

The geographical and seasonal distributions of many infectious diseases, particularly vector borne, as well as those of many parasites and pests of various kinds are affected by climate. Pathogens, vectors, and intermediate and final hosts can all be affected both directly by the climate (e.g. temperature and humidity) and by the effects of climate on other aspects of their habitats (e.g. vegetation). If the climate changes, hosts and pathogens may be brought together in new locations and contexts, bringing new threats to animal (and in some cases human) health and new challenges for livestock management and policy. However, it is difficult to segregate out epidemiological changes that can be attributed unambiguously to climate change. Climate is characterized not merely by averages, but also by short-term fluctuations, seasonal oscillations, sudden discontinuities and long term variations, all of which can influence disease distribution and impacts. Rapid spread of pathogens, even small spatial or seasonal changes in disease distribution, may expose livestock populations to new disease challenges. Disease-related threats can be acute or chronic and can be caused by the direct effects of disease or indirectly by the measures used to control disease. The most severe recent epidemics in India in terms of the numbers of livestock lost have involved quite a narrow range of diseases: most notably foot-and-mouth disease, avian influenza, Blue tongue, African swine fever, classical swine fever and contagious bovine pleuropneumonia.

16.8.3 Mitigation strategies

- Culling of unproductive animals and thereby reducing animals numbers significantly reduced methane emission.

- Increasing the productivity per animal.

- Increased green fodder in the animal diet decreased methane production.

- Feeding of more concentrate more efficiently reduced the green house gas emission per unit of product.

- Feeding of methane inhibitors like bromo-chloromethane (BCM), 2-bromo-ethane sulfonate (BES), chloroform, cyclodetrine found to reduced methane production by up to 50% in cattle and small ruminants.

- Feeding of ioniphore antibiotics like monensin are found to decrease methane production.

- Improved animal waste management such as burning, burring, covered storage facilities found to reduce the methane emission.

16.9 Adaptation and Mitigation Strategies to Climate Change

16.9.1 Genetic approach

Many local breeds are having valuable adaptive traits that have developed over a long period of time which includes tolerance to extreme temperature, humidity, adaptation to survive in low/ poor management conditions and feeding regimes. Hence, Genetic approach to adapt the climate change should include measures such as :

- Identifying and strengthening the local genetic groups which are resilient to climatic stress/ extremes .

- Genetic selection for heat tolerance or bringing in types of animals that already have good heat tolerance and crossbreeding the local genetic population with heat and disease tolerant breeds.

- Identifying the genes responsible for unique characteristics like disease tolerance, heat tolerance, ability to survive in low input conditions and using it as basis for selection of future breeding stock will help in mitigating the adverse effect of climate stress.

Breeding management strategies

Changing the breeding animal for every 2-3 years (exchange from other district herd) or artificial insemination with proven breed semen will help in enhancing the productivity. This may be supplemented with supply of superior males through formation of nucleus herd at block level. Synchronization of breeding period depending on the availability of feed and fodder resources results in healthy offsprings and better weight gain. Local climate resilient breeds of moderate productivity should be promoted over susceptible crossbreds. In India, with small flock sizes, large fluctuations in rearing conditions and management between flocks, and over time within a flock,lack of systematic livestock identification, inadequate recording of livestock performances and pedigrees, and constraints related to the subsistence nature of livestock rearing (where monetary profit is not the most important consideration), the accuracy of selection will be much lower, resulting in even lower rates of genetic gain. However, locally adapted breeds are likely to be highly variable and the highest performing animals of such breeds can have great productive potential. Therefore, the screening of livestock populations previously not subjected to systematic selection is likely to give quicker results to provide high genetic merit foundation stock for nucleus flocks.

16.9.2 Nutritional adjustments

The feed intake by the livestock during thermal stress is significantly lower than those in comfort zone. Hence, the care should be directed towards providing more nutrient dense diet while will help to minimize production losses due to the high temperatures as well as those feed which generates less heat during digestion. This can be achieved by following measures:

- Feeding dietary fat remains an effective strategy of providing extra energy during the time of negative energy balance. Incorporation of dietary fat at level of 2-6 % will increase dietary energy density in summer to compensate for lower feed intake.

- Adjusting animals' diets to minimize diet-induced thermo genesis (low fibre and low protein diets). High-fiber diets generate more heat during digestion than lower fiber diets.

- Using more synthetic amino acids to reduce dietary crude protein levels. Excessive dietary protein or amino acids generate more heat during digestion and metabolism.

- Feeding of antioxidant (Vitamin A, C & E, selenium, Zinc) reduces the heat stress and optimize feed intake.

- Addition of feed additives/vitamins and mineral supplementations that helps in increasing feed intake, modify gut microbial population and gut integrity and maintain proper cation and anion balance.

- During lean/drought periods, shepherds migrate along with their animals in search of fodder. This migration sometimes creates social conflicts with local people for available scarce fodder resources. Further, this could invite new diseases and parasites which pose health problems in small ruminants. Protein is the first limiting nutrient in many grazing forages and protein availability declines in forages as the plant matures towards the end of winter season. When daytime temperatures and humidity are elevated, special precautions must be taken to keep livestock comfortable and avoid heat stress. Allow for grazing early in the morning or later in the evening to minimize stress.

- Concentrate mixture (18% DCP and 70% TDN) prepared with locally available feed ingredients should be supplemented to all categories of animals. When no green fodder is available, addition of vitamin supplement in concentrate mixture helps in mitigating heat stress.

- Further, in extreme conditions, energy intake becomes less compared to expenditure as the animal has to walk more distance in search of grazing resources which are poor in available nutrients. Hence, all the animals should be maintained under intensive system with cut and carry of available fodder. The concept of complete feed using crop residues (60%) and concentrate ingredients should be promoted for efficient utilization of crop residues like red gram stalk, etc. Further, productivity and profitability from ruminants can be increased by strengthening feed and fodder base both at village and household level with the following possible fodder production options.

16.9.3 Management interventions

- *Water supply:* Animals must have access to large quantities of water during periods of high environmental temperatures. Much of the water is needed for evaporative heat loss via respiration to help them cool off. Hence, provision has to be made for supply of continuous clean, fresh and cool water to the animals. Cleaning the feeding trough frequently and providing fresh feed will encourage the animals to take more feed. Splashing the cool water over the animals at regular intervals during the hot period will reduce the heat stress.

- *Feeding time:* Providing feed to the animals during cool period i.e. evening or night will improve the feed intake by the animals.

- *Stocking density:* Reducing the stocking density during hot weather will help the animals in dissipating the body heat more efficiently through manifestation of behavioural adaptation.

- *Shade*: The use of shades is an effective method in helping to cool animals. Shades can cut the radiant heat load from the sun by as much as 40%. Shades with straw roofs are best because they have a high insulation value and a reflective surface. Uninsulatedaluminum or bright galvanized steel roofs are also good. The best shades have white or reflective upper surfaces. Provision of trees at certain distance from the shed which will provide shade to the animals.Shifting the animals to cool shaded area during the hot climatic conditions.

- *Provision of vegetative cover over* the surrounding area will reduce the radiative heat from the ground. The surface covered with green grass cover will reflect back 5-11% of solar radiation as compared to 10-25% by dry bare ground and 18-30% by surface covered by dry sand adding to thermal stress.

- *Provision of elongated eaves* or overhang will provide shade as well as prevent rain water from entering the sheds during rainy season.

- *Ventilation*: increasing the ventilation or air circulation in the animal sheds will aid the animals in effective dissipation the heat. The air circulation inside the shed can be increased by keeping half side wall i.e., open housing system, use of fan, increasing the height of the building etc.

- *Roof material*: the roof material to be used should be bad conductor of heat. i.e., it should prevent radiative heat from entering into the shed. Thatch along with bamboo mat is excellent roofing material for tropical conditions. However, it is prone for fire hazards as well as its longevity is less. The outer surface of the roof should be painted white so that the white surface will reflect the solar radiation back. Some materials such as aluminium reflect heat well as long as they are not too oxidized.

16.9.4 Other interventions

- *Revival of Common Property Resources* (CPRs) Majority of the total feed requirements of ruminants are met by the CPRs. There is no control over the number of animals allowed to be grazed, causing severe damage on the re-growth of number of favourable herbaceous species in grazing lands. Thus causing severe impact not only on herbage availability from CPRs but also quality of herbage affecting the productivity of animals adversely; hence there should be some restriction on number and species of animals to be grazed in any CPR as a social regulation. CPRs need to be reseeded with high producing legume and non-legume fodder varieties at every 2-3 years intervals as a community activity. Further, grazing restriction till the fodder grows to a proper stage and rotational grazing as community decision would improve the carrying capacity of CPRs.

- *Intensive Fodder Production Systems:* Growing of two or more annual fodder crops as sole crops in mixed strands of legume (Stylo or cowpea or hedge Lucerne, *etc*) and cereal fodder crops like sorghum, ragi in rainy season followed by berseem or Lucerne etc., in rabi season in order to increase nutritious forage production round the year. Fodder crops like Stylohamata and Cenchrusciliaris can be sown inthe inter spaces between the tree rows in orchards or plantations as hortipastoral and silvopastoral systems for fodder production .

- *Use of Unconventional Resources as Feed:* The available waste products form food industries like palm press fibre, fruit pulp waste, vegetable waste, brewers' grain waste and all the cakes after expelling oil *etc.*, and thorn-less cactus should be used as feed to meet the nutritional requirements of animals.

16.10 Conclusion

The growing human population and its increasing affluence will increase the global demand for livestock products. But the expected big changes in the climate globally will affect directly or indirectly the natural resource base, the animal productivity andhealth and the sustainability of livestock-based production systems. Global warming is expected to introduce an additional level of pressure for livestock production systems in dry areas. Livestock production system is sensitive to climate change and at the same time itself a contributor to the phenomenon, climate change has the potential to be an increasingly formidable challenge to the development of the livestock sector. Responding to the challenge of climate change requires formulation of appropriate adaptation and mitigation options for the sector. Although the reduction in GHG emissions from livestock industries are seen as high priorities, strategies for reducing emissions should not reduce the economic viability of enterprises if they are to find industry acceptability. As the numbers of farm animals reared for meat, egg, and dairy production increase, so do emissions from their production. By 2050, global farm animal production is expected to double from present levels. The environmental impacts of animal agriculture require that governments, international organizations, producers, and consumers focus more attention on the role played by meat, egg, and dairy production. Mitigating and preventing the environmental harms caused by this sector require immediate and substantial changes in regulation, production practices, and consumption patterns. The livestock development strategy in the changing climate scenario should essentially focus on minimization of potential production losses resulting from climate change, on one hand, and on the other, intensify efforts for methane abatement from this sector as this would also be instrumental in increasing production of milk by reducing energy loss from the animals through methane emissions. Constant research, education and sensitization are needed in order to adapt to and combat the possible effects of climate change.

References

Hand Book of Animal Husbandry, ICAR, New Delhi.

19th Livestock census 2012. All India report, Ministry of Agriculture, Department of Animal Husbandry, dairying and fisheries, New Delhi.

Pankaj, P.K., D.B.V. Ramana, R. Pourouchottamane and and S. Naskar, 2013. Livestock Management under Changing Climate Scenario in India. World Journal of Veterinary Science, 2013, 1: 25-32 .

Sandeep Reddy, S. and Ananda Rao, K. 2014.Effects of climate change on livestock production and mitigation strategies –a review .International Journal of Innovative Research and Review ISSN: 2347-4424 (Online).

State action plan on climate change, Government of Tamil Nadu, 2013.

17

Futuristic Strategies for Fodder and Waste Management in Climate Change Scenario

Francis Xavier and Deepak Mathew

Kerala Veterinary and Animal Sciences University, Mannuthy, Thrissur Kerala – 680651, India

"Terra"*per se* is an extraordinarily complex, messy geophysical system with dozens of variables, most of which change in response to one another. The basic proposition behind the science of climate change is also rooted in the floral and faunal cohabiting with the *Homo sapiens*. Livestock and related fodder cultivation and the green waste generated in a farming enterprise are of extreme significance when we ponder over the climate change. Confusion and complacency reigns the thinking process and planning process related to climate change. Confusion in the sense, that risk communication in this regard is so placid. Complacency, we say due to the lackadaisical attitude of our policy makers. To make it very plain, fodder production sector as well as farm sector will suffer the most in Kerala for want of scientific thinking, planning and execution. Drinking water becomes a question mark when the coastal aquifers and coastal belt is ingresses by saline water. Along with this the irrigation water source will also play it's role. The higher temperatures are going to dewater the topsoil putting greater stress on the plants whether wild or domesticated. Anthropogenic emissions related to fodder and livestock waste may pose problems in future as the population increases fast. The thought on luxury emissions and survival emissions started way ahead with changing climate thoughts. We would also like to impress upon the "survival emissions" by the livestock farming sector directly and indirectly on the climatic elements. Climate change may influence land use patterns and degradation of cultivable fodder land, irrigation and drinking water pattern for fodder and livestock respectively. Fodder cultivation and Livestock green waste has a significant role in green house gas emissions and ecology (Fig. 17.1).

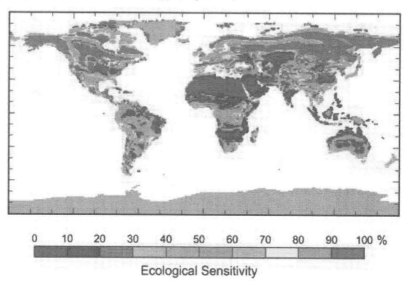

Ecological Sensitivity

(*Courtesy*: NASA, 2013)
Fig. 17.1: Climate change and ecological sensitivity an overview.

17.1 Fodder Production

Out of the 143 crore hectares of agriculture land in the whole world per capita land for food production is estimated to be 0.5 ha. The land availability for food production is getting reduced and hence the aggressive agricultural steps in. In 1960 when the population was 300 crore on this globe the 0.5 ha. land for producing food existed. The population explosion (680 crores in 2010) reduced the per head availability of the land for food production which became 0.21 ha. This ratio will naturally reach drastic levels as the population grabs the way of Malthusian thinking. Reduction in the land utilisation will naturally be reflected in the fodder crops. Indian population increases at 1.7% .At national level also we have to face the grim situation soon. In the state of Kerala, only 12% of the available agricultural land is utilised for food production, Acute shortage of land for food production whether it is human food or animal food exists. In this scenario the reduced fodder land availability to feed livestock poses problems. In the developing countries, more and more people are going to move out of agriculture into the industrial and service sectors. And this should be facilitated if we really are serious about poverty reduction. The land use also may come down globally. All these mean a shift from labour intensive to techno-intensive cultivation is in the anvil in a state like Kerala. Increasing inputs like fertilizers and pesticides, which beyond a point is non- sustainable and eco-damaging will also have to creep in, for this we need to strike a middle path between the Nemesis and Hubris approaches. In case of fodder production in Kerala the immediate needs may be listed as follows:

- Increase publicly funded research in indigenous fodder that may survive and sustain in the harshness of the climate change. (Emphasis on Public, because the private players frequently distort perspectives for example marketing a new gene pool or technology that would require as input, some product manufactured by them or by a related company.

- Ban patenting of fodder genes, trans-genes and DNA or RNA sequences.

- Strictly enforce green Bio-safety protocols.

- Strictly monitor cross pollination/hybridization events and destroy hybrids between wild and artificial fodder crops if found.

- Uundergo all the phases of evaluating nutritional value.

- Periodic review of each fodder crop and the cultivation protocol for a mandatory minimum period.

The stock taking on the indigenous fodder verities will sure project us as paupers. Input has not gone in this field seriously. Fodder production in Kerala fits well into that famous 'Lifeboat ethics". 'Lifeboat ethics' is a term to describe a very interesting metaphor for the age-old philosophical problem of deciding to what degree the rich and the wealthy are ethically obliged to help the poor and needy. The proponents of lifeboat ethics argue that the implications of modern farming must be examined from the perspective of human population growth. Increased food production in poor countries due to the adoption of modern farming methods allowed many people- who would otherwise have starved- to survive and to have children. Garrett Hardin, a renowned American human ecology Professor, in his 1974 essay "Living on a lifeboat" used this metaphor. Let us quote Hardin, (1974). (Hardin, G. 1974. "Living on a lifeboat" *Bioscience* 24 (10): 561-568).

Consider a lifeboat with a capacity of 50 people, with 40 people already on board. The lifeboat is in an ocean surrounded by 100 unfortunate drowning people. The "ethics" of the situation put you in a dilemma of whether the hapless drowning people should be taken aboard the lifeboat. Hardin compared individual lifeboats as rich nations and the swimmers as poor nations. He argued that if a lifeboat's carrying capacity is exceeded, everyone dies. Therefore, it is suicidal for the people on board to allow the unfortunate drowning people to get into the lifeboat as helping the drowning people threatens the people already on board. Moreover, it is not going to aid the drowning in the end. In the above example, is it feasible to allow anymore individuals to board the boat when there are already 40 people on board? If all the 100 unfortunate drowning people are allowed to get in, he argues, we get "complete justice, but complete disaster," that is, everybody will be drowned. There is a possibility to let 10 aboard, but the choice from 100 is difficult. Further, it is not proper to fill all the berths as a safety factor in case of possible emergencies. In short, the people already on the lifeboat should not allow any drowning individuals to board the boat! So true if you think on the land availability in the small state of Kerala. So the prime dilemma is who will make use of the land for fodder. Human resource, technical inputs and fodder caring inputs surely add to the emissions. The positive side we see here is most organisms can do little to change the carrying capacity of their environments, but humans can! If food (for humans and animals) is not available in a region, we can purchase it from another region by paying cash or kind. The historical pattern of human population growth shows the effect of breakthroughs. Here comes, the sustainable emissions tagged to transport, technology and ever increasing population. We are capable of intensive fodder production coupled with technological advances, increasing the effective carrying capacity of fodder cultivable lands. A nation's land has a limited capacity to support a population and in most of the developing countries, the carrying capacity of the land has already been exceeded. So the life boat ethics in Fodder production is there in Kerala. The fodder gene stock in the indigenous granary is meagre. Somehow, the native varieties have been ignored for long. The adaptability of the native fodder in the emerging changes has to be researched. Also an indirect impact is there on the livestock population on the climatic elements. The UN's Food and Agriculture Organisation has estimated that meat production accounts for nearly a fifth of global greenhouse gas emissions. These are generated during the production of animal feeds, for example, while ruminants, particularly cows, emit methane, which is 23 times more effective as a global warming agent than carbon dioxide. The agency has also warned that meat consumption is set to double by the middle of the century. The world is undergoing an extinction crisis – the most rapid loss of biodiversity in the planet's history (IUCN, 2013) http://portals.iucn.org/memberscommunication – and this loss is likely to accelerate as the climate changes. It has been estimated that 20–30 percent of plant and animal species will be at higher risk of extinction due to global warming and that a significant proportion of endemic species may become extinct by 2050 as a consequence. Some taxa are more susceptible than others. When climate change disrupts ecosystems that provide global services, the implications are even more serious. Ecosystems are exposed to the effects of changing climates in different measures. Although the impacts of climate change may be difficult to detect since they are often combined with the effects of other activities, such as land use changes, the most recent Global Biodiversity Outlook report (Secretariat of the Convention on Biological Diversity, 2010) identifies climate change as one of the main factors responsible for the current loss of biodiversity. Fodder gene biodiversity loss is also an important point. Due to their high productivity, many land holdings have been converted to croplands over the centuries or used as pasture for domestic livestock. Many apparently natural grasslands have been altered more subtly. Grasslands are amongst the least protected ecosystems on the planet. One good example is Vagamon Grass lands (Fig. 17.2) of Idukky District of Kerala (Felix *et al.*, 2010). The pasture land has changed so profoundly over time that in many cases scientists and planners remain unsure about their ecological histories.

Fig. 17.2: Vagamon grassland ecology; breach of commons

Possible suggestions are

- Maintaining current ecosystems wherever possible as, healthy and intact ecosystems are best able to withstand climate change.

- Adapting fodder and livestock management tools to address climate change.

- Restoring damaged or changing ecosystems. Highly diverse ecosystems are likely to be most resilient in the face of rapid environmental changes.

To sum up; counter degradation of pasture and fodder lands, un scientific alterationsand restore native species and also counter invasions both floral and faunal.

17.1.1 Thumburmuzhy fodder

Cattle Breeding Farm Thumburmuzhy under KVASU hand picked and propagated a fodder variety (Fig.17.3) that thrives and yieldswell in extreme summer and Monsoon times of Kerala.Thumburmuzhy fodder has a thin stem

Fig. 17.3: Thumburmuzhy fodder

abundant growth and violet inflorescence and on an average they yield 20 kg from a single planted stem cutting. Apart from reducing feed wastage this has more protein content (Table 17.1) when compared to all other cultivated fodder in the farm. The ability to withstand changing environmental factors like humidity, rainfall, hot and humid climate and changing irrigation factors may make this an ideal fodder for the changing climate. The nutritional analysis of the different fodder crops in the farm to compare Thumburmuzhy fodder is also appended. The soil evaporation was also less in Thumburmuzhy planted area as they have a thick biomass. It is assumed to be a natural cross of Napier and a wild fodder of the forest near the village. So far the performance of the fodder in Farmers plots was excellent in different climatic conditions.

Table 17.1 Analysis Report of Fodder crops at CBF Thumburmuzhy Felix, F. *et al.* (2010) Vagamon Eco terrorism A case study 2nd Int Responsible Tourism Conference. Acad Paper. pages 25-30

Nutritional analysis of samples at Dept of Nutrition with lab registration number:

Fodder	Co1	Co3	Co6	Killikulam	African	Australin.	*THUM BURMUZHY1	Congo signal	Guinea grass	Para Grass	Hybrid Guinea grass
Lab Reg.no	51/08	52/08	44/08	53/08	45/08	46/08	43/08	54/08	55/08	56/08	57/08
Moisture	77.5	89.1	81.7	78.5	78.8	85.2	80.1	82.7	84.3	84.6	89.6
Crude protein	16.6	16.0	14.0	16.0	15.8	14.8	17.5	15.9	15.5	16.2	15.0
Ether Extract%	1.5	0.9	1.4	0.8	1.7	1.4	1.3	0.6	1.3	0.7	1.5
Crude fibre %	25.6	22.1	23.3	31.3	20.0	22.2	19.9	28.4	38.6	25.0	23.5
NFE %	42.5	46.2	52.2	40.3	52.1	49.2	50.2	44.4	29.4	45.0	46.1
Total Ash %	13.8	14.8	9.1	11.6	10.4	12.4	11.1	10.7	15.2	13.1	13.9
AIA %	2.0	4.5	2.3	7.0	4.4	4.6	5.6	6.4	3.9	4.1	3.5

17.2 Futuristic Strategies for Livestock Waste in Climate Change Scenario

Livestock sector developed with agriculture where animals were primarily used as a source of manure and draught power. Further development in agriculture found utility in cattle in the form of milk and meat. The increase in population and per capita income resulted in increased demand for livestock products which in turn led to intensification and industrialisation of the sector with concentrate based feeding. The increase in human population and dwindling land holdings increased the density of livestock. The manure and waste became surplus than that could be applied over the land where the stock had been reared. This necessitated storage and transportation of waste. The release of Carbon and Nitrogen from livestock and manure operations in the form of Carbon dioxide, Methane and Nitrous oxide is considered to be a major contributor to the GHG emissions. It generates 65 percent of human-related nitrous oxide, which has 296 times the Global Warming Potential (GWP) of CO_2, most of which comes from manure. Livestock also accounts for 37 percent of all human-induced methane (23 times as warming as CO_2), largely produced by the ruminant digestive system and degradation of manure, and 64 percent of ammonia, which contributes significantly to acid rain (FAO, 2006). The dwindling water resources, contamination of these sources with animal wastes and manure, resulting eutrophication, all pose a threat to global warming.

The intensified agriculture system has resulted in trading of nutrients. Nutrient imbalance in the form of nutrient depletion and soil degradation in the regions where the feed is produced, and nutrient accumulation and environmental pollution in the regions where livestock production is concentrated occur as a result of intensive and Industrialised livestock farming. The dung of animals contain nutrients which were not utilised by the animals. The death of the animal returns part of the nutrients to the soil but part is released into the atmosphere. The Carbon dioxide, Methane and Nitrous oxide which are released as part of livestock activities and forming a part of the Carbon and Nitrogen cycle are thought to have a significant role in global warming. Manure surplus formed in regions of high cattle density makes it necessary for proper removal and transport of manure. The excess nutrient release into water bodies and ensuing eutrophication is also a threat. For a farmer the emissions are loss in nutrients, energy and soil organic matter. The brighter side is that with proper manure management and waste recycling sequesters most of these minerals. The increased foliage that could be subsequently produced by utilising the manure increases the carbon sequestration in plants. Biogas when trapped and utilised for energy production decreases the methane to lesser potent Carbon dioxide. The livestock waste could grow into a key fuel and power source changing the outlook on waste as a menace to a valuable resource which replaces part of other polluting methods of energy generation. The utility of composting to sequester animal nutrients also has significant positive impacts on global warming.

17.2.1 Animal waste and global warming

Animal waste include manure and urine produced in farms as well as during grazing, dead animals, offal produced in meat industry and waste water from farms and processing units.

17.3 Animal Species and Emissions

Kinsman *et al.* (1995)found that stored manure of Dairy cows contributed to 5.8 and 6.1% of methane and CO_2 respectively. Manure storage and processing are the second largest source of emissions, in swine representing 27.4 percent emissions. Most manure emissions are in the form of CH_4 (19.2 per cent, predominantly from anaerobic storage systems in warm climates); the rest is in the form of N_2O (8.2 per cent). Manure storage and processing account for 10% of emissions. Backyard systems of rearing are considered to have more volatile solid and N excretion due to poor conversion of low quality feed. Manure emissions account for 20 per cent of emissions in layers but only 6 per cent in broilers. This is due to different management systems; most of the manure from specialized meat production is managed in dry, aerobic conditions whereas that from hens is often managed in liquid systems with long-term pit storage. In backyard systems there is higher emission due to lower feed conversion (poor breeds, low quality feed and wastage of energy for scavenging), higher mortality (disease and predation) and poor manure management.

17.3.1 Manure

The manure is naturally produced by animals as part of their digestive process and contains two chemical components that can lead to GHG emissions during storage and processing: organic matter that can be converted into CH_4 and N that leads to nitrous oxide emissions. The carbon dioxide produced during the microbial action is negligible. The manure and urine produced during grazing is not usually recoverable, except where cattle dung is collected dried and burnt as a fuel or sold as manure. The amount of manure produced and degree of processing varies with production system and species reared. A large portion of the detrimental effects of intensive livestock production relate to poor management of livestock excreta, which contains large amounts of organic matter and mineral nutrients (Sims *et al.*, 2005).

17.3.2 Manure and methane

Intensive livestock rearing systems show a concentration of livestock production close to urban centres as well as a shift from ruminant to monogastric production. Also the large and specialised units have showed interest in slurry based systems which require less labour. (Harald M. *et al.*, 2010). The methane production potential of manure depends on the specific composition of the manure, which in turn depends on the composition and digestibility of the animal diet. Methane is released from the anaerobic decomposition of organic material. This occurs mostly when manure is managed in liquid form, such as in deep lagoons or holding tanks during storage and processing. The amount of methane produced during decomposition is also influenced by the climate, management system, contact with oxygen, water content, pH, and nutrient availability. Climate factors include temperature and rainfall.

17.3.3 Nitrous oxide and ammonia

Nitrogen is mostly released in the atmosphere as ammonia (NH_3) that can be later transformed into N_2O (indirect emissions). Ammonia volatilization contributes strongly to the high rates of atmospheric N deposition. A large part of the remaining ammonia reacts in the atmosphere with SO_2 and Nox and is transported over a distance of 5 to about 1000 km.

Because N_2O production requires an initial aerobic reaction and then an anaerobic process, it is theorized that dry, aerobic management systems may provide an environment more conducive for N_2O production. Nitrous oxide is produced from the combined nitrification-denitrification process that occurs on the nitrogen in manure. The majority of nitrogen in manure is in ammonia (NH_3) form. Fresh dung and slurry is highly anoxic and well-buffered with near neutral pH, N_2O production is expected to increase with increasing aeration. However, the denitrification process that produces N_2O requires an anaerobic environment. In grazed pasture urine patches are the main source of nitrous oxide emissions. Nitrous oxide emissions from manure storage and processing, and from the application of manure on crops and pasture, represent about 3 million tonnes of N. This is about 15 percent of the mineral N fertilizer use that can be ascribed to feed (crop and pasture) production for the livestock sector (FAO, 2006). Additional losses of N take place in the form of NH_3 and Nox emissions into the atmosphere and leaching of soluble forms of N into ground water. Ammonia has a short life time in atmosphere oxidising to form NO and N_2O and thereby having global warming effects. In animal houses that do not use bedding materials the dung, slurry and urine remain in a predominantly anaerobic state with little opportunity for NH_4^+ to be nitrified and therefore little or no N_2O emissions are likely to occur. Solid manure provides aerobic and anaerobic conditions in close proximity and could be a source of N_2O production/consumption and emission. There is slight emission of N_2O during aerobic composting after the initial thermophilic phase. Intensive aeration of slurry to remove excess N has been shown to increase N_2O emissions. During manure application there is a delayed N_2O emission.

17.3.4 Manure management systems

There are basically two systems "liquid" and "dry" systems. Dry systems include solid storage, dry feedlots, deep pit stacks, and daily spreading of the manure. In addition, unmanaged manure from animals grazing on pasture falls into this category. Liquid management systems often use water to facilitate manure handling. These systems include tanks and lagoons which store manure until it is applied to cropland. Liquid systems create the ideal anaerobic environment for methane production. Greenhouse gases are associated with storage and application of animal manure. CO_2 generated from manure storage and management has been very less studied. Nitrous oxide generation within manure is a resultant of nitrification/denitrification process that occurs in manure storage and application. Methane production within manure in earthen storage systems is attributed to the solid liquid interface with the methane bacteria present at this interface. Anaerobic storage systems of manure would convert non-lignin organic matter to methane under warm, moist, anaerobic conditions. The variation in the CH_4 production for different manure can be attributed to the solids and bacterial populations in the manure systems. Methane emissions from manure can be effectively controlled by shortening storage duration, ensuring aerobic conditions or capturing the biogas emitted in anaerobic conditions. However, direct and indirect N_2O emissions are much more difficult to prevent once N is excreted. Techniques that prevent emissions during initial stages of management preserve N in manure that is often emitted at later stages. Thus, effective mitigation of N losses in one form (e.g. NH_3) is often offset by N losses in other forms (e.g. N_2O or NO_3). These transference effects must be considered when designing mitigation practices. Numerous interactions also occur among techniques for mitigating CH_4 and N_2O emissions from manure (Fig. 17.4).

Fig. 17.4: Gas arising from stored manure

17.3.4.1 Dry lot storage

If urine is not collected and bedding is sparsely used, losses of N and K in particular will be high as most urine is lost. Depending on the storage facilities and storage time of the faeces part of the nutrients in faeces will also be lost through leaching and surface runoff, in the case of a precipitation surplus and uncovered manure heaps. Urine collection will minimize K losses but N losses will often remain high as volatilization will increase, though this is dependent on climatic conditions, storage time and storage method. Using bedding, with sufficient absorption capacity to capture urine, might reduce N losses with ca. 15% of the mineral N. Due to increasing pressure to minimize water quantity and odor problems, some producers are evaluating dry systems that may result in fewer liquid-based manure management systems. Dry screw press and other mechanisations help in separation of solid organic contents from manure, but is expensive and energy dependent.

17.3.4.2 Liquid manure storage

Liquid waste is mainly stored in liquid/slurry tanks, pit storage and anaerobic lagoons. The cattle farming operations mainly use the first two options, whereas swine farmers utilise the anaerobic lagoons to store and handle the diluted waste.This the main system in intensive livestock systems, except for broilers. Volatilization losses are dependent on the level of ventilation, depth of storage tanks and storage time, but often range between 5 and 35% of the total N excreted. With the use of confined and intensive livestock production systems continuing to increase, the use of liquid-based manure management systems will probably increase. Such systems are often preferred for large-scale livestock production systems because they allow for the efficient collection, storage, and, in some cases, treatment, of livestock manure. This shift towards liquid systems would result in significant increases in emissions because liquid systems produce considerably more methane than dry systems. All three systems can be considered as non-optimised anaerobic reactors, operated in a fed-batch mode. Fed-batch means that the storage/reactor is filled in time, until it is completely full. Two general options exist for reducing emissions from liquid systems: (1) switching from liquid management systems to dry systems; or (2) recovering methane and utilizing it to produce electricity, heat or hot water. The former one is having several difficulties including cost and labour. On a weight basis, manure in dry systems produces significantly less methane than liquid systems. Even with increased manure production, shifts toward dry systems could decrease methane emissions dramatically. Based on default emission factors, such shifts would presumably have an opposite effect on nitrous oxide emissions.

17.3.4.3 Dead and fallen animals

Burial, incineration, rendering and composting are the options available for carcass and offal management. Burial is the method of choice for safe disposal in deaths due to highly contagious diseases. Here there is sequestration of minerals. Incineration is also resorted to but has an impact on global warming with GHG emissions. Rendering though safe with limited emissions is energy dependent and costly. Composting is an economical option of disposal and is especially practiced in poultry farms (Fig. 17.5).

Fig. 17.5: Aerobic Composting of dead birds

17.4 Mitigation Strategies

17.4.1 Improvement of feed quality

Higher quality and digestibility of feed result in reduced manure emissions. Diets affect manure emissions, by altering the content of manure: ration composition and additives have an influence on the form and amount of N in urine and faeces, as well as on the amount of fermentable organic matter in faeces. Steps need to taken to optimize the utilisation of feed resources, thereby decreasing the need to feed excess and to manage excess manure.

17.4.2 Diet management

Reducing crude protein content decreases N excretion and ammonia emissions.Most of the N in the urine (NH_3 emission) originates from an imbalance between the amount and the quality of digestible protein and animal requirements, it can be reduced by adjusting the feed ration. Feeds with high rumen degradable protein are excreted in urine when not enough rumen fermentable energy is present; when the degradable carbohydrate suffices, microorganism utilize a large fraction of this protein and keep the nitrogen in organic form. An increase in dietary fermentable energy content at similar nitrogen intake level results in a large amount of nitrogen excretion in the faeces and a pronounced decline of nitrogen excreted in urine. It was reported that a 27% increase in the total slurry-nitrogen loss as the intake energy was increased, and a 56% increase in nitrogen lost due to an increase in crude protein (van der Stelt *et al*. 2008).In intensive ruminant production systems, balancing the intake of rumen degradable protein and rumen degradable energy can produce major effects. In most European countries this would require a substantial reduction in the quantities of grass fed, as grass with a reasonable productivity per hectare, almost inevitably has a high surplus of rumen degradable protein. However, this would imply a significant shift in feeding practices and even in the whole livestock system. In intensive monogastric production systems, reductions in N losses will be mainly achieved by adjusting the protein content of the feed to the variable requirements of animals of various age and productivity (phase-feeding) and by balancing amino acid requirements and digestible amino acids offered. For the last-mentioned option, addition of synthetic amino acids is an important and increasingly popular strategy. Many commercial compound feeds for monogastrics contain added synthetic amino acids, mainly lysine and methionine as these are often the first limiting amino acids. High costs may prevent broader application of other limiting amino acids (Schutte and Tamminga, 1992). Tannins and other phenolic compounds appear to have major effect on the amount and pathway of nitrogen excretion. Ruminant consuming less tannins excreted high amount of nitrogen compared to those fed high tannins feed. Whereas, faecal nitrogen was higher in animals consuming tannin fortified feeds against those fed with tannin free diets.(Powell *et al.*, 2009).

17.4.3 Composting

Composting is a method to treat dry manure and waste to sequester the nutrients and to reduce emissions as well as the pathogenic effects. Good composting practices that balance the carbon: nitrogen ratio and provide adequate aeration and moisture will minimize GHG emissions. Greenhouse gas savings associated with the application of composted products to soil, including:

- Carbon sequestration in soil

- Avoidance of chemical fertilisers and other chemical plant/soil additives

- Improved soil properties and related plant growth

- Rehabilitation of degraded land and mitigation of land degradation (ROU, 2007)

The additional advantage is that composting also addresses a number of other environmental concerns such as pathogens, surface and groundwater quality and ammonia emissions. The composting process does not require as much capital investment as some of the other organic waste solutions. Composted materials have gained a wide acceptance as organic amendments in sustainable agriculture, as they have been shown to provide numerous benefits whereby they increase soil organic matter levels, improve soil physical properties (increased porosity and aggregate stability and reduced bulk density) and modify soil microbial communities (Knapp *et al.*, 2010)

17.4.4 Aerobic composting

Livestock food venture, hatchery waste, broiler waste and crop residues does not have a scientific waste disposal system practically implemented. Biodegrading in aerobic composting system is successfully used by the above waste generators in different forms (Murphy and Handwerker,1988). The advantage of aerobic composting over anaerobic composting is that the smell is limited and also there is temperature rise in the heap to the tune of 65 to 70 degree Celsius during the thermogenic phase that helps in destroying most pathogenic bacteria as well as the viability of weed seeds. Almost any biodegradable waste could be converted including bones and feathers. The usual period of composting varies with climatic conditions and the system utilised. Burton (1992), found that shifting the anaerobic manure storage to an aerobic storage reduced the potential loss of NH_3 to the atmosphere, but lead to N_2O production.

17.4.5 Thumburmuzhy composting

Thumburmuzhy composting is an aerobic system developed at Kerala Veterinary ad Animal Sciences University. In this system Carbon is sequestered from waste, manure and dried leaves to form value added manure. The dried leaves which are otherwise burned add Carbon dioxide to the atmosphere is converted in the organic manure. The cattle manure which lie idle until utilised result in methane release along with ammonia. In aerobic composting the methane release from dung is minimized. The waste if untreated result in emission of volatile fractions including ammonia along with detrimental health and environment effects. Thus the utility is threefold reducing the environmental impact. Thumburmuzhy composting is done in 4 ft x 4ft x 4ft ferro-cement bin (Fig.17.6) with airspace and grooves utilising bacterial consortium from cow dung and carbon source, from dry leaves, hay, straw and dry paper bits, worked well in all Kerala climatic zones with a roof to prevent rain water during monsoons. The layering system had also been modified so that labour need is minimised, the core temperature maintained at 70 degree Celsius had a self limiting cycle after the composting process was over (Francis *et al.*, 2013).

Fig. 17.6: Thumburmuzhy model aerobic composting unit

Climate Change Community

Technologies Suggested for Compendium

- Excel Industries Limited Delhi has suggested the Organic Waste Converter (OWC).
- Kerala Agricultural University has fine tuned THUMBURMUZHY or Gandhian Waste Management System, a cost effective, convenient composting system.
- Sakthi Surabhi development by Vivekananda Kendra Natural Resources Development Project (VK NARDEP) generates bio-gas from kitchen waste.
- ARTI-Appropriate Rural Technological Institute, Pune has developed household cooking appliances & fuels that utilize Clean Technology.
- The System of Rice Intensification (SRI) developed by the Development Research, Communication and Services Centre in West Bengal.

Fig. 17.7: Thumburmuzhy composting suggested in UN Climate Change Compendium

17.4.6 Vermicompost

Vermicomposting involves the bio-oxidation and stabilization of organic material by the joint action of earthworms and microorganisms. Although it is the microorganisms that biochemically degrade the organic matter, earthworms are the crucial drivers of the process, as they aerate, condition and fragment the substrate, thereby drastically altering the microbial activity and gradually reducing the ratio of C:N and increasing the surface area exposed to microorganisms – thus making it much more favourable for microbial activity and further decomposition (Domínguez *et al.*, 1997). The two phases can also be distinguished here, (i) an active phase where the earthworms process the waste modifying its physical state and microbial composition (Lores *et al.*, 2006), and (ii) a maturation-like phase marked by the displacement of the earthworms towards fresher layers of undigested waste, where the microbes take over in the decomposition of the waste. The duration will depend on the species and density of earthworms, the main drivers of the process, and their ability to ingest the waste (ingestion rate). Vermicomposting is not fully adapted to the industrial scale

(Domínguez *et al.*, 1997). Since the temperature is always in the mesophilic range, pathogen removal is not ensured, although some studies have provided evidence of suppression of pathogens (Monroy *et al.*, 2008). In some cases, organic residues require pretreatment before being vermicomposted as they may contain substances that are toxic for earthworms, such as acidic compounds (Nair *et al.*, 2006). The combination of composting and vermicomposting has recently been considered as a way of achieving stabilized substrates (Tognetti *et al.*, 2007). Composting enables sanitization of the waste and elimination of toxic compounds, and the subsequent vermicomposting reduces particle size and increases nutrient availability; in addition, inoculation of the material resulting from the thermophilic phase of composting with earthworms reduces the expense and duration of the treatment process (Ndegwa and Thompson, 2001).

17.4.7 Biogas

Energy harvesting from manure in the form of Biogas has been popular in China and India but the cost of the digester and technical difficulties still limits it's widespread adoption. The resultant slurry is difficult to handle and N loses via volatilisation may be high as most Nitrogen is in mineral form. Biogas technology utilises the microbial consortium under anaerobic conditions to act on feedlot, which usually contains manure or other wastes producing biogas and end product called slurry. The biogas contains about 60% Methane along with other gases like CO_2, H_2S etc. The amount of feedlot available in the form of manure determines the size of the biogas plant. The amount of manure potentially available as feedstock for biodigesters may be estimated at over 1 billion tonnes a year (FAO, 2009). Methane generation takes place in the volatile solids portion (VS) of the manure. The VS portion depends on livestock type and diet. Animal type and diet also affect the quantity of methane that can be produced per kilogram of VS in the manure. This quantity is commonly referred to as "Bo" and is measured in units of cubic meters of methane per kilogram of VS (m^3 CH_4/kg VS). Optimal conditions for methane production include an anaerobic, water-based environment, a high level of nutrients for bacterial growth, a neutral pH (close to 7.0), warm temperatures, and a moist climate.

Size of plant according to number of cattle

Plant size	No of cattle's	Dung & Water mixture/day	Food prepared for persons/day
1 m^3	2-3	25 kg x 25 L	3-4
2 m^3	4-6	50 kg x 50 L	5-8
3 m^3	7-9	75 kg x 75 L	8-12
4 m^3	10-12	100 kg x 100 L	12-16

Source: http://www.rcsdin.org/Biogas%20tech%20manual.pdf

The dimensions of digesters volume corresponded to treatment of manure may be 5-10 pigs per m^3.

Gas Production Potential of Various Types of Dung

Types of Dung	Gas Production Per Kg dung (mJ)
Cattle (cows and buffaloes)	0.023-0.040
Pig	0.040-0.059
Poultry (Ciuckens)	0.065-0.116

Source: Updated Guidebook on Biogas Development, 1984

The combustion of biogas also releases CO_2. However, the main difference, when compared to fossil fuels, is that the carbon in biogas was recently up taken from the atmosphere, by photosynthetic activity of the plants. The carbon cycle of biogas is thus closed within a very short time (between one and several years). Biogas production by AD reduces also emissions of methane (CH_4) and nitrous oxide (N_2O) from storage and utilisation of untreated animal manure as fertiliser. The GHG potential of methane is higher than of carbon dioxide by 23 fold and of nitrous oxide by 296 fold. When biogas displaces fossil fuels from energy production and transport, a reduction of emissions of CO_2, CH_4 and N_2O will occur, contributing to mitigate global warming. (Teodorita A. S. 2008)

Biogas and power generation unit at University Livestock Farm Mannuthy, Kerala

17.4.7.1 Large Sized Biodigesters

For larger intensive production systems three types of methane recovery technologies are available. Covered anaerobic digesters may be used at farms that have engineered ponds for holding liquid waste. Complete-mix and plug-flow digesters can be used for other farms.

17.4.7.2 Covered Anaerobic Digesters

Covered anaerobic digesters are the simplest type of recovery system and can be used at dairy or swine farms in temperate or warm climates. Here manure flows into the primary lagoon where it decomposes and generates methane which is collected under the cover and used to power an engine-generator. Waste heat from the generator is used for on farm heating needs. The digested wastewater flows into the secondary lagoon where it is stored until it can be applied to cropland. A two-lagoon system also provides added environmental benefits, including odor and pathogen reduction. This technology is often preferred in warmer climates and/or when manure must be flushed as part of on-going operations.

17.4.7.3 Complete-Mix Digesters

Complete-mix digesters are tanks into which manure and water are added and an equal amount of digested material is removed and transferred to a lagoon. The digesters are mixed mechanically on an intermittent basis to ensure uniform digestion. The average retention time for wastewater in the tanks is 15 to 20 days. To speed decomposition, waste heat from the utilization equipment heats the digesters. Complete-mix digesters are typically used at swine farms in colder regions and are not recommended for use at dairy farms because of the high solids content of dairy manure.

17.4.7.4 Plug-Flow Digesters

Plug-flow digesters consist of a long concrete-lined tank where manure flows through in batches, or "plugs." As new manure is added daily at the front of the digesters, an equal amount of digested manure is pushed out the far end. One day's manure plug takes about 15 to 20 days to travel the length of the digesters. Plug-flow digesters are almost always used at dairies where the consistency of the cow manure allows for the formation of "plugs." Manure digestion using plug-flow digesters also provide the added benefit of digested solids, which can be recovered and used as a soil amendment or bedding for cows. Plug-flow digesters are generally used in colder climates or at newly constructed dairies instead of lagoons. Because manure handled as a solid produces very little methane, the emission reduction from plugflow digesters can be minimal, depending on climate and waste systems. Electricity generation for on-farm use can be a cost effective way to reduce farm operating costs. The generated electricity displaces purchased electricity. The economic feasibility of electricity generation usually depends on the farm's ability to use the electricity generated on-site. Selling the electricity to an electric power company has seldom been economically beneficial because the utility buy-back rates are generally very low.

17.5 Improvement of Manure Management

The timing of manure application is very important to decrease the loss of nutrients. Application of manure at the time of optimal utilisation by the plant and also at the time when excess precipitation occurs which may favour the bacterial GHG production from manure. Urease inhibitors and Nitrification inhibitors- slow down nitrogen turnover by slowing down oxidation of N to NO_3^- causing N to stay more in the immobile form of NH_4^+ for usually 3-4 weeks. But the effects are still debatable?

17.5.1 Manure heap management

Mixing of cattle manure and straw 50% (v/v) at start of storage, reduced methane production by 45%. Air tight covering of manure heap inhibiting aerobic organisms and associated increase in temperature that stimulates CH_4 emissions from anaerobic microenvironments. Frequent turning of heaps helped to reduce anaerobic zones decreasing methane production by 0.5%. initial C content.

17.5.2 Slurry management

Slurry stores are source of CH_4 emission, allowing a crust formation can result in formation of a Methane sink as a result of methane oxidation. (Eg. Covering of slurry surface with porous surface of straw, recycled polythene etc). Frequent removal of slurry and lower temperatures reduce methane formation. Reducing organic content by

17.5.3 Improvement of intermediate and backyard systems

At the low end of the productivity spectrum, in backyard systems, emission intensity is low. The feed ration is mostly made up of wastes and by-products with low emission intensity which compensate for the high manure emissions per unit of product due to poor nutrient balancing and low digestibility. The possibility to increase backyard production is limited by the availability of the feed materials these systems rely on. There is, however, a strong mitigation potential in upgrading intermediate systems to improve herd efficiency. Furthermore, independent of the production system, manure storage, processing and application practices can be altered to mitigate emissions. For chicken, the broiler and layer systems display lower levels of emission intensity than backyard systems for meat and eggs. Feed represents about 75 percent of emissions in intensive systems, so the type and origin of feed materials explain most of the emission intensity variability within these systems. Biogas as part of the backyard system can mitigate the GHG emissions to a large extent.

17.6 Policy & Regulations

The stocking density on land should be regulated to such that the waste generated can be safely handled and processed reducing environment impacts. There should be initiation and support for development of clean bio energy strategies from livestock waste. Regulations should ensure efficient and reduced use of water for cleaning operations. Farmers should be compensated for the cost borne on methods to mitigate GHG emissions. Intermediate systems must be optimised to ensure maximum productivity at minimum emissions. The use of inorganic fertilizers should be discouraged and utilisation of farmyard manure could be promoted. Extension and education of farmers for better feeding and management strategies to reduce emissions of GHG should be initiated. Optimal use of fertilizer at the correct time following soil testing should be ensured.

References

Burton, C.H. 1992. A review of the strategies in aerobic treatment of pig slurry; purpose, theory and method. J. Agi. Eng. Res. 53, 249-272.

Donald, C.R., Jerry, L. H and Ronald, L.S. 2000. Agricultural contributions to green house gas emission in Climate Change And Global Crop Productivity eds. Reddy, K. R. and H. F. Hodges CABI Publishing, New York, 37-55

Francis X., D. Girija, M.O.Kurien, D.K M. Deepak, 2013. 'Thumburmuzhy' a new model developed for livestock waste aerobic composting, International Conference on Waste, Wealth and Health, Organised by IIWM, VigyanBharati, & MPCST, Bhopal in association with MPPCB, giz and NSWAI at Bhopal on Feb 15-17th 2013 pp.

Gerber, P.J., Steinfeld, H., Henderson, B., Mottet, A., Opio, C., Dijkman, J., Falcucci, A. &Tempio, G. .2013.*Tackling climate change through livestock – A global assessment of emissions and mitigationopportunities.* Food and Agriculture Organization of the United Nations (FAO), Rome.

Harald M, O. Oenema, C. Burton, O. Shipin, P. Gerber T. Robinson and G. Franceschini.2010, *Impacts of intensive livestock production and manure management on the environment* In H. Steinfeld *et al.*, eds. Livestock in a changing landscape: drivers, consequences, and responses, volume 1, pp 144-145.

Kinsman R, F.D. Sauer, H.A. Jackson, M.S. Wolynetz .1995..Methane and carbon dioxide emissions from dairy cows in full lactation monitored over a six month period. J. Dairy Sci., 78 (12): 2760-2766.

Lessard, R., Rochette, P., Gregorich, E.G., Pattey, E. and Desjardins, R.L., 1996. Nitrous oxide fluxes from manure- amended soil under maize. J. Env. Quality 25, 1371-1377.

Murphy, D.W. and Handwerker T.S.,1988., Preliminary Investigations of composting as a method of dead bird disposal, Procd. National Poultry Waste Management Symposium.Ohio State University, Columbus, Ohio.pp.65-71.

Powell, J.M., G. A. Broderick, J. H. Grabber, and U. C. Hymes-Fecht, 2009.Technical note: Effects of forage protein-binding polyphenols on chemistry of dairy excreta. Journal of Dairy Science, 92(4): p. 1765-1769.

Rufino, M.C., E.C. Rowe, R. J. Delve, and K. E. Giller. 2006. Nitrogen cycling efficiencies through resource-poor African crop- livestock systems. *Agriculture, Ecosystems and Environment*112 : 261-282.

Sims, J. T., Bergström, B.T. Bowman, and O. Oenema. 2005. Nutrient management for intensive animal agriculture : Policies and practices for sustainability. Soil Use and Management 21:141-151.

Teodorita Al Seadi 2008, Biogas Handbook, University of Southern Denmark Esbjerg, NielsBohrs, Denmark, p 11.

Updated Guidebook on Biogas Development-Energy Resources Development Series.1984. No. 27.United Nations. New York, USA.

van der Stelt, B., P. C. J. Van Vliet, J. W. Reijs, E. J. M. Temminghoff, and W. H. Vanriemsdijk, 2008. Effects of Dietary Protein and Energy Levels on Cow Manure Excretion and Ammonia Volatilization. Journal of Dairy Science, 91(12): p. 4811-4821.

http://www.fao.org/ag/againfo/home/en/news_archive/2009_IGG.html accessed on 2/11/13

http://www.recycledorganics.com/infosheets/ghg/ghg.pdf accessed on 14/11/13

http://www.epa.gov/methane/reports/05-manure.pdf accessed on 14/11/13

http://unfccc.int/files/meetings/workshops/other_meetings/application/pdf/rys.pdfaccessed on 14/11/13.

18

Conceptual Design of Future Farms in the Context of Changing Climate

Stephen Mathew, Marykutty Thomas, Abhilash, R.S. and Ajith, K.S

Livestock Research Station, Thiruvazhamkunnu, Kerala Veterinary and Animal Sciences University, Kerala, India

> *Mother earth never attempts to farm without livestock*
> *– Howard (1940)*

Livestock have the potential to be transformative; by enhancing food and nutrition security and providing the income to pay for education and for other needs, livestock can transform lives of many (Jimmy Smith *et al.*, 2013). The projected growth of population is 9.6 billion in the year 2050 from the current population of 7.2billion (Gerber *et al*, FAO, 2013). The world demand for milk and meat in 2050 is expected to expand 58 and 73% respectively from their demand levels in 2010 (FAO, 2011). Nevertheless, the growth animal agriculture is not on par with the demand. Today, in a world of limited resources, in the face of climate change and a rising world population, a major issue is global food security (Augustine *et al* 2013). The key inferences of the latest approved summary for policymakers by the IPCC (2013) on climate change are 1). Warming of climate system is unequivocal and will continue beyond 2100. Global surface temperature change for the end of the 21st century is expected to exceed 1.5°C relative to 1850 to 1900 for almost all RCP scenarios 2). It is very likely that frequency of heat waves and heavy precipitation events have increased 3). The atmospheric concentrations of carbon dioxide, methane, and nitrous oxide have increased to levels unprecedented in at least the last 8, 00,000 years. Intergovernmental Panel on Climate change (IPCC) cautioned that over the next half-century, it is very likely that climate change will make it more difficult for nations to achieve the Millennium Development Goals in general and in particular goal[-1] namely food security (IPCC, 2007). Climate change will certainly have adverse effects on livestock productivity (IPCC, 2007). Jane Kabubo-Mariara(2009) reported that the climatic variables in general and summer temperature in particular are having a significant impact on stocking and net revenue from livestock in Kenya. His results revealed that a change in temperature from 2.5°C to 5°C shifts the impact on net value of livestock from a loss of 8% to 43%. Across the United States heat stress is estimated to be responsible for a loss of between $1.69 and $2.36 billion to livestock industries (St-Pierre *et al*., 2003). Climate change distress livestock directly and indirectly. Direct effects from heat stress influence animal production and reproduction, metabolic and health status and immune response. The indirect effects of global warming such as soil infertility, water scarcity, grain yield and quality and diffusion of pathogens may impair the industrialised livestock farming systems more than the direct effects (Nardone *et al*., 2010). In the current era of climate change, in order to match the genotypes of their livestock with the environment, farmers can follow two alternate strategies: adapt the environment to the need of the animals as is the case in industrial animal production systems or keep animals that are adapted to the respective environment as in the case in low input small holder or organic production systems (Mirkena *et al*., 2010).

18.1 Kerala Scenario

Kerala is situated in the southern peninsular region of India is bordered in the east by Western Ghats and in the west by the Arabian Sea. To its south is Kanyakumari district of Tamil Nadu while towards north is the state of Karnataka. The state is blessed with 44 rivers. Kerala is proud of having a highest literacy rate of 93.9% compared to other Indian states. Kerala is an agro-bio diversity paradise and having an agrarian economy. It is rich in rice, banana, jack fruit, tubers, coconut and rubber plantations and also spices. Kerala is the abode of Vechur cattle, world's smallest dairy cattle breed, Kuttandu buffaloes, Malabari and Attapady-black goat breeds and Chara and Chamballi duck breeds. The state have a population of 3.3 crore people with majority of them are non-vegetarians. It has a cattle population of 1.62 million of which more than 93% are crossbreds. While the cattle and buffalo populations are dwindling over the past decade, goat, poultry and duck population are showing an upward trend. The milk and egg production in the state are also registering a slight upward trend (Economic Review, 2010). The state is experiencing a huge gap between demand and production of all livestock and poultry products except chevon. In Kerala, majority of livestock farmers are small and marginal or even landless. Poor land availability, high population density, changing agricultural patterns, high cost of production, climate change and social issues are the major drivers in the change of livestock production in the state. Increase in mean atmospheric temperature (0.47°C in 54 years) and decline in annual rainfall are the major the climatic shifts in Kerala (Prasada Rao *et al.*, 2013). Climate change presents a major threat to Kerala's food and water security systems as well as the lives and livelihoods of coastal communities. Nevertheless, an unprecedented interest is noticed from several pockets of the state to evolve sustainable livestock ventures from small scale farming system prevailing in the state. Designing a self-supporting, sustainable and climate resilient livestock farming system suitable to the state in this climate change poses a serious challenge.

18.1.1 Thiruvazhamkunnu - A treasure trove

Future farm project is proposed at Livestock Research station, Thiruvazhamkunnu situated in Kottoppadamgrama panchayat of Mannarkkad taluk in Palakkad District- the granary of Kerala. The station is having a land area of 163 hectors. Currently the station is having a dairy farm with 300 cattle and a Murrah buffalo farm with 61 heads of Murrah buffaloes- the world's best dairy buffalo breed — and a Conservation unit of Attappady goats. Apart from this, the LRST is having fodder plots, agro-forestry plots, cashew, coconut and other agricultural crops and some natural forest area. LRST is bestowed with demographic location in Palakkad- the district with the highest cattle population and highest paddy cultivation in the state. Palakkad district lies between north latitude 10° 46′ and 10° 59′ and east longitude 76° 28′ and 76° 39′. The climate here is hot and humid and perhaps the hottest place in the state by virtue of its special geographical features.

18.2 Adaptation Strategies

The AR4 notes that a wide array of adaptation options is available, but more extensive adaptation than is currently occurring is needed to reduce vulnerability to future climate change. There are barriers, limits, and costs, but these are not fully understood, let alone quantified (IPCC, 2007). There is a great variety of possible adaptive responses available, including technological (such as more drought-tolerant crops), behavioural (such as changes in dietary choice), managerial (such as different farm management practices), and policy options (such as planning regulations and infrastructural development) (Thornton *et al.*, 2009). Adaptation options to climate change has been summarised by Kurukulasuriya and Rosenthal (2003), who define a typology of adaptation options that includes the following:

- Micro-level adaptation options, including farm production adjustments such as diversification and intensification of crop and livestock production; changing land use and irrigation; and altering the timing of operations.

- Income-related responses that are potentially effective adaptation measures to climate change, such as crop, livestock and flood insurance schemes, credit schemes, and income diversification opportunities.

- Institutional changes, including pricing policy adjustments such as the removal or putting in place of subsidies, the development of income stabilization options, agricultural policy including agricultural support and insurance programmes; improvements in (particularly local) agricultural markets, and the promotion of inter-regional trade in agriculture.

- Technological developments, such as the development and promotion of new crop improved animal health technology.

18.2.1 Future livestock

The most productive exotic breeds of livestock are suited to the highly industrialised, specialised production systems that require highly efficient management options. Though, the direct effects of heat stress on these animals can be ameliorated with the adoption suitable managerial measures, the indirect effects are difficult to curtail. The genetic studies on adaptation of animals to different extraneous stresses revealed that there is enormous amount of variation in adaptive performances at three genetic levels: species breed and among individual animals within breed. Unique genetic attributes of zebu cattle like coat, hide, skin, haematological characteristics and physiological attributes make them more suitable to hot climate than Bostaurus cattle. In the warmer tropical areas, where pathogens are epidemic diseases are widespread, climatic conditions are stressful, and feed and water are scarce, locally adapted autochthonous breeds display far greater levels of adaptation and resistance due to their evolutionary roots as compared to imported breeds (Mirkena *et al.*, 2010). Appropriate strategy for any breeding program would therefore be to set suitable selection goals, which match the production system rather than ambitious performance objectives that cannot be reached under the prevailing environment. The selection goals for selecting animals that up to now has been oriented toward productive traits, from now on, must be oriented toward robustness, and above all adaptability to heat stress.

Scientists are of the view that there will be a change in the species composition as a result of climate change. Goat population is likely to increase owing to its high adaptability to climate change and high production potential. In designing future animals, three features should be considered as constituting the core of trait development focus: feed efficiency, adaptability, and product quality (P. Mormede*, E. Terenina , 2012). Increasing feed efficiency is a main avenue to reducing the cost of production and the environmental problems caused by animal production systems (reduction of effluents, nitrogen and phosphorus excretion, methane emission). This goal can also be reached using alternative sources of food, such as by-products of agro food and bio-fuel industries, which would also reduce the competition between animals and humans for noble feedstuff. Adaptability denotes the key instrument to improve sustainability of livestock production systems under the pressure of climate and other  weaker factors (Nardone *et al.* 2010). Increasing intensification of dairy systems in the developing world through the use of temperate-breed genetic stock could lead to greater vulnerability to increasing temperatures (Thornton *et al.*, 2009). Though, the direct effects of heat stress on these animals can be ameliorated with the adoption suitable managerial measures, the indirect effects are difficult to curtail. Nardone *et al.* (2010) emphasised that selection of animals must be oriented towards robustness and above all adaptability to heat stress rather than towards productive traits.

Adaptability is a global measure of the sensitivity of the animal to the environment and to the metabolic load of its genetic potential for production traits. Adaptability also includes traits that are sensitive to inadequate environmental conditions, such as disease resistance and mortality in various stages (e.g. neonatal), altogether known as "functional traits." Such traits are important not only for performance levels but also for animal health and welfare. Their improvement

also has a positive impact on the environment by reducing production losses as well as the need for the preventive and therapeutic use of drugs (Mormede, E. Terenina, 2012].

The genetic studies on adaptation of animals to different extraneous stresses revealed that there is enormous amount of variation in adaptive performances at three genetic levels: species breed and among individual animals within breed. Unique genetic attributes of zebu cattle like coat, hide, skin, haematological characteristics and physiological attributes make them more suitable to hot climate than Bos taurus cattle. In the warmer tropical areas, where pathogens are epidemic diseases are widespread, climatic conditions are stressful, and feed and water are scarce, locally adapted autochthonous breeds display far greater levels of adaptation and resistance due to their evolutionary roots as compared to imported breeds (Mirkena *et al.*, 2010). Appropriate strategy for any breeding program would therefore be to set suitable selection goals, which match the production system rather than ambitious performance objectives that cannot be reached under the prevailing environment. The quality of animal products is the third category of traits to take into consideration in designing the animals of tomorrow. Food safety is of primary importance for products of animal origin, together with their sensory and technological qualities. However, there is a growing demand for food products that improve consumer health status (e.g. specific fatty acid composition (P. Mormede, E. Terenina, 2012).

18.2.2 Animal housing

A suitable housing system has an immense role to play in sustaining production potentiality of animals. Animal shelter should be designed by giving due considerations to animal health, feed cost, labour requirements and the

environmental attributes. It is essential to adopt micro climate modification strategies in highly intensified production systems. The plan of animal shelter must ensure maximum comfort and welfare of the animal. Nevertheless, minimalistic approach for the infrastructure is well appreciated. It should guarantee maximum energy efficiency. Environmental modifications of animal housing that helps to alleviate heat stress problems are structure orientation, ventilation, use of shades, and use of cooling systems. Climatic data can be used as a guide for the type of housing, insulation ventilation etc. For the best outcome in animal housing, the producers, building suppliers, and livestock specialists must adopt a rational and integrated approach for animal housing that provides for health management and husbandry, material and construction technology, energy and pollution, and economic considerations.

18.2.3 Animal welfare

Farm animals are sentient beings that should not be treated simply as 'commodities'. There has been increasing public awareness of farm animal welfare issues. World Organisation for Animal health (OIE) approved its global guidelines for animal welfare in 2005 (D. Fraser, 2008).The passing of the guidelines by 167 countries, some of which did not have national animal protection legislation of their own, signalled that animal welfare was no longer a concern only of certain (generally prosperous) nations, but had become an issue for official attention at a global level (Bayvel *et al.*, 2005, D. Fraser, 2008). Farm animal welfare implies 'natural' behaviour in a 'natural 'environment. The welfare of a farm animal depends on its ability to sustain fitness and avoid suffering. The responsibility of the farmer is to make provision for good welfare through good husbandry. The five freedoms and provisions as depicted below are used as an indicator of animal welfare in a farm.

The Five Freedoms and Provisions (Webster, A.J.F. 2001)

1. Freedom from thirst, hunger and malnutrition – by ready access to fresh water and a diet to maintain full health and vigour.

2. Freedom from discomfort – by providing a suitable environment including shelter and a comfortable resting area.

3. Freedom from pain, injury and disease – by prevention or rapid diagnosis and treatment.

4. Freedom to express normal behaviour – by providing sufficient space, proper facilities and company of the animal's own kind.

5. Freedom from fear and distress – by ensuring conditions which avoid mental suffering.

Another key feature that will be important in the future is the traceability of livestock products. The primary reasons for improving traceability of food supply chains have traditionally been for safety and quality assurance. However, the increasing social concerns regarding animal welfare, ethical production methods and environmental issues have meant that the sustainability of the milk supply will become another important driver for traceability. Consumers require information about the traceability of the food they eat so that they can make an informed decision about purchasing a product. This is influencing milk procurement decisions by milk processors, requiring the development of traceability systems and education of suppliers.

18.2.4 Water conservation strategies

Global warming present challenges regarding water supply and water quality. The phenomenon of water salination is spreading in many areas of the World. Other than salination, water may contain chemical contaminants, either organic or inorganic, high concentrations of heavy metals and biological contaminants. Animals exposed to hot environments drinking an amount of water 2–3 times more than those in thermo-neutral conditions can run many risks. Indeed, altered water pH may affect metabolism, fertility and digestion; the excess of nitrite content can impair both cardiovascular and respiratory systems; excess of heavy metals can impair the hygienic and sanitary quality of production, and the excretory, skeletal and nervous systems of animals (Nardone *et al.,* 2010). Water scarcity will impose serious threats in all farming systems. There is a need to manage the both surface and ground water resources in a way so that maximum water use efficiency is attained. Cost effective water conservation must be adopted in the farms. There is a need to recharge the aquifers and

conserve rain water through rain water harvesting structures. All the water sources should be used economically and judiciously so that no water is wasted. Water management strategies like water harvesting through water shed management, recycle & reuse of water are to be adopted. Various techniques in the proper water shed management include in situ and inter-basin rain water harvesting, water harvesting in ponds, stream flow harvesting through check dams and earth fill dams and roof- top water harvesting.

Water conservation and its effective utilisation will be one of the top priorities of future farms. Above, 5-million litre capacity artificial water pond at the Cashew Research Station, Anakkayam.

18.2.5 Water foot print

The rational and judicious use of water in the livestock production process is an adaptation mechanism in the current scenario of climate change. The total volume of fresh water required to produce a livestock product is known as the water foot print of that product. The water foot print concept is closely linked to the virtual water concept. The volume of water that is used to produce a livestock product, measured at the place where the product is produced, is called the virtual water content of that commodity. Reducing the water foot print or virtual water content of livestock products is important for coping with the water stress associated with the climate change. The farms in those regions with adverse water scarcity must concentrate on production of livestock products with low water foot print. Likely, all global warming effects on water availability could force the livestock sector to establish a new priority in producing animal products that need less water. For estimating the water foot print of animal product, we can compare water consumption of different animal species, referring the total need of water (for feed, drinking, cleaning, *etc.*) to produce 1 unit of protein. We can assume one unit of protein corresponds to 30 g of proteins, which is about the daily animal protein requirement for humans. To produce 30g of protein from beef, about 3.7 tons of water is needed. This value is about 6 times the amount of water required to produce the same quantity of protein from pigs. The differences are related to the water source and utilization for producing feed for beef cattle or pigs: i.e. pasture for beef cattle in grazing systems or in rain fed mixed systems versus irrigated farming producing grain to feed pigs in industrial livestock systems (Nardone *et al.,* 2010). Global warming present challenges regarding water supply and water quality. Water scarcity will impose serious threats in all farming systems. The phenomenon of water salination is spreading in many parts of the world. Other than salination, water may contain high concentrations of chemical and biological contaminants. Animals exposed to hot environments drinking an amount of water 2-3 times more than those in thermo-neutral conditions can run many risks

(Nardone *et al.*, 2010). Future farms have to develop a sustainable water security system with four pronged approaches viz. supply augmentation, demand management, adoption of new technologies and anticipatory action to mitigate the impact of global warming. To augment the supplies, we must harvest rain water and store it carefully, both above and below ground. Rain water harvesting should become a way of life. We should ensure that every drop of farm effluent is purified and if not, recycled. We should show equal interest in getting the best out of available water by emphasising the economy and efficiency of water use. Among the new technologies, genetic selection and development of fodder resistant to salinity, drought and flood seems to be most promising. Annual rain fall in the state dwindles every year, though we are getting 300 mm of annual rain fall. What we lack is a proper water security system. A reliable source of water should be developed in the future farms. Let the future farms be the forerunner in the water literacy movement in the state. Water scarcity under the scenario of global warming calls for the production of livestock products that require less water. The beef production needs 6 times more water than for the same quantity of pork. A soil conservation policy for enhancing the macro and micro nutrients and soil carbon should be envisaged in the future farms.

18.2.6 Farm management information system

The managerial tasks in agriculture are currently shifting to a new paradigm, requiring more attention on the interaction with the surroundings, namely environmental impact, terms of delivery, and documentation of quality and growing conditions. Information science and information technology are having a rapidly growing impact on the methods used in livestock production. Among other things, this managerial change is caused by external entities (government, public) applying increasing pressure on the livestock sector to change production from a focus on quantity to an alternate focus on quality and sustainability. This change has been enforced by provisions and restrictions in the use of production input and with a change of emphasis for subsidies to an incentive for the farmer to engage in a sustainable production rather than based solely on production. In general, this change of conditions for the managerial tasks on the farm has necessitated the introduction of more advanced activities monitoring systems and information systems to secure compliance with the restrictions and standards in terms of specific production guidelines, provisions for environmental compliance and management standards as prerequisites for subsidies.

Until now, farmers most often have dealt with this increased managerial load by trying to handle manually, the mass of information in order to make correct decisions. The increasing use of computers and the dramatic increase in the use of the internet have to some degree improved and eased the task of handling and processing of internal information as well as acquiring external information. However, the acquisition and analysis of information still proves a demanding task, since information is produced from many sources and may be located over many sites and is not necessarily interrelated and collaborated. The potential of using these data will reach its full extent when suitable information systems are developed to achieve beneficial management. Farm management Information System (FMIS) is defined as a planned system for the collecting, processing, storing and disseminating of data in the form of information needed to carry out the operations functions of the farm. Precision livestock farming (PLF), which is based on these concepts, is a leading example. Using electronic information transfer, PLF applies principles of control engineering in optimising production and management processes. A combined application of new communication technology, sensor systems, more powerful computing power, positioning systems (GPS) and geographical information systems (GIS) have enabled the development of new systems for cultivating and harvesting crops and to improve indoor animal feeding management and milking systems. PLF consists of *measuring* variables on the animals, *modelling* these data to select information, and then using these models in real time for *monitoring and control* purposes. Thereby, PLF is currently regarded as the heart of the engineering endeavour towards sustainability in (primary) food production. Its application allows making optimal use of knowledge and information in the monitoring and control of processes. A first step in PLF is monitoring, collecting and evaluating data from on-going processes. Collection of data from animals and their environment, by innovative, simple and low-cost techniques, is followed by evaluation of the data by using knowledge-based computer models. Currently, considerable PLF research is directed toward development and validation of various techniques for data measuring and registration on livestock farms. The research scope ranges from monitoring feeding times, feed intake, and performance parameters to real time analysis of sounds, images, live weight assessment, condition scoring, on-line milk analysis and more.

The final aim is to achieve a full picture of the state of the animals (cows, pigs, chicken, *etc*) and their environment on a continuous basis, PLF has a great potential in developing the technology for continuous automatic monitoring and improvement of animal health, animal welfare, quality assurance at farm and chain level, and for improved risk analysis and risk management.

18.2.7 Farm tourism

One last benefit of livestock, perhaps least understood and quantified, is the subtle but powerful attachment of people to the animals themselves – the almost mystical bond of the "ancient contact". Aside from their own intrinsic appeal, animals enhance the aesthetic value of meadows and pastures that enthral residents and visitors to pastoral lands worldwide. Cummins (2003), pondering the rewards of looking after livestock, wrote: "But when I start thinking about how our animals and crops and fields and woods and gardens sort of all fit together, then I get that good feeling inside. Such examples, which presumably occur in countless ways worldwide, imply that humans and societies derive benefits from animals beyond mere monetary value. Farm tourism defined most broadly, involves any farm-based operation or activity that brings visitors to a farm. Farm tourism includes a wide variety of activities, including buying produce direct from a farm stand, feeding animals, or staying at a farm. It is one alternative for improving the incomes and potential economic viability of future farms. Goulding *et al.* (2008), for example, note that in rural areas of the UK, "the income from tourism is perhaps 10 times that from farming" presumably, in part, from the livestock enhanced appeal of the countryside. Farm tourism is a form of niche tourism that is considered a growth industry in many parts of the world, including Australia, Canada, the United States, and the Philippines. People have become more interested in how their food is produced. They want to meet farmers and processors and talk with them about what goes into food production. Farmers can use this interest to develop traffic at their farm, and interest in the quality of their products, as well as awareness of their products

18.2.8 A Fistful of soil

In the era of land rush and climate change, future of food security will depend upon the sustainable management of land resources. There should be a commitment on the part of future farm to manage the land and soil for long term advantage rather than short term expediency. Indiscriminate use of pesticides and fertilisers will hamper the soil ecosystem. Soil testing and precise use of fertilisers should be the routine of future farms. Soil carbon banks represent a win- win situation for both food security and climate change mitigation. Aforestation and planting fodder trees will enhance carbon sequestration. Utmost attention should be paid to prevent the soil erosion. Repeated soil over-applications of manure, above crop requirements, lead to the accumulation of not only macro nutrients such as N, P and K, but also heavy metals particularly Cu and Zn, impacting animal health through grazing and crop feeding. The main consequence of nutrient overloaded soils is related to the interaction between soils and its water and air fractions causing water and atmospheric pollution. Water pollution occurs mainly through the leaching of nitrates applied in excess of plant uptake, while air pollution is the consequence of complex processes including nitrification /denitrification and also the breakdown and transformation of organic matter in soils. Composting systems or related technologies producing a useful solid product; biological systems for liquids that effectively breakdown some of the organic load and separation systems removing solids for the clarification and/or concentration of manure nutrients are three broad options available for manure treatment.

18.2.9 Precision livestock farming

Information science and information technology are having a rapidly growing impact on the methods used in livestock production. Precision livestock farming (PLF), which is based on these concepts, is a leading example. Using electronic information transfer, PLF applies principles of control engineering in optimising production and management processes. A combined application of new communication technology, sensor systems, more powerful computing power, positioning systems (GPS) and geographical information systems (GIS) have enabled the development of new systems for cultivating and harvesting crops andto improve indoor animal feeding management and milking systems. PLF consists of measuring variables on the animals, modelling these data to select information, and then using these models in real time for monitoring and control purposes. Thereby, PLF is currently regarded as the heart of the engineering endeavour towards sustainability in (primary) food production. Its application allows making optimal use of knowledge and information in the monitoring and control of processes. A first step in PLF is monitoring, collecting and evaluating data from on-going processes. Collection of data from animals and their environment, by innovative, simple and low-cost techniques, is followed by evaluation of the data by using knowledge-based computer models. Currently, considerable PLF research is directed toward development and validation of various techniques for data measuring and registration on livestock farms. The research scope ranges from monitoring feeding times, feed intake, and performance parameters to real time analysis of sounds, images, live weight assessment, condition scoring, on-line milk analysis and more. The final aim is to achieve a full picture of the state of the animals (cows, pigs, chicken, etc.) and their environment on a continuous basis, PLF has a great potential in developing the technology for continuous automatic monitoring and improvement of animal health, animal welfare, quality assurance at farm and chain level, and for improved risk analysis and risk management.

18.3 Mitigation Strategies in the Future Farms

Both bottom-up (specific mitigation options) and top-down (economy-wide) studies indicate that there is substantial economic potential for the mitigation of global greenhouse-gas emissions over the coming decades, that could offset the projected growth of global emissions or even reduce emissions below current levels (IPCC, 2007). For the medium term, various technologies are listed as being currently available and promising for their mitigation potential, including improved crop and grazing land management to increase soil carbon storage, and improved livestock and manure management to reduce methane emissions. The increase in demand for livestock products will be met partly from increased productivity of livestock but also through increases in livestock populations. Considerable amounts of carbon can be sequestered from improved management in grasslands. Such management would include conversion of cropland to grassland, reduction in grazing intensity and biomass burning, improving degraded lands and reducing erosion, and changes in species mix. Big gains could result from converting the wetter grasslands back to woodland or forest (Thornton *et al.*, 2009). In terms of methane mitigation in pastoral systems, probably the only effective way is through reducing livestock numbers. It is not very likely to happen. There is also considerable potential for some agro forestry options that can mitigate emissions while at the same time providing opportunities for increasing the resilience of agricultural systems. As for impact assessments of adaptation options, there is a real need for analytical frameworks that can examine the trade-offs involved in mitigation options (Thornton *et al.*, 2009).

18.3.1 Clean development mechanism

Though developing countries are not intended to adopt any GHG abatement targets as per the Kyoto protocol of United Nations Framework Convention on climate change (UNFCC), India is used to participate in Clean Development mechanism (CDM) projects on GHG mitigation measures for the benefit of certified Emission Reductions (CERs). Various managemental and nutritional measures can be adopted in farming systems as GHG mitigation measures. Some of the issues associated with the livestock farming that influence the production of CH_4 low digestibility of animal feeds and exposed treatment plants used for the anaerobic decomposition of animal manure. Supplementation of cow with molasses, multi-nutrient feed blocks and bypass protein may overcome the problem of their low digestibility. Currently, the emphasis world over is on the development of clean- green technology at an affordable price. Methane digester and natural compost are examples of waste management and clean technology that can be adopted in future farms. Methane digester technology is used worldwide to use animal waste and waste water for the production of methane. This fuels a gas turbine to create electricity. The effluent can be used to grow grass for cows and provide greenery. It is well known that methane is a highly potent GHG than carbon dioxide. However, it is a clean fuel when burnt. Its combustion produces no sulphur dioxide or particulates. Vermi-compost, a manure rich in nitrogen can recondition the soil, thus increasing agriculture produce by 20% to 30%. This compost is a very good replacement for chemical fertilizers. Use of natural compost in farming yields organic fodder and hence chemical free livestock products. In order to reduce the emissions into the atmosphere, less emphasis should be given on the use of energy that is based on non fossil fuels in future farms. The electric power production industry and transportation contribute substantially to GHG emission. Nevertheless, future farm plans emphasize on the energy conservation and the measures taken up for efficient utilization of energy. There are several non-CO_2 energy options are available like wind, solar, hydropower and *gobar* gas. Clean livestock production also envisages the production of clean, germ free, chemical free, livestock products.

18.3.2 Greenhouse gas mitigation strategies

Ruminant livestock systems contribute to global warming through the emission of nitrous oxide (N_2O), methane (CH_4) and carbon dioxide (CO_2). Greenhouse gas production from livestock can be ascribed to fermentation of feed by ruminants and animal waste. Though developing countries are not intended to adopt any GHG abatement targets as per the Kyoto protocol of United Nations Framework Convention on Climate Change (UNFCC), India is currently participating in Clean Development mechanism (CDM) projects on GHG mitigation measures for the benefit of certified Emission Reductions (CERs). Various managementl and nutritional measures can be adopted in farming systems as GHG mitigation measures. Nitrous oxide emissions from soils can be reduced by implementing practices aimed at enhancing the ability of the crop to compete with processes that lead to the escape of N fromthe soil-plant system. This can be done by using slow release/controlled fertilizers, nitrification inhibitors (NIs) precision management techniques targeting both mineral and organic fertilizers timing and rate, breeding for more efficient plants or animals and dietary-based methods, *e.g.* the manipulation of the urine hippuric acid content Other management strategies for reducing N_2O emissions include reducing the grazing time during wet periods, manipulation of drainage and irrigation systems, use of dietary amendments for animals (e.g. salt supplementation), and using different manure application techniques. Methane emissions can be reduced by: (i) adopting dietary methods, including; e.g. increasing the level of starch or rapidly fermentable carbohydrates, di *et al.,* teration resulting in improved animal productivity, addition of fat in the diet or stimulation of acetogenic bacteria and reduction of methanogens or removal of protozoa through additives or probiotics (ii) housing and manure storage methods, including; deep cooling of manure or reduction of manure pH, removal of the gas source, generation of biogas from waste and careful management of the bedding and manure heaps.

18.3.3 Organic pathway

There has been a tremendous increase in the number of organic farms worldwide over the last few years. Livestock production, and especially ruminant livestock production forms an integral part of many organic farms due to its role in nutrient recycling on farms. In Switzerland, Finland, Austria and Denmark milk products are the most important organic products (Hermansen, 2003). The future development of organic farming can be attributed to increased consumer interests in organic products throughout the world. Organic livestock farming is defined by basic guidelines that has been formulated and are further being developed by the International Federation of Organic Agriculture Movements (IFOAM) in association with FAO. The guidelines include specifications for housing conditions, animal nutrition, and animal breeding, as well as animal care, disease prevention and veterinary treatment. An important key principle is reliance on the management of internal farm resources rather than on external input and, in relation to health management, reliance on preventive measures rather than on treatment. There are, however, some well-identified areas, like parasite control and balanced ration formulation, where efforts are needed to find solutions that meet with organic standard requirements and guarantee high levels of health and welfare. There is an unprecedented interest among public in the state regarding pesticide, hormone and chemical free organic food products. This is the driving force for the many livestock farmers to switch over to organic livestock production. Livestock production especially goat and poultry production in the state based on the low inputs are in a way 'organic by default" not by choice. Translating them into organic system is comparatively an easier task.

18.3.4 Mixed farming systems

Mixed farming systems, in which crops and animals are integrated on the same farm, produce 92% of the global milk supply, all of the buffalo meat and approximately 70% of small ruminant meat. The waste products (crop residues) of one enterprise (crop production) can be used by another enterprise (animal production), which returns its own waste (manure) back to the first enterprise. As a way of diversifying the sources of income and employment for resource-poor farmers, mixed farming offers considerable potential for poverty alleviation in rural areas. There can be little doubt that small-scale mixed farming systems will continue to play a pivotal role in animal production in the state in the foreseeable future. The main crops that can be associated with mixed farming systems in the future are paddy, coconut, fish, honey bee, rubber and various spices.

18.3.5 Small scale farming systems

In the scenario of land rush and climate change, small scale/ low input/ subsistence farming systems will continue to play a substantial role in food security the state. Majority of livestock farmers in the state are marginal or small scale farmers and their drain from livestock sector will have serious implications. Design of economically viable and sustainable small farming systems is imperative for the future of livestock sector in Kerala.

18.3.6 Remembering our humanity

Any unilateral progress without social commitment will turn out to be hollow. The future farm should be able to address the hidden hunger and malnutrition among the tribal community, the original guardians of biodiversity. It must be able to provide the food and livelihood for the poorest of the poor, backward and the forgotten people in its neighbourhood. It can initiate a project on alleviating poverty and undernourishment among the tribal people in Attapady in collaboration with government departments and NGOs.

18.3.7 Human resource development

The institution building must be concentrate on brains, not merely on bricks. Future farms must impart trainings on livestock farming, marketing, self employment, entrepreneurship and technology transfer. It ought to be the model farm for the whole state. Consultation services may be provided to the farmers in the design and implementation of farms, and in tackling the challenges ahead. The future farm must be the forerunner in imparting climate literacy, mitigation and adaptation strategies, climate resilient farming systems, and water shed management to the public in the State. The future farms must generate process and analyse the data regarding livestock production and economics from time to time and play a key role in policy decisions for tomorrow.

18.3.8 Pro-women: Strengthening the role of women in livestock farming

There is a growing feminisation in livestock farming in Kerala and more than 60% of livestock farmers are women. The gender specific needs of women farmers, both as women and as farmers, will have to be met, if women are to play their rightful role in the state's agricultural progress. Empowering the women livestock farmer in areas related to enhancing productivity, profitability and sustainability of small- scale livestock farming will be one of the priorities of future farms. The empowerment measures incorporated access to technology, credit, inputs and market. It will help women farmers to contribute more effectively to food security of the State. Women self help groups and Kudumbasrees can contribute significantly to various aspects of future farms at Thiruvazhamkunnu right from fodder production to marketing of farm produce. Though the direct effects of climate change is relatively easy to conquer as in the case of highly industrialised production, there is little control over indirect effects of climate change. Paradoxically, the livestock sector contributes directly and indirectly to climate change. The FAO's description of livestock's long shadow (FAO, 2006) invited fervent debate all around the world. Livestock is responsible for the 18% of anthropogenic emission of green house gases as a result of enteric fermentation, manure management and deforestation (Hoffemann 2011). Beef and cattle milk production account for the majority of emissions, respectively contributing 41 and 20 percent of the sector's emissions (Gerber *et al.*, 2013). Furthermore, currently, livestock supply 13% of energy to the world's diet but consume one-half the world's production of grains to do so (Jimmy Smith *et al.*, 2013). Drawing the architecture of the future farms is a great challenge in the present scenario. Global perspective of future farms is footed on self-supporting, sustainable, clean, green and ethical livestock production. The WUN portrays future farming systems as one which ensure sustainable and responsible production of healthy food from healthy animals. One must perceive the future livestock farming system as a unified whole – not as bits and pieces such as animals, crops, soil, air or water – along with their interactions. Productivity, sustainability and commerce are the pillars on which this edifice must be built.

18.3.9 Conservation of biodiversity

Genetic erosion and species extinction are now occurring at an accelerated pace due to habitat destruction, demographic explosion, enlarging ecological foot print, and climate change and changing farming system. Thiruvazhamkunnu, Livestock Research station is a treasure trove of bio diversity. It rests in the lap of the Silent Valley national park an evergreen forest with wide range of flora and fauna. The bio diversity conservation and sustainable management should become the future farm ethic. There is an urgent need to identify, characterise and document the vast genetic bio diversity of livestock research station. Future farms need to be built without hampering bio diversity. The future farm should be a paradise where butterflies flutter, birds twitter and the cattle roam freely amongst themselves. The issue of man-animal conflict should be taken into serious consideration and an amicable solution reached in each case which ought to be a win-win situation for both the parties.

18.3.10 Think globally, assess regionally, and act locally

The core concept of future farm can be interpreted thus: Think globally, assess regionally and act locally. The author finds the new expression 'glocal' most appropriate in the present context. There is a glut of knowledge regarding the future farm concepts, adaptation and mitigation strategies under the changing climatic scenario. The climate change and its impact on livestock and livestock farming systems ought to be accessed on a meso-scale range. Besides, adaptation strategies need to be tailored to local availability of input and impacts of climate change. Mitigation strategies and targets are already set by Kyoto protocol globally. Local governments must decide how to implement them.

Conclusions

Climate change will impair production, reproduction, health and immune status of livestock. Modification of existing regimes of precipitation and the increase of aridity will have repercussions on the availability of feedstuff for animals. The difficulty in livestock production will correspond to the increasing needs in animal products. Scientific research can help the livestock sector in the battle against climate change. All animal scientists must collaborate closely with colleagues of other disciplines, first with agronomists then, physicists, meteorologists, engineers, economists, etc. in the design of future farms. Productivity, sustainability and commerce are the pillars of future farms. The effort in selecting animals that up to now has been primarily oriented toward productive traits, from now on, must be oriented toward robustness, and above all adaptability to heat stress. Research must continue developing new techniques of cooling systems such as thermo-isolation, concentrating more than in the past on techniques requiring low energy expenditure. Farm management information systems and precision farming are the basic tools available for the farmers for the effective management of farm resources. Mixed farming, organic farming, farm tourism are some of the avenues open to farmers to overcome loss due to climate change. Regional climate change need to assessed and time-bound weather forecast reports developed to inform the farmers in advance. Above all to beat the climate change or in any case not to let the climate beat livestock systems, researchers must be very aware of technologies of water conservation. In the future we can profit, more than in the past, from the years of experience of the people living in arid zones by applying our scientific knowledge to useful traditional practices (Nardone *et al.,* 2010).

References

Augustin. M.A., Udabage, P., Juliano,P., Clarke, P.T., 2013. Towards a more sustainable dairy industry: Integration acrossthe farm-factory interface and the dairy factory of the future. International Dairy Journal 31: 2-11.

Bayvel, A.C.D., Rahman, S.A., Gavinelli, A. (Eds.), 2005. Animal Welfare: Global Issues, Trends and Challenges. Revue Scientifique et Technique de l'Office International des Epizooties, 24: 463–813.

Bassoc, B. and Blackmore S.B 2010Conceptual model of a future farm management information systemComputers and Electronics in Agriculture 72.

David Fraser. 2008. Toward a global perspective on farm animal welfare. Applied Animal Behaviour Science 113 (2008) 330–339.

Erber, P.J., Steinfeld, H., Henderson, B., Mottet, A., Opio, C., Dijkman, J., Falcucci, A. and Tempio, G. 2013. Tackling climate change through livestock – A global assessment of emissions and mitigation opportunities. Food and Agriculture Organization of the United Nations (FAO), Rome.

FAO. 2006. Livestock's long shadow – Environmental issues and options, by H. Steinfeld, P. J. Gerber, T. Wassenaar, V. Castel, M. Rosales & C. de Haan. Rome.

FAO, 2013 Tackling climate change through livestock – A global assessment of emissions and mitigation opportunities. Food and Agriculture Organization of the United Nations , Rome. By Gerber, P.J., Steinfeld, H., Henderson, B., Mottet, A., Opio, C., Dijkman, J., Falcucci, A. & Tempio, G.

FAO. 2011. World Livestock 2011 – Livestock in food security. Rome.

FAO. 2007. Global Plan of Action for Animal Genetic Resources and the Interlaken declaration, Rome.

FAOSTAT. 2010. Food and Agricultural Commodities Production. Rome, Italy.

IPCC (Intergovernmental Panel On Climate Change). Climate change. 2007: Impacts, Adaptation and Vulnerability.Summary for policy makers.http://www. ipcc.cg /spm13apr07.pdf

IPCC, 2007. Summary for policymakers. In: Solomon S, Qin D, Manning M, Chen Z and others (eds.), Climate change 2007: the physical science basis. Contribution of Working Group I to the Fourth Assessment Report of the Intergovernmental Panel on Climate Change. Cambridge University Press, Cambridge.

IPCC, 2013: Summary for Policymakers. In: Climate Change 2013: The Physical Science Basis. Contribution of Working Group I to the Fifth Assessment Report of the Intergovernmental Panel on Climate Change [Stocker, T.F., D. Qin, G.-K. Plattner, M. Tignor, S. K. Allen, J. Boschung, A. Nauels, Y. Xia, V. Bex and P.M. Midgley (eds.)]. Cambridge University Press, Cambridge, United Kingdom and New York, NY, USA.

Jane Kabubo-Mariara. 2009. Global warming and livestock husbandry in Kenya: Impacts and adaptations. Ecological Economics. 68. 1915–1924.

Jimmy Smith, Keith Sones, Delia Grace, Susan MacMillan, Shirley Tarawali, and Mario Herrero. 2013. Beyond milk, meat, and eggs: Role of livestock in food and nutrition security. Animal Frontiers. vol. 3 no. 1 6-13.

Janzen,H.H. 2011. What place for livestock on a re-greening earth? Animal Feed Science .and Technology 166-167 (2011):783-796.

Kurukulasuriya, P., Rosenthal, S., 2003. Climate change and agriculture: a review of impacts and adaptations. Climate Change Series Paper No. 91, World Bank, Washington, DC.

Leaver, J. D. (2011). Global food supply: a challenge for sustainable agriculture.Nutrition Bulletin, 36: 416-421.

Mirkena, T., Duguma, G., Haile, A.,Tibbo, M., Okeyo, A.M., Wurzinger, M., Solkner. J. 2010 Genetics of Adaptation In Farm Animals: A review. Livestock Science,132: 1-12.

Mormede, P. and E. Terenina, E. 2012 Molecular genetics of the adrenocortical axis and breeding for robustness. Domestic Animal Endocrinology 43. 116–131.

Nardone, A., Ronchi, B., Lacetera, N., Ranieri, M.S. and Bernabucci, U. 2010. Effects of climate changes on animal production and sustainability of livestock systems.Livestock science 130: 57-69.

Rao, G.S.L.H.V.P., Varma, G. G., Prasad, A., Sankaralingam, S., Usha, A. P., Rajeev, T. S., Deepa, A. and Unnikrishnan, T. 2013. Climate change and animal agriculture, In: Proceedings of preliminary workshop on ensuring sustainable and responsible production of healthy food from healthy animals, 1-13.

Schinckel AP. Modeling, management and selection of genetics for optimal commercial performance. In: German Society for Animal Science. Giessen, Germany: 9th World Congress on Genetics Applied to Livestock Production; 2010: paper 0045.

Sejian.V., Naqvi., S.M.K., Ezeji., Lakriz, J., Lal. R. 2012. Environmental Stress and Amelioration in Livestock Production, Springer Publications.

Thorton, P.K., Van de Steeg, J., Notenbaert, A., Herrero, M., 2009. The impacts of climate change on livestock and livestock systems in developing countries: A review of what we know and what we need to know. Agricultural systems. (2010) 37–47.

UNDP. 2010. World population prospects, the 2010 revision. New York, NY, USA:United Nations Population Division.

Webster, A.J.F. 2001. Farm Animal welfare: the five Freedoms and the free market. The Veterinary Journal 2001, 161: 229–237.

19

Climate Change and Agriculture in North East India: A Retrospective Analysis

Anjumoni Mech, Sejian Veerasamy, Arindam Dhali and R.U. Suganthi

ICAR-National Institute of Animal Nutrition and Physiology, Adugodi Bangalore – 560030, India

Climate change has become a global threat as its impact is evident on the Earth. In the coming decades,the severity of climate change is expected to aggravate the ecosystem. Due to its vulnerable ecosystem and geographical location, the North eastern region (NER)of India that is home to about 40 million people is considered as one of the most climate sensitive areas in the country. The NER is among India's least developed region andscarce information is available on the impact of climate change in this region. However, initial scientific research and field observations confirm that the region is suffering from the impacts of climate change already. The region comprising 0.26 million km²area occupying 8% of nation's territoryconsists of the states of Assam, Arunachal Pradesh, Manipur, Meghalaya, Mizoram, Nagaland, Tripura and Sikkim. These states consist of a part of the east Himalayan region, which extends eastwards from Arunachal Pradesh to the Darjeeling hills of West Bengal. The entire region is a part of Indo-Burma and Himalayan hotspots *i.e.* characterized by rich biodiversity, heavy precipitation and high seismicity. The climate is predominantly humid subtropical with hot, humid summers, severe monsoons and mild winters. Along with the west coast of India, this region has some of the Indian subcontinent's last remaining rain forests. The region's lowland and moist to wet tropical evergreen forests are considered to be the northern most limits of true tropical rainforests in the world (Procter *et al.*, 1998). Northeast (NE) India has 61% of the total geographical area under forest cover. The rich forest area that accounts for approximately 25% of total forest cover in India covering around 66.8% of its geographical area is much higher than the national average of 21%.The state wise forest cover area in the NER India is shown in Table 19.1.

Table 19.1: Forest cover in North East India, 2001 and 2003

N.E States	Geographical area of the state	Total Forest cover (in sq. km) in 2003	Percentage Forest cover in the State	Forest cover in 2001
Arunachal Pradesh	83,743	68,019	81.22	68,045
Assam	78,438	27,826	24.04	27,714
Manipur	22,327	17,219	77.12	16,926
Meghalaya	22,429	16,839	75.08	15,584
Mizoram	21,081	18,430	87.42	17,494
Nagaland	16,579	13,609	82.09	13,345
Sikkim	7,096	3,262	45.97	3,193
Tripura	10,486	8,093	77.18	7,065

Source: FSI 2003

19.1 Geographic Location and Climate

The Northeast India is situated at the confluence of Indo-Malayan, Indo-Chinese and Indian biogeographical territories. Geographically it is located between 21.5° to 29.5° North latitude and 85.5° to 97.5° East longitudes. Due to its unique location and topography NER has distinct precipitation and drainage patterns. The predominating climate in the region is in compliance with humid sub-tropical climate that is characterized with hot and humid summers, severe monsoons and mild winters. The mighty Brahmaputra, Barak River and their tributaries surrounds the territory. The average rainfall is very high (10000 mm per annum) and flood and seismic waves are the most common natural calamities. The greater part of evergreen rain forest of India is found to be located in the foothills of eastern Himalaya, i.e. Assam valley, lower parts of Nagaland, Meghalaya, Mizoram and Manipur where the annual rainfall exceeds above 2300 mm per annum. These rain forests are well known niche of natural biodiversity. During last 50 years the rainforests of the region are depleting at a very alarming rate due to increasing human population, negligence and ignorance of the state authorities and the local communities. The adverse effect of such activities is reflected on climatic condition with consequential severe and frequent incidences of natural disasters like flood, cloudburst, earthquake, etc in the recent days. Flood damages bio-diversity and ecology, Sedimentation and deposition of silt and sand affects the bills, wetland, natural drainage, fish migratory birds, crops, forests, animals, air pollution resulted from dead bodies of livestock etc. All this in a body damages the bio-diversity and harms the balanced ecology.

Moreover, flood also damaged the grazing fields and creates problems for the animals just after the flood. Looking into the increase intensity and occurrences of such incidences it is considered that the issues related to climate change impacts in the region need to be addressed at the earliest.

19.2 Evidences of Climate change in Northeast India

Climate Change in the Eastern Himalaya is expected to cause significantly warmer temperatures in the North East India and its neighborhood region imposing more stress on existing ecosystem and living being including human, livestock, wildlife, fish, water resources and agriculture.These environmental changes in turn are showing its impact on both economy and social life of the people in the region. The impact of climate change is significantly affecting thesocial life of ruralinhabitants residing in the hilly terrains of north east India as they arestill dependent on local natural resources for their day to day living. The local people and the livestock used to survive mostly on unique flora with edible wild plants and fruits that were found in abundance in the forest area of the region in the earlier days. However, with changing weather patterns the availability of such edible wild plants and fruits are diminishing in the region. The animal fodder that was available in abundance in the wild is now scarce and collection of these edible floras from wild is becoming more and more difficult for local people.Deforestation and jhum cultivation have led to extensive loss of forest cover in the last few decades.Forest area is squeezing making the life of both people and livestock much tougher. The farmers are facing difficulty to adapt to the frequent climatic changes as there have been occurrences of unpredictable and frequent changes in weather (droughts and floods). Particularly unusual pattern of rain, hailstorms, flood is battering the crops in a massive way. Assam is considered as the most flood prone areas in NER (Fig. 19.1). Besideswith the rise in global temperature there are reports of new species of pests infecting the crops in the region like elsewhere in the Country.

Fig. 19.1: Flood prone areas of NER

According to an analysisconducted by North East climate change adaptation programme (NECCAP, 2011) the long-term fluctuations (year 1901-2007) of seasonal rainfall time series shows distinct periods of above and below-normal rainfall. The variability is very high in West and East Khasi Hills of Meghalaya, while the variability is high in Changlang, Tirap, Dibang Valley, Lohit, East Siang, West Siang (Arunachal Pradesh) and Tinsukia, Dhemaji and Dibrugarh (Assam). The same study has confirmed that the Northeast Indian region has 'warmed' significantly during the last decade due to rapid increases in both, the maximum as well as minimum temperatures, with the minimum temperature increasing more rapidly (NECCAP, 2011).

19.3 Estimate of Climate Change Impact

In India, over the period of last 100 years (year 1901 to 2000) 0.4°C increase in temperature has been recorded. While in north east India the spatial distribution of temperature changes result comparatively warmer winter and post monsoon seasons. According to analyses done by Indian Meteorological Department and Indian Institute of Tropical Meterology, Pune, monsoon rainfall has been found to decrease by -6% to -8% of normal in North East along with several other states of the country over this period of 100 years (Dogra and Srivastava, 2012). This type of climatechange is already affecting the livestock industry through various direct or indirect effects on vector born diseases, soil fertility and pasture availability. Further, increase atmospheric CO_2 has direct impact on fertilizationof soils. One of the most alarming facts is that global climate change will have significant impact on the ecology of Himalayan glaciers which may result in rapid melting of glaciers forming large numbers of glacier lake as well as glacier lake outburst floods in the valleys downstream causing flash floods in the Himalayan region. There has been predictions saying that in the short term of 20-to-40 years, the states of Assam, Meghalaya, Sikkim and Arunachal Pradesh is expected to receive "much less monsoon precipitation" while rainfall is likely to increase in Tripura, Mizoram, Manipur and Nagaland (The Herald of India, July 2013).The graphical representation of the rainfall data recorded at the meterological department,Tocklai Experimental Station, Assam showed that that there has been a steady decline in annual rainfall in the state since 1950 till 2010 (Fig. 19.2). Further monsoon rainfall deficient years in NER has been growing and the Assam-Meghalaya subdivisions of the region have already recorded six monsoon rainfall deficient years since 2001 (Borah et al., 2012). The same report cites that in the year 2011, Assam-Meghalaya subdivisions recorded 35% deficit monsoon rainfall while Nagaland-Manipur-Mizoram Tripura subdivision recorded 41% deficit rainfall leading to an alarming situation. Similarly the graphical representation of minimum temperature in Assam since 1950 to 2010 reveals a steady rise and there has been a rise of 1°C to 1.5°C within a period of last 90 years (Fig. 19.3).

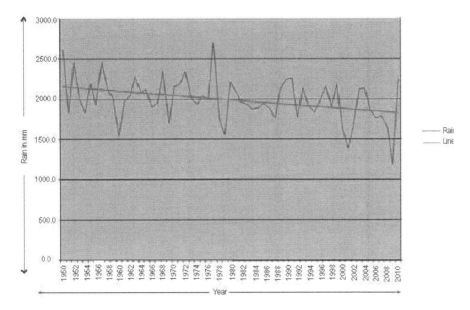

Fig. 19.2: Rainfall trend in Assam since 1950 to 2010
(Borah *et al.*, 2012)

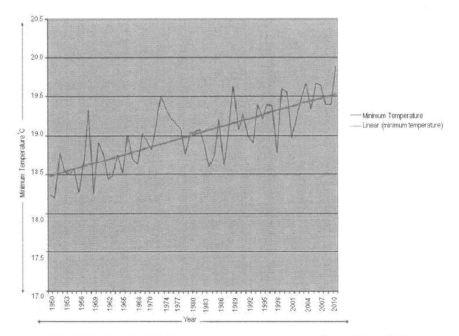

Fig. 19.3: Trend in Minimum temperature over Assam from 1950 to 2010
(Borah *et al.*, 2012)

According to a recent report by Indian Network for Climate Change Assessment (INCCA) India is going to experience an increase in precipitation in the 2030s and the increase is highest in the Himalayan region and lowest in the North Eastern region. While the temperature in the region is expected to increase between 1.8°C to 2.1°C. The extreme precipitation events may also increase by 5-10 days throughout the region and the number of rainy days is likely to increase by 1-10 days with an increase intensity of rainfall by 1-6 mm/day. This is alarming as the region is prone to landslides and floods which may get aggravated by heavy rainfall (Shimray, 2011).The northeastern systems of Mahanadi and Baitarni rivers are predicted to experience extreme precipitation events, with increased frequency and intensity, thus causing enhanced flood risk (GOI, 2011). Increase of frequency and severity of floods and droughts will have implications on the functioning of the ecosystems.

As indicated by the IPCC,drought like condition has been reported in as many as 15 districts in Assam for two consecutive years in 2005 and 2006, signifying the evidence of climate change (IPCC, 2007). The intense drought-like conditions that occurred during the monsoon months of 2006 due to below-normal (nearly 40%) rainfall in the region. Consequently, more than 75% of the 26 million people associated with agriculture suffered due to crop failure and other associated secondary effects. Usually such fluctuations are being observed due to inter-annual variability in the monsoons. However, climate change has increased the variability of the southwest monsoon beyond the regular level. According to IMD records among the seven northeastern states, Assam has suffered the most deficit rainfall and high temperatures during 2006 (Table 19. 2). There has been only four years of normal or above-normal rainfall in the region betweenthe year 1991 and 2000. As most of the agriculture in the region is carried out under rain fed condition henceboth heavy and continuous showersimposes difficulty for the farmers to accomplish agricultural operationsin time and intense heat further aggravates the situation (Chhabra *et al.*, 2009).

Table 19.2: Irregular rainfalls recorded in the year 2005 and 2006 in NER India

State	Excess (+) Year 2005	Deficiency (-) Year 2006
Assam and Meghalaya	-23%	-32%
Arunachal Pradesh	Normal	-25%
Manipur	-22%	-25%
Mizoram	-22%	-25%
Nagaland	-22%	-25%
Tripura	-22%	-25%

Source: IMD, Guwahati, India (Chhabra *et al.*, 2009)

Modeling studies on the impacts of climate change on forests for short to midterm (2021–2050) could reveal that the forests in the northern part of the North East are mainly affected by climate change. However, the net primary productivity is projected to increase by 23% in this region that will be followed by increase in biomass and soil carbon, leading to probable changes in vegetation type (Ravindranath *et al.*, 2011).Besides the projected vegetation change in the forestry sector will be 8% (Sharma and Chauhan, 2011). The Himalayan region is expected to experience moderate to extreme drought severity in the 2030s (INNCA, 2010); whereas the water yield in region is expected to decline to certain extent (Fig. 19.4).

Another study was conducted by Ravindranath *et al.* (2011) with an index-based approachwherein a set of indicators that represent keysectors of vulnerability (agriculture, forest, water) were selected using the statistical technique principal component analysis. The impacts of climate change on key sectors were derived from impact assessment models and it was represented by changes in the indicators which were further utilized for the calculation of future vulnerability to climate change. The results of the study demonstrated that the majority of the districts in North East India are subjected to climate induced vulnerability during the study period and in the near future.

19.4 Impact on Agriculture

Eighty six per cent of the population is dependent on agriculture for their livelihood and over 60% of the crop area is under rained agriculture that are extremely susceptible to any kind of climate change. Rice is the predominant crop that covers about 84% of cultivated area. The whole region is under diverse climatic regime and is highly dependent on the southwest monsoon during June to September. The total area affected by agricultural drought during the south-west monsoon period constitutes an average of 40% of the geographical area of the region,of which 39% area is under rice. Shifting cultivation is predominant in the eastern and north-eastern regions on hill slopes and in forest areas of Meghalaya, Nagaland, Manipur, Tripura, Mizoram and Arunachal Pradesh (Fig. 19. 5). Shifting cultivation is a type of subsistence farming where a plot of land is cultivated for a few years until the crop yield declines due to soil exhaustion and the effects of pests and weeds. Once crop yield declines, the plot of land is deserted and the ground is cleared by slash and burn methods, allowing the land to replenish. Crops such as rain fed rice, maize, wheat, small millets, root crops as well as vegetables are being grown in this system.In this system of Slash-and-burn cultivation which is also known as jhum,clearing of sloping land is being done and that may lead to increased soil run-off following disappearance of the protective vegetative cover. The soil run-off and redeposit ion affects soil fertility and spatial patterns of fertility parameters in a field. Soil erosion is

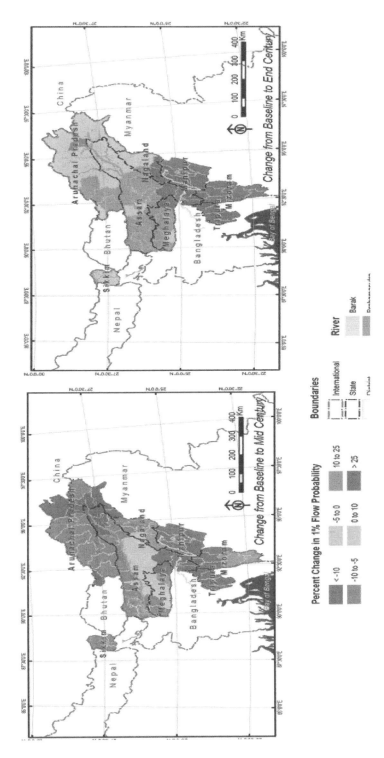

Fig. 19.4: Spatial variation in Change in stream discharge at 99th percentile for IPCC SRES A1B BL, MC and EC scenarios for North Eastern Region (INCCA, 2010)

an irreversible phenomenon causing land degradation and deterioration of surface water quality which may be caused due to inappropriate land use and poor management (Yadav, 2011).

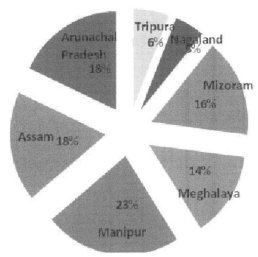

Fig. 19.5: Annual area under shifting cultivation in the North East India (RTFSC, 1983 and Basic Statistics of NER, 2002, GOI)

19.5 Contribution of Livestock to Climate Change

The economic contribution of livestock sector is enormous to Indian economy as it contributes 26% of total agricultural GDP and provides employment to 18 million people in principal or subsidiary status. Consequently maintaining a sustainable livestock production system with changing climatic scenario is a tough challenge to the livestock producers and entrepreneurs. Agriculture is the prime source of livelihood for a majority of rural population in the North-Eastern region (NER) of India and it is significantly supported by livestock as an alternative source of income. Methane is considered to be most important greenhouse gas which contributes significantly to global warming and livestock is a major anthropogenic source of methane emission for agriculture. Sources of methane emission from livestock has been categorized into two *i.e.* emission from enteric fermentation and manure management. According to a report (Chhabra *et al.*, 2009b) using remote sensing derived livestock feed and fodder area the total methane flux from Indian livestock was computed as 74.4 kg/ha whereas the total methane emission including enteric fermentation and manure management was calculated as 11.75 tg for the year 2003. The report further cites that considering the total CH_4 emission of 3.16 Tg and total milk yield of 88.1 Mt for the year 2003 in India, the CH_4 emission per kg milk produced amounts to 35.9 $gmCH_4$/kg milk. However, among the milching livestock, CH_4

emission is less in high yielding exotic/cross-bred cows (23.8gmCH$_4$/kg milk) as compared to low milk yielder indigenous cows (44.6 gm CH$_4$/kg milk).Further, category-wise CH$_4$ per unit milk produced is suggested to be in the order of cross-breed cow<dairy buffalo<goat<indigenous cow. In the NER India the CH$_4$ emission from livestock is highest in Assam (77.6 gmCH$_4$/kg milk produced) followed by Arunachal Pradesh, Nagaland, Meghalaya, and Tripura with 51-75 gmCH$_4$/kg milk produced, then Manipur and Sikkim with 25-50 gm CH$_4$/kg milk produced and least was recorded for Mizoram with <25 gm CH$_4$/kg milk produced (Fig. 19.6). In contrast the gap between requirement and production of milk in the region is as wide as 51% as it produces only 1.06 Mt milk against the requirement of 2.15 Mt milk per annum and the per capita availability of milk is only 81g/capita/day i.e. 1/3rd of national average (Paul and Chandal, 2010).The spatial distribution of methane emission from livestock in NER has been shown in the Fig.19.7. Analysis of performance and factors influencing

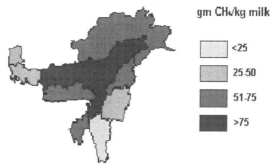

Fig. 19.6: Distribution of methane emission per kg milk produced by the milching livestock in NER India (2003) (*Source*: Chhabra *et al*., 2009a)

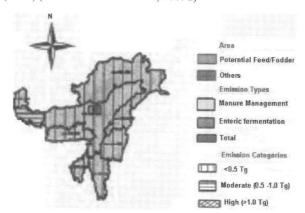

Fig. 19. 7: Spatial distribution of methane emission from livestock in NER India for 2003 (Potential feed/fodder area implies both area under crops and grasslands available for livestock. Other area include land cover classes like forest, ice or snow, water bodies, wetlands, bare soils urban- built- up land etc. (Chhabra *et al*., 2009^b)

development of the livestock sector in NER could reveal that the growth of livestock sector is slower in the NER than at the national level (Kumar *et al.*, 2007). Nevertheless, meeting the escalating demand for livestock products in a sustainable manner is a big challenge for the people of the region.

Evaluation of state wise detailed estimates of methane emission for the year 2003 in the NER India (Table 19. 3) showed that Assam possessing the highest number of livestock in the region contributes maximum to emission from enteric fermentation and manure management of livestock and contributes 2.97% of total contribution of methane emission from livestock in the country.

Table 19.3: Sate wise detailed estimates of methane emissions (Tg) for 2003 in North East India

Region	Livestock population (millions)	Enteric fermentation emission (Tg)	Manure management emission (Tg)	Total emission (Tg)	Percentage contribution
Assam	13.83	0.27	0.0791	0.349	2.97
Arunachal Pradesh	1.26	0.01	0.0025	0.012	0.10
Manipur	0.97	0.02	0.0034	0.023	0.19
Meghalaya	1.55	0.02	0.0044	0.024	0.20
Mizoram	0.28	0.00	0.0011	0.001	0.01
Nagaland	1.35	0.02	0.0047	0.025	0.21
Sikkim	0.34	0.01	0.0007	0.011	0.09
Tripura	1.46	0.02	0.0034	0.023	0.19

(Shhabra *et al.*, 2009b)

19. 6 Impact of shifting cultivation on climate change

The NER is being designated a typical case of economic underdeveloped region in the country. The region occupies 3.4% of the agricultural land of the country but it contributes only 1.5% to that total food grain production providing livelihood support to 70% of the population (Karmakar, 2008; Misra and Misra, 2006). Eighty-five per cent of the total cultivation in northeast India is by shifting cultivation. Due to increasing requirement for cultivation of land, the cycle of cultivation followed by leaving land fallow has reduced from 25-30 years to 2-3 years. This significant drop in uncultivated land does not give the land enough time to return to its natural condition. Because of this, the resilience of the ecosystem has broken down and the land is increasingly deteriorating. Frequent shifting from one land to the other has affected the ecology of the region. Shifting cultivation was assessed by the FAO to be one of the causes of deforestation.The Forest Survey of India (FSI) is condemns shifting cultivation as the main reason for decrease of forest cover in North east India. Degradation of forest causes ecological imbalance, rapid drying up of small water sources and loss of productivity of land causing major economic loss for people who are solely dependent on agricultural income and are without any other sources of subsidiary income.The results of successive studies carried out by the Forest Survey of India, *i.e.* State of Forest Report, 1995 and 1997 states that the loss in forest cover in the northeastern states was mainly due to the shifting cultivation. The total forest area affected by the shifting cultivation in the region is shown in the (Fig. 19.8). The extent of area under shifting cultivation is maximum (0.39 m ha) in Nagaland followed by Mizoram (0.38 m ha) and Manipur (0.36 m ha). These states together account for about 65% of the total area under shifting cultivation in the N-E. According to a satellite record in Tripura, comparison of the current forest cover (Satellite data of Nov 2006-Jan 207) with the previous assessment (Satellite data of Nov 2004) showed a loss of 100sq. km of forest cover. In Tripura comparative analysis of satellite data between Nov 2004 to Nov 2006-Jan 2007 could reveal a decrease of 2 km^2 in the very dense forest, 46 km^2 in moderately dense forest and 52 km^2 in open forest (INCCA, 2011). However, in the NER only about 8% of the 73 forest grids are projected to undergo changes in the 2030s. The region is projected to be an increase of 23% in NDP on average (NECCAP, 2011). The cumulative area in the seven NE states affected by shifting cultivation during 10 years was found to be 1.73 m ha (Table 19.4).

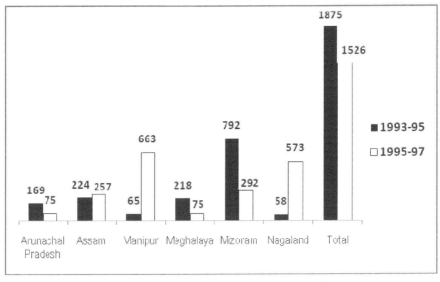

Fig. 19.8: Loss of forest Area in NE region due to shifting cultivation
Source: State Forest Report 1995, 1997

Table 19.4: Area under shifting cultivation during year 1987-1997

States	Cumulative area of Shifting Cultivation(1987-89 to 1995-97)*	Annual area under shifting cultivation	Fallow period (Years)
Assam	0.13	69,600	2-10
Arunachal Pradesh	0.23	70,000	3-10
Manipur	0.36	90,000	4-7
Meghalaya	0.18	53,000	5-7
Mizoram	0.38	63,000	3-4
Nagaland	0.39	19,000	5-8
Tripura	0.06	22,300	5-9
Total	1.73	386900**	**

Source: FSI report on shifting cultivation (Between 1987 to 1997) **Yadav ,2013

19.7 Loss of Nutrients and Top Soil due to Shifting Cultivation

With reduction in jhum cycle from 20-30 years to 2-3 years, the land under shifting cultivation looses its nutrients and the top soil. So long as the jhum cycle has duration of 10 years or more this type of cultivation did not pose any threat to the ecological stability and soils of the largely forested hill area. While studying jhum ecology in Meghalaya, it was reported that water and nutrient losses in shifting-cultivation areas were far greater than in the virgin areas, and areas left for 50 years after jhuming. Thus, reduction in the cycles of jhuming, adversely affects the recovery of soil fertility, and the nutrient conservation by the ecosystem. Repeated short-cycle jhuming has created forest-canopy gaps which are evident from the barren hills.

19.8 Other Ecological Consequences of Shifting Cultivation

As ecological consequences of shifting cultivationthe area under natural forest has been declining, the habitat is becoming fragmented, native species are disappearing and there is invasion by exotic weeds and other plants. The area having jhum cycle of 5 and 10 years is more vulnerable to weed invasion compared to jhum cycle of 15 years. The area with fifteen-year jhum cycle has more soil nutrients, larger number of species, and higher agronomic yield with ratio of energy output to input as 25.6 compared to jhum cycle of 10 and 5 years (4.6–9.8).

Although jhuming has many benefits from livelihood point of view, but in long run it destroys the ecosystem balance because one inch soil formation in nature takes about 1000 years. But several inches of soil are washed out each year due to jhuming. Heavy siltation of Brahmaputra River from its tributaries and frequent breaking of embankments are caused by heavy soil erosion from hills of North East India. A study by ICAR institute of NEH region, Barapani, reported that the shifting cultivation on steep slopes (44-53%) has soil loss value of about 50 tones/ ha/ year with corresponding nutrient losses of 703 kg of organic carbon, 144 kg of phosphorus and 7 kg of potash annually. Major adverse effect of shifting cultivation apart from soil and nutrient losses are rapid siltation of river beds causing floods, denudation of forests which hardly gets regeneration time because after 3 to 5 years they have to be burnt for another jhuming due to population pressure. Previously this period was for 7 to 10 years.

Photo 1: A view of land prepartion for jhum cultivation by burning jungles at Pfeutseromi village, Phek district, Nagaland

19.9 Impact of Climate Change on Livestock

Increase incidence of flood and land erosion due to climate change is one of the main reasons of reducing grazing land. In addition another reason that reduces grazing land is rehabilitation of erosion and flood-affected people in the grazing land. Already there has been severe shortage of fodder due to sharp decline in area under permanent pasture and other grazing land in the region. The illegal encroachment of village grazing land for cultivation is also another reason behind declining area under pasture and other grazing land. Besides, there is hardly fodder grown for livestock. According to a study conducted in the high altitude rangelands of Indian Transhimalaya, to investigate the ecology of endangered Tibetan argali (*Ovis ammon hodgsoni*) the wild sheep species and related issues regarding its conservation and potential conflict with the local nomadic pastoralists the dramatic changes in the nomadic pastoralism that has occurred in the past five decades in the area has led to loss of pastures (-25 to -33%) of the nomads, consequent readjustment in traditional patterns of pasture use, intensified grazing pressures (25to 70%) and rangeland degradation. Consequently such changes would have serious effect on the survival of local wildlife, as tested with a study of the effects on argali of livestock presence and resource exploitation (Singh, 2008). In Assam , during earlier days in 1950 to 1970s there used to be large herd of buffaloes kept primarily for production of milk everywhere, which was locally known as khuti' or 'Bathan' system. However, with the shrinkages of swamp, marshy area and grazing land, a large number of these Khutis and Bathans is disappearing rapidly. Moreover, according to INCCA (2010), the Temperature Humidity Index (THI) i.e. the index used to define losses due to thermal stress is likely to increase in the region during April-October with THI>80. It is a matter of great concern as most of the animals come under severe stress with THI>80.

19.10 Impact on Wild Life

Much of the dense forests of Assam, Nagaland and Arunachal Pradesh are part of Himalayan biodiversity hotspot.The impact of climate change is likely to result in large scale changes in this Himalayan biodiversity hotspot.The natural landscape of the region has been extensively modified in the recent past due to pressure on land. The one-horned rhino, living on floodplain grasslands in Assam and Nepal, relies on annual monsoon rains to replenish the vegetation it feeds on. There have been worries that climate change could disrupt that through frequent occurrences of droughts and floods.

19.11 Population and Urbanisation

At present the region is home to 38 million people i.e. 3.8% of the country's total population. The changes in demographic and socioeconomic profile of the region have brought about alteration in its resource base in the last three decades. Over the time the region has witnessed intense land usewith increase in the demand for urbanization, grazing, agricultural land, increased demand for fodder, fuel wood and timber production. Due to all these factors it is estimated that approximately 30 % of the total forest cover in the region is under pressure. According to a study, sponsored by The Energy and Resource Institute (TERI), change in temperature, quantum and intensity of rainfall coupled with extreme weather conditions would have a long-term impact, particularly on the structure and composition of forests in the region. Severe impact is expected in the areas where stability of the natural systems affected due to socio-economic pressures such as encroachment on forest areas, over-grazing, felling of trees for jhum cultivation, etc (The Hindu, 2011).

19.12 Deploying Mitigation Strategies

19.12.1 Preserving biodiversity and indigenous breeds

By adhering to climate-friendly farming practices, agriculture can continue to provide food for the present and future population and will continue to be a source of livelihood for large number of local people who rely on farming for income and sustenance. If agriculture is to play a positive role in the combat against climate change, the agricultural practices that mitigate or adapt to climate change has to be studied, modified and validated.Growing diverse and locally adapted indigenous crops of the region, can provide a source of income and improve farmers' chances of withstanding the effects of climate change, such as heat stress, drought, and the expansion of disease and pest populations. Preserving plant and animal biodiversity will reduce farmers' dependency on small number of commodity crops that make them vulnerable to frequentshifts in global markets.

19.12.2 Sustainable approaches to land and water use

The undulating nature of the terrain surrounding the foothills and the high density of rainfall require proper water management and the prevention of soil erosion. Organized water source and an irrigation management planare needed to combat occurrence of drought and flood, and the enormous economic loss to the agriculture in the region. Raising Livestock in pasture is a key cultural practice in Northeast instead of usage of chemical fertilizers in agriculture. A healthy pasture ecosystem always requires all key mineral element and nutrients for the livestock as well as for growing crops. Rearing or allowing to graze livestock on the land between two cropping,reducesoverall emissions and the total amount of energy expended to produce a crop or animal.

19.12.3 Scientific Jhum cultivation

The shifting cultivation is helpful in the organic management of land if followed in scientific manner by following an ideal jhum cycle of 20- 30 years. The use of controlled fire in shifting cultivation enables to manage soil fertility and control weeds and pests in a labour efficient manner and also enables one to avoid use of agrochemicals. The per hectare consumption of chemical fertilizers (N, P, K and N+P+K) for cropped area during the year 1999/2000, was only 2.8 for Nagaland, and 2.0 for Arunachal Pradesh, as compared to 95.6 in India as a whole (Yadav, 2013). The ash deposits after burning, helps to fertilize the soil by immediate release of the occluded mineral nutrients-Mg, Ca, and available P, for crop use (Scheuner *et al.*, 2004; Niemeyer *et al.*, 2005).

19.12.4 Improving productivity and reducing emissions in milk animal

As milk production in India is primarily based on feeding crop residues to milk animals it results in more acetic acid and methane production. Technology for enrichment of crop residues with urea, molasses, mineral etc., which would result in reduction of methane, should be promoted in the region. In addition the farmer of the region should be taught about the benefits of providing balanced ration to livestock with appropriate energy, protein and minerals that will help in production of more microbial protein and less methane. Using SF_6 tracer technique 10-15% reduction in methane emission has been recorded in lactating animals in UP. In addition significant ($P<0.05$) reduction in gross energy lost as methane has been observed. Therefore, it is interpreted that ration balancing has the potential to improve production and reduce methane emission from lactating cattle and buffaloes (Kannan *et al*,. 2010). Feeding of bypass protein in ruminants also enables efficient use of protein meals, results in reduction of methane and improves milk yield at the same time. Programmes for improving the genetic merit of the milch animals should also be undertaken that may lead to improved feed conversion efficiency along with reduction in methane emission per litre of milk. In addition other initiatives that should be undertaken to safeguard the livestock industry from the impact of climate change are:

- Selection of livestock breeds that can cope to the changing environmental condition calls for better utilization of traditional/indigenous breeds.

- Systematic analysis of breed characteristics should be undertaken under variable environment.

- Study should be designed to understand the genetic basis of adaptation by linking genomic analysis with spatial analysis of the landscape.

- Controlled crossbreeding or upgrading of indigenous livestock with improved breed which can thrive well under poor feeding and management condition and can grow faster than indigenous ones should be identified and popularized via different schemes. Some examples of the livestock breed that has been found to perform very well throughout the region are Vanaraja birds as dual purpose poultry- breed and Jersey cross cattle for milk production.

19.12.5 Increasing agroforestry

Increasing agroforestry can assist in removing atmospheric carbon dioxide and hence planting trees in livestock farms whenever possible can help mitigate climate change. The trees would also provide shade for livestock and create habitats for animals and insects, such as bees, that pollinate many crops.

19.12.6 Urban agriculture

The practice of cultivating, processing and distributing food in or around a city or peri urban area that may involve animal husbandry, aquaculture, agroforestry

Photo 2. Roof top vegetable cultivation in Assam year 1997 shifting cultivation during (*Courtesy*: Post jagoran.com, 18 July 2013)

and horticulture is known as urban agriculture. Urban agriculture contributes to food security and food safety as it increases the amount of fresh vegetables, fruits and meat products available to people living in the cities. A common and efficient form of urban agriculture is the biointensive method that promotes energy-saving local food production. The another aspect of urban agriculture is that it can contribute in mitigating the greenhouse gas emissions released from the transport, processing, and storage of food destined for urban populations and it is also consider as a sustainable agriculture. In this context Assam government is promoting roof top vegetable gardening in urban areas. The state Directorate of Horticulture and Food Processing has been entrusted to implement this programme and several workshop and training programme related to urban agriculture.

19.12.7 Green manure and hedgerows

Cover cropping, also known as green manuring, is the practice of strategically planting crops that delivers a range of benefits to a farming system. In this system the crops are often plowed into the soil instead of harvesting. Planting cover crops improves soil fertility and moisture by making soil less vulnerable to drought or heat waves. Cover crops also serve as a critical deterrent against pests and diseases that affect crops or livestock. Hedgerows farming system is another simplest and best control practices for sloping land. In this system nitrogen fixing woody plants, grasses, fruit trees and other crops are planted in rows only in the contour. This farming system is an improved soil fertility system accepted by the farmers and adopted in the mountain areas in many countries (Shabba and Ghosh, 2003).

19.12.8 Major existing schemes for formulating futuristic strategies

The ecosystem of northeastern states of India is expected to be most susceptible to the increasing severity of climate change in the coming decades. The probable impacts of climate change are expected to increase the incidences of flood and extend the drought periods still further. Simultaneously, increase population growth and rising prosperity in the region has an adverse impact on the natural resources as well. Initiatives should be taken up to design and develop collaborative, interactive programmes by the scholars and institutions in partnership with the institutes working on climate change issues in the country. However, initiative has been taken up by the GOI in collaboration with Germany's KfW Development Bank for assessing the extent of impact of climatic change in a scientific approach. The highlight of this programme was to deal with the susceptible areas and then sketch the possible adaptation measures on the basis of concrete data (*By Marcus Stewen and Nand Kishor Agrawal*). There is also a need to properly channelize funds to disseminate information on adaptation and coping opportunities to the changing environment extensively and comprehensively to vulnerable communities of the region. Such information should be developed and conveyed in local language. As most local governments and media are not adequately aware of climate change issues dissemination of knowledge and information on climate change should be carried out through effective partnerships between stakeholders, institutions, community organizations, civil society groups and the media. There is also a need to tackle climate change by involving gender experts who can analyze the situation and address the respective issues accordingly. The ever-increasing gap between the demand and supply of livestock products can be met through bringing out changes in the production structure or opening up the international trade.Emphasis should be given on region specific vulnerability assessment through modeling of climate change scenarios and vulnerability Index development by using suitable and precise impact assessment model. Accordingly suitable strategies should be developed for better preparedness towards impacts of climate change.

19.13 Conclusion

The NER India has not been put in the priority list of nation's development agenda. Being one of the ecological hotspots in the country and considering its sensitiveness to climatic change NER region need specific attention for protection from climatic change and further development process of its rich flora and fauna. Like other parts of the country the per cent contribution from human developmental activities to climate change in the NER has been observed to be quite substantial. The issues of climate change are closely linked to other environmental concerns and hence are the focal challenge for overall sustainable development of the region. Moreover, any impact of climate change at the very outset disproportionately affects the poorest and weakest section in the society. All sectors including rapid urbanization, industry, energy production, transport, forestry, sustainable agriculture and livestock farming and waste management can contribute to the overall mitigation efforts of climate change. However, a significant proportion of landless labourers, small and marginal farmers have access to livestock resources and acceleration in the growth of livestock in NER offers significant opportunities for intensification of household income and employment generation. Hence concentration should be imparted for technical, institutional and policy initiatives for the improving breeds, feed availability, disease control and food safety of livestock to combat the ever-increasing climatic change in the region. More studies should be conducted for identifying the sectors that are most vulnerable to climate change and accordingly the adaptation interventions should be prioritized. More information should be generated for developing vulnerability profiles via climate change and impact assessment from multiple models for the region. Such type of information will give the basis for planning policies in combating climate change at administrative level.

References

Basic Statistics of NER, 2002, Government of India, North Eastern Secretariat, Shillong. p. 42.

Borah A, Devi S, Medhi M. 2012. A Report on "Impact of climate change on marginalized women: An Exploratory study across 6 six districts in Assam."Published by INECC.

Borthakur, D. N., Singh, A., Awasthi, R. P. and Rai, R. N., in Proceedings of National Seminar on Resources, Development and Environment in Himalayan Region, D. S. J., New Delhi, 1980, pp. 330–342.

Chhabra A, Manjunath KR Panigrahy S. 2009[a]. Assessing the Role of Indian Livestock in Climate Change.ISPRS Archives XXXVIII-8/W3 Workshop Proceedings: Impact of Climate Change on Agriculture, 359.

Chhabra A, Manjunath KR, Panigrahy S and Parihar JS. 2009[b]. Spatial pattern of methane emission from Indian livestock. Current Science, 96(5): 683-689.

Dabral, DD, Baithury N, Pandey A. 2008. Soil Erosion Assessment in a Hilly Catchment of NorthEastern India Using USLE, GIS and Remote Sensing. Water Resource Management. 1783-1798.

Dogra N and Srivastava S. 2012. Climate change and disease dynamics in India. Published by TERI, New Delhi, India.

FSI, State of Forest Report, Forest Survey of India, Dehra Dun, 1995.

FSI, State of Forest Report, Forest Survey of India, Dehra Dun, 1997.

FSI, State of Forest Report, Forest Survey of India, Dehra Dun, 2003.

Government of India. 2011. Climate change and 12th five year plan report of sub-group on climate change. Planning commission,GOI, New Delhi ,October, 2011.

INCCA. 2010. Climate Change in India: A 4×4 Assessment, A sectoral and regional analysis for 2030s. Ministry of Environment and Forests, Government of India.

Indian-German Financial Cooperation North East Climate Change Adaptation Programme (NECCAP), Ministry of Development of North Eastern Region (DoNER,Government of India KfW Development Bank. 2011. Project Document, pages: 999.

IPCC. 2007. Climate Change 2007: The Physical Science basis, Summary for Policy Makers. Contribution of Working Group Ito the Fourth Assessment Report of Intergovernmental Panel on Climate Change, http://www.ipcc.ch

Jain SK, Kumar V, and Saharia M. 2012. Analysis of rainfall and temperature trends in northeast India. International journal of climatology. 33: 968-978.

Kannan A, Garg MR and Singh P. 2010. Effect of Ration Balancing on Methane Emission and Milk Production in Lactating Animals under Field Conditions in Raebareli District of Uttar Pradesh. Indian Journal of Animal Nutrition 27(2): 103-108.

Karmakar KG. 2008. Agriculture and rural development in North Eastern India, The Role of NABARD. ASCI Journal of Management 37 (2): 89-108.

Kumar A, Staal S, Elumalai K and Singh DK. 2007. Livestock Sector in North-Eastern Region of India: An Appraisal of Performance. Agricultural Economics Review, 20:255-272.

Misra AK and Misra JP. 2006. Sustainable development of agriculture in North Eastern India: A quest for more economical and resources sustainable alternatives. ENVIS Bulletin, Himalayan Ecology 14 (2): 3-14.

NECCAP. 2011.Project Document by North East Climate Change Adaptation Programme (NECCAP). Government of India.

Niemeyer T, Niemeyer M, Mohamad A, Fottner S, Hardtle W. 2005. Impact of prescribed burning on the nutrient balance of heathlands with particularreference to nitrogen and phosphorus. Applied Vegetation Science, 8: 183-192.

North east will get warmer as climate changes. The Herald of India. 27th July 2013.

Paul D and Chandal BS. 2010. Improving milk yield performance of crossbred cattle in North Eastern States of India. Agricultural Economics Review, 23:69-75.

Ravindranath NH, Rao S, Sharma N, Nair M,Gopalakrishnan R, Rao AS, Malaviya S, Tiwari R,Sagadevan A Munsi M, Krishna N andBala G. 2011. Climate change vulnerability profiles forNorth East India.Current Science, 101 (3): 384-394.

RTFSC, Report of the task force on Shifting Cultivation, 1983, Ministry of Agriculture. Government of India.

Scheuner ET, Makeshin F, Wells ED, Catrer PQ. 2004. Shorth term impacts of harvesting and burning disturbances on physical on physical and chemical characteristics of forest soils in western Newfoundland, Canada. European Journal of Forest Research, 123: 321-330.

Sharma SK and Chauhan R. 2011.Climate change research initiative: Indian Network for Climate Change Assessment.Current Science, 101 (3): 308-311.

Shimray N.2010.New Report on Climate Change Impact Assessment for India Highlights Key Concerns for Northeast. Available on: http://negreens.com/2010/11/18/new-report-on-climate-change-impact-assessment-for-india-highlights-key-concerns-for-northeast/

Singh, J. S. et al., in Eco-Development Guidelines and Model of Development of the Central Himalaya, Department of Botany, Kumaun University, Nainital, 1986, p. 48.

Singh NJ. 2008. Animal-habitat relationship in high altitude rangeland. PhD Thesis , University of TromsØ

Subba TB and Ghosh G.C. 2003. The Antrhopology of North East India. Published by Orient Longman Private Limited, New Delhi.

The Hindu. Climate change to have large-scale effect on NE biodiversity. 10 June 2011.

Yadav PK. 2013. Slash-and-Burn Agriculture in North-East. Expert Opinion on Environmental Biology,2:1Available online: http://dx.doi.org/10.4172/2325-9655.1000102

20

One Health and Climate Change

C. Latha and K. Vrinda Menon

Department of Veterinary Public Health, College of Veterinary and Animal Sciences, Mannuthy, Kerala Veterinary and Animal Sciences University Kerala, India

One Health is defined as "the collaborative effort of multiple disciplines working locally, nationally, and globally to attain optimal health for people, animals and the environment." The concept of 'One Health' dates back to the Greek physician Hippocrates (ca. 460 BCE–a.370) who in his text "On Airs, Waters and Places" stated:

Whoever wishes to investigate medicine properly, should proceed thus: in the first place to consider the seasons of the year, and what effects each of them produces, for they are not all alike, but differ much from themselves in regard to their changes.

Rudolf Virchow a German physician and pathologist formally recognized the connections between human and animal health, thus; *"Between animal and human medicine there is no dividing line, nor should there be. The object is different, but the experience obtained constitutes the basis of all medicine"*. In the 21st century, emerging disease outbreaks triggered the realization that animal, human and environment health are interrelated. Thus the concept of 'One Health' is an evolving, interdisciplinary way of approaching complex health issues by recognizing the inter-connectedness of human health, animal health and the environment. It encourages people to move beyond narrow, professional perspectives toward a more holistic view of health.

One Health is an approach for addressing complex public health issues mainly:

- Diseases especially vector borne
- Pandemics e.g. influenza
- Food safety
- Antimicrobial resistance (AMR).

This approach is integrated and holistic and considers challenges to health through a trans-disciplinary lens. The factors such as growth in human and livestock populations, climate change, and the globalization of trade in animals and animal products will likely intensify, creating favorable conditions for the emergence of new and more complex public health threats. Consequently, public health threats will continue to be a significant economic, social and public health burden worldwide. The concept has gained importance because:

- Over 75% of emerging worldwide infectious diseases are zoonotic, meaning they pass between humans and animals. Most of these originate in wildlife.

- We live in a changing environment in which humans and animals share the same environment. Animals support us through social and economic means This intimate relationship creates a heightening risk of disease transmission and pose complex challenges that call for cross-disciplinary leadership skills.

- The increasing global population creates a growing demand for sufficient and safe food supplies. Future food production will require broad based, interdisciplinary collaborations, sharing of information, and cooperative technology development which is the essence of One Health.

- Local communities are affected by global issues such as rabies control, vector borne diseases, and emergency preparedness, among many others. One Health recognizes the importance of 'thinking globally while acting locally.'

- Regardless of which of the many definitions of One Health is used, the common theme is collaboration across sectors. Collaborating across sectors that have a direct or indirect impact on health involves thinking and working across silos and optimizing resources and efforts while respecting the autonomy of the various sectors. To improve the effectiveness of the One Health approach, there is a need to establish a better sectoral balance among existing groups and networks, especially between veterinarians and physicians, and to increase the participation of environmental and wildlife health practitioners, as well as social scientists and development actors. In the past few years, One Health has gained significant momentum. It is now a movement and it is moving fast. The approach has been formally endorsed by most of the developed countries ,World Bank, World Health Organization (WHO), Food and Agriculture Organization (FAO), World Organization for Animal Health (OIE), United Nations System Influenza Coordination (UNSIC), various Universities. The current One Health movement is an unexpected positive development that emerged following the unprecedented Global Response to the Highly Pathogenic Avian Influenza. Since the end of 2005, there has been increasing interest in new international political and cross-sectoral collaborations on serious health risks.

20.1 Impact of Climate Change on Health

Over the last 50 years, human activities, particularly the burning of fossil fuels have released sufficient quantities of carbon dioxide and other greenhouse gases to trap additional heat in the lower atmosphere and affect the global climate. In the last 100 years, the world has warmed by approximately 0.75°C. Over the last 25 years, the rate of global warming has accelerated, at over 0.18°C per decade (Hadley research centre 2008). Sea levels are rising, glaciers are melting and precipitation patterns are changing. Extreme weather events are becoming more intense and frequent. Although global warming may bring some localized benefits, such as fewer winter deaths in temperate climates and increased food production in certain areas, the overall health effects of a changing climate are likely to be overwhelmingly negative. Climate change affects social determinants of health – clean air, safe drinking water, sufficient food and secure shelter.

20.2 Patterns of Infection

Climatic conditions strongly affect water-borne diseases and diseases transmitted through insects, snails or other cold blooded animals. Changes in climate are likely to lengthen the transmission seasons of important vector-borne diseases and to alter their geographic range. For example, climate change is projected to widen significantly the area of China where the snail-borne disease schistosomiasis occurs (Zhao, 2008). Malaria is strongly influenced by climate. The Aedes mosquito vector of dengue is also highly sensitive to climate conditions. Studies suggest that climate change could expose an additional 2 billion people to dengue transmission by the 2080s (Hales, 2008). The disease pattern is true in the Indian context also with the incidence of vector borne diseases increasing every year. The events in the last decade have taught us that we are vulnerable to fatal zoonotic diseases such as those caused by haemorrhagic fever viruses, influenza, rabies and BSE/vCJD. Future research activities should focus on solutions to these problems arising at the interface between animals and humans.

The resulting holistic approach to emerging infections links microbiology, veterinary medicine, human medicine, ecology, public health and epidemiology. As emerging 'new' respiratory viruses are identified in many wild and domestic animals, issues of interspecies transmission have become of increasing concern. The development of safe and effective human and veterinary vaccines is a priority. For example, the spread of different influenza viruses has stimulated influenza vaccine development, just as the spread of Ebola and Marburg viruses has led to new approaches to filovirus vaccines. Interdisciplinary collaboration has become essential because of the convergence of human disease, animal disease and a common approach to biosecurity. High containment pathogens pose a significant threat to public health systems, as well as a major research challenge, because of limited experience in case management, lack of appropriate resources in affected areas and a limited number of animal research facilities in developed countries. Animal models that mimic certain diseases are key elements for understanding the underlying mechanisms of disease pathogenesis, as well as for the development and efficacy testing of therapeutics and vaccines. An updated veterinary curriculum is essential to empower future graduates to work in an international environment, applying international standards for disease surveillance, veterinary public health, food safety and animal welfare. (Kahn *et al.,*2009). The work initiated in the different departments of Kerala Veterinary and Animal Sciences University is cited below:

20.3 Department of Veterinary Public Health

The department has undertaken four externally aided projects and around 30 post graduate thesis mainly focusing on food safety. The food borne pathogens and its isolations from different sources *viz.* animal and animal products, humans and environmental samples across the state has been carried out by the department through the various research projects. The main focus was to understand the potential sources of contamination and pathogens involved in foodborne diseases but effect with respect to seasonal variation has been studied to a lesser extent. The modules for HACCP implementation in the meat and milk processing industry to ensure food safety has been formulated by the department. The study on the incidence of brucellosis and some parasitic diseases had been undertaken by the department.

The salient observations of the studies are as cited below:

20.3.1 Isolation of *Listeria* spp from different sources

Samples	Strain	Year	Place	Percentage of positive samples
Chicken	*Listeria monocytogenes*	2011-12	Kerala (Thrissur)	1.01
Beef	*Listeria monocytogenes*	2011-12	Kerala (Thrissur)	0.816
Fish	*Listeria monocytogenes*	2011-12	Kerala (Thrissur)	0.819
Milk	*Listeria grayi*	2010-12	Kerala (Thrissur)	0.33
Soil	*L. welshimerrii*	2010-12	Kerala (Thrissur)	0.380
	L. innocua	2010-12		4.95
	L. monocytogenes	2010-12		0.380
Vegetables	*L. welshimerrii*	2010-12	Kerala (Thrissur)	1.960.7155
	L. innocua	2010-12		

20.3.2 Diagnosis of parasitic diseases and its public health significance

- Of the 300 samples collected from laboratories, six were positive for ascarid ova. The 30 samples from students revealed no parasitic ova.

- Of the 112 raw vegetable samples two samples of cabbage and one onion sample was positive for Ascarid ova.

- A survey was conducted on the socio-economic status, food habits, awareness of taeniosis/ cysticercosis, health problems and pig rearing practices in households distributed in three districts of Kerala state, viz., Kottayam, Ernakulam and Palakkad and the data was analysed.

20.3.3 Bacterial quality assurance of meat in processing plant

- The detailed microbiological analysis of animal carcasses i.e. Beef, pork and chicken was undertaken.

- The bacteriological quality was determined and critical control points were assessed.

- Various sanitizers were used on the carcasses at different concentrations to reduce the microbial load.

- A cost effective sanitizer was chosen and recommended for use.

- A shelf life study of meat and meat products were also undertaken and ways to improve the quality of these products was evolved.

20.4 Evaluation of Quality of Market Milk in Kerala

- A detailed survey of production, transportation, and distribution pattern of milk in co-operative societies in three districts viz. Palakkad, Ernakulam and Thrissur was carried out.

- Microbiological quality of 200 pasteurized milk samples collected from retail shops of Thrissur and Palakkad Districts were analysed.

- The samples belonging to Thrissur district had an overall mean TVC, CC, ECC, Psychrotrophic count, faecal streptococcal count and yeast and mould count of 5.08 ± 0.05, 2.89 ± 0.09, 0.53 ± 0.11, $5.30\pm.01$; 3.40 ± 0.14 and $1.89\pm0.08 \log_{10}$cfu/ml, respectively.

- *Escherichia coli*was isolated from 10 % of the samples belonging to Thrissur and 11% samples collected from Palakkad district. The isolates consisted of serotype O4, O60 rough strains, and untypable strains.

- Pseudomonas organism was isolated from 4 and 6 % of the samples from Thrissur and Palakkad. The isolates identified were *Pseudomonas aeruginosa, P. fluorescens,P.cepacia* and *P.putida.*

- *Bacillus cereus* was isolated from three samples obtained from Thrissur district and two samples belonging to Palakkad district.

20.5 Studies on Isolation of *E. coli* O157 in Meat

- Out of 105 beef samples analyzed, 62 samples were found positive for the presence of *E. coli* isolates.

- All the isolates were subjected to identification techniques for the detection of presence of Enterohaemorrhagic *E. coli* and among the 64 samples, eight samples were found positive for the *E. coli* O 157: H7.

- Forty six percentage of human stool samples analysed, were found to be positive for the presence of *E. coli* isolates.

- But none of the samples were found positive for Enterohaemorrhagic *E. coli.*

- Out of eighteen beef cutlet samples analysed, none was containing *E. coli* isolates.

20.6 Post Graduate Research since 1979

• **Prevalence studies of bacterial pathogens in foods, clinical samples (human and animal) and environment viz.**

a) *Escherichia coli*

b) *Listeria monocytogenes*

c) *Staphylococcus aureus*

d) *Bacillus cereus*

e) *Pseudomonas*

f) *Bacillus cereus*

g) *Yersinia*

h) *Salmonella*

i) *Aeromonas*

j) *Vibrio parahaemolyticus*

k) *Brucella* spp

• **Prevalence studies of parasitic agents *viz.***

a) Hydatidosis

b) Taeniosis

c) Cryptosporidiosis

d) Giardiosis

• **Comparative study of the water quality in industrial and non industrial areas in Kerala with seasonal variation.**

Water samples each from Eloor and Ollurkara panchayat well water samples in four seasons was assessed:

- Coliform count 1.32 ± 0.07 $\log_{10}$cfu/ml (Eloor) No seasonal difference 0.98 ± 0.07 log 10 cfu/ml (Ollurkara)- Highest during monsoons.

- E coli 0.54 ± 0.08 log 10 cfu/ml and 0.29 ± 0.05 log 10 cfu/ml at Eloor and Ollurkara respectively.

- Lead 0.56 ± 0.07 mg/l (Eloor) 0.09 ± 0.01 mg/ml (Ollurkkara) –highest during monsoons.

- Zinc Eloor higher in summer Ollurkkara – post monsoon.

- Cadmium high in monsoons in both places.

- Nitrate high during post monsoon in both places.
- Fluoride higher in summer and pre monsoon in Ollurkara.

Assessment of duck egg quality- seasonal variation:

- E coli higher during post monsoon in eggs.
- Coliform higher in post monsoon in eggs.
- APC both in shell and egg higher count.

Department of Veterinary Pathology

Rabies Prevalence- 2012-13

- Total of 331 suspected cases of animals have been presented (Dog-232, Cat-42, Cattle-17 and Goat-10)
- 135 cases were found to be positive (Dog-121, Cat-5, Cattle-6, Goat-3)

20.7 Conclusion

Planning for changes in the array of zoonotic diseases faced under global climate change is a challenge. A continued partnership between veterinary and medical together with public health professionals, biologists, and epidemiologists offers the best chance to make early determinations of emerging health threats especially due to climate change. No country is safe from invasion by pathogens unless an international effort is made to continually take action, share experiences and assist poor countries in controlling the impact resulting from climate changes and disease epidemics. The *One Health* approach with strong coordination and collaboration among all tiers of animal health and government agencies should come together to overcome this problem. The academia must focus their research to study the interactions between environmental health, animal health, and human health. We may not be able to stop climate change, but we can pay heed towards these climatic changes while planning control measures of any particular disease.

References

Hales, S. 2002. Potential effect of population and climate changes on global distribution of dengue fever: an empirical model. The Lancet, 360:830–834.

Hippocrates. 1978. Airs, waters and places. An essay on the influence of climate, water supply and situation on health. In: Hippocratic Writings. Lloyd G.E.R. ed. London, UK, Penguin p 138

Kahn, R.E, Clouser D.F, Richt J.A.2009. Emerging infections: a tribute to the one medicine, one health concept. ZoonosesPublic Health, 56:407-28.

Zhou, X.N. 2008. Potential impact of climate change on schistosomiasis transmission in China.American Journal of Tropical Medicine and Hygiene,78:188-194.

21

Advanced Biotechnological Tools with Potential Applications in Climate Change Studies in Farm Animals

T.V. Aravindakshan

Centre for Advanced Studies in Animal Genetics & Breeding
Kerala Veterinary and Animal Sciences University, Mannuthy
Thrissur – 680651, Kerala, India

The climatic changes and global warming may impact the economic viability of livestock production systems in a significant manner. The possible consequences of increased ambient temperature on the livestock include reduced feed intake, reduced weight gains, lower milk production and lower conception rates during summer periods. Though animals can, to some extent, adapt to higher temperatures with prolonged exposure, the production losses are likely to be exhibited in response to higher temperature events. However, potential direct and indirect impacts of climate change on livestock production have not yet been thoroughly explored. Changes in availability and quality of fodder have also been identified as one of the major impacts of climate changes leading to lowered productivity of livestock species. Climate change could also affect the distribution of vector-borne livestock diseases. These changes occur as a result of shifts in the geographical ranges of ticks, mosquitoes, flies and other vectors.

21.1 Contribution of Livestock Sector to Climate Change

The livestock sector plays an important role in climate change as 14.5 % of human-induced green house gas (GHG) emissions is contributed by livestock species whose GHG emissions has been estimated at 7.1 giga tonnes CO_2-eq per annum. Of the total emissions of this sector, beef production accounts for 41%, milk 20%, pig meat 9% and poultry egg and meat production for 8%

(FAO, 2013). In addition to the contribution of GHG, there is significant losses of energy, nitrogen and organic matter to the livestock through emissions of nitrous oxide (N_2O), methane (CH_4) and carbon dioxide (CO_2), the three main GHG emitted by the sector, resulting in reduced efficiency and productivity. It is, therefore, essential to reduce emissions through technologies and practices in order to achieve high production efficiency at animal and herd levels. They include the use of better quality feed and feed balancing to lower enteric and manure emissions, improved breeding and genetic selection, methods to alter rumen microbial profiles etc.

21.2 Biotechnological Tools Applicable to Climatic Change Studies

The novel 'omics' technologies such as genomics, transcriptomics, proteomics, and metabolomics are now allowing researchers to identify the genetics behind animals' response to stress on account of climatic changes. These 'omics' technologies enable a direct and unbiased monitoring of the factors affecting animal growth, production and reproduction and provide data that can be directly used to investigate the complex interplay between the animal, its metabolism and also the stress caused by the environment. Animal responses to stress are mediated via profound changes in gene expression which result in changes in composition of transcriptome, proteome and metabolome.

21.2.1 Genomics

Genome represents the complete set of instructions for making an organism. It contains the master blueprint for all cellular structures and activities for lifetime of the cell or organism. Genomics is a branch of science that studies the structure and function of genome and it includes determination of the entire DNA sequence of organisms and fine-scale genetic mapping. Genome studies have been greatly favoured by recent developments in molecular biology such as high throughput DNA sequencing, computing capabilities (infrastructures, databases, and algorithms), data generation and analysis. The number of livestock and fish genomes sequenced is increasing every year. The list includes cow, pig, horse, chicken, turkey, cod and tilapia with many species in the process of annotation. The rapid decline in the cost of whole genome sequencing through various next generation sequencing (NGS) platforms is an impetus for sequencing more agriculturally important species in the years to come. Such genome sequence data generated have a tremendous impact in livestock breeding and research, including production, reproduction, stress tolerance, resistance to diseases and better utilization of plant fiber in ruminants.

21.2.1.1 Reference Genome Assemblies

Currently, all the genome sequences available for animals are draft versions, which need further refining in order to develop fully annotated high quality reference genomes of these species. Such genome sequence data, annotated for functional elements and sequence variants, helps in the improvement of animal production to a great extent. The reference genome assembly serves as a framework to connect diverse sets of data related to genes, regulatory elements, polymorphisms, genotypes, and phenotypes. Initially, reference genome assembly was developed based on single consensus representation of the sequence. Subsequently, realizing the importance of single nucleotide polymorphisms (SNPs) and copy number variations (CNVs) among individuals of a species, such large scale variations are also being catalogued in reference genome assemblies. High quality genome assemblies are needed to:

- Generate complete and correct annotations of all functional elements.

- Align next generation sequence (NGS) information from individuals to discover single nucleotidepolymorphisms (SNPs), which can be used to characterize linkage disequilibrium, map genetic defects or quantitative trait loci (QTL) and identify signatures of selection.

- Correctly order SNPs to enable haplotype based imputation of genotypes.

- Map NGS transcriptome information to reliably annotated genes within the reference genometo correctly quantify gene expression.

- Map epigenetic information to understand the relationship between epigenetic modification and genes.

- Discover complex genomic variation within species.

- Accurately compare genomes of agricultural animal species with genomes of model organisms and human.

21.2.1.2 Functional Genomics

Functional genomics allows large scale gene function analysis with high throughput technology and incorporates interaction of gene products at cellular and organism level. The information coming from sequencing programs is providing enormous input about genes to be analysed. The availability of many animal genomes nowadays facilitates studying the functions of genes on a genome wide scale. The lack of information from other animal genomes will also be compensated in part by the availability of large collection of expressed sequence tags (ESTs)and complementary DNA (cDNA) sequences. The basic interest behind this EST projects is to identify genes responsible for critical functions. ESTs, cDNA libraries, microarray, serial analysis of gene expression

(SAGE) and high throughput transcriptome profiling using NGS are currently employed analyse global gene expression profiles in functional genomics programme. The understanding of the complexity of stress signaling and animal adaptive processes would require the analysis of function of number of genes involved in stress response.

21.2.2 Transcriptomics

Transcriptome is the set of all RNA molecules, including mRNA, rRNA, tRNA, and other non-coding RNA produced in a tissue at a given time and usually refer to the amount or concentration of each RNA molecule in addition to the molecular identities. It varies with external environmental conditions and reflects the genes that are being actively expressed at any given time. The study of *transcriptomics*, also referred to as gene expression profiling, examines the expression level of mRNAs in a given cell population, often using high-throughput techniques such as DNA microarray technology and next generation sequencing (NGS) followed by bioinformatics analysis of the sequence data. The use of next-generation sequencing technology to study the transcriptome at the nucleotide level is known as RNA-Seq. Numerous investigations show that heat response genes undergo transcriptional reprogramming during heat stress. It has been described that induction of specific stress genes, in the response against heat stress, also elicit defense against certain pathogens suggesting the existence of a complex signaling network that allows the animals to recognize and protect itself against pathogens and environmental stress. Elucidating the molecular mechanism that mediates the complex stress responses in animals is an important step to identify this cross tolerance and to develop improved variety of stress tolerant breeds.

21.2.2.1 Gene Expression Studies Using Microarray Analysis

Microarrays have been widely used for studying gene expression profiles. This technique can be adapted to simultaneously compare the expression profiles of genes involved in stress response among different genetic groups of livestock species in an attempt to identify and catalogue novel and differentially expressed genes under climatic stress. In the past, blotting techniques have used hybridization between complementary nucleic acid molecules for the detection of DNA and RNA. Microarrays can be considered as high throughput variants of the conventional Southern/Northern hybridization. To understand how different genes express in a particular situation such as in stress conditions, a transcriptome analysis need to be carried out. A transcriptome is the collection of all mRNAs produced in a cell under a given condition. It gives an indication of how many genes are expressed in a cell and how much each gene is expressed in a given condition. A microarray chip contains thousands of short oligonucleotides or cloned cDNA representing the mRNA sequences, spotted on a silicon chip which acts as the probe in the hybridization and the total mRNA isolated from the cells/tissues is labeled with fluorescent dyes and hybridized to the chips. The level of gene expression is determined by the intensity of fluorescence from each spot. It means that if a gene is expressed more, its mRNA copies will be more in the sample, which in turn results in a more intense hybridization with its probe on the chip resulting in increased intensity of fluorescence. Usually, the gene expression profiles of two samples are compared such as control versus treatment with a drug, normal versus exposure to stress etc. Total RNA from both control and test samples are isolated, cDNA prepared from both samples labeled using different fluorescent dyes (eg. red and green) and are allowed to hybridize to the probe on the array. Each spot on the array, corresponding to a gene, would hybridize to cDNA molecules tagged with red and green dye. The ratio of red:green fluorescence is a measure of the relative levels of expression of the gene in control and test samples.

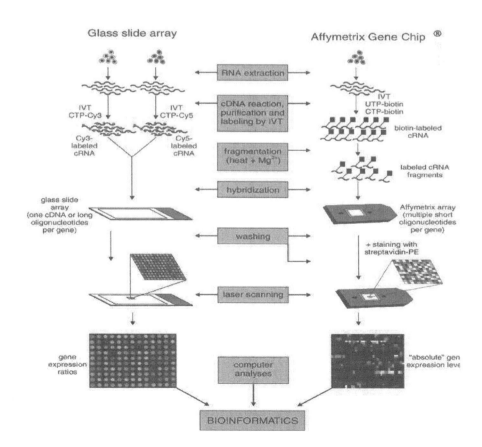

21.2.2.2 Gene Expression Studies Using Quantitative Real Time PCR (qPCR)

Real time PCR is a laboratory technique based on the PCR, which is used to amplify and simultaneously quantify a targeted DNA molecule. It enables both detection and quantification of one or more specific sequences in a DNA sample. The procedure follows the general principle of polymerase chain reaction; its key feature is that the amplified DNA is detected as the reaction progresses in *real time*, a new approach compared to standard PCR, where the product of the reaction is detected at its end. Two common methods for detection of products in real-time PCR are: (1) non-specific fluorescent dyes that intercalate with any double-stranded DNA, and (2) sequence-specific DNA probes consisting of oligonucleotides that are labeled with a fluorescent reporter which permits detection only after hybridization of the probe with its complementary DNA target.

21.2.2.3 Real-time PCR with Double-Stranded DNA-Binding Dyes as Reporters

A DNA-binding dye binds to all double-stranded (ds)DNA in PCR, causing fluorescence of the dye. An increase in DNA product during PCR therefore leads to an increase in fluorescence intensity and is measured at each cycle, thus allowing DNA concentrations to be quantified. However, dsDNA dyes such as SYBR Green will bind to all dsDNA PCR products, including non-specific PCR products (such as primer dimer). This can potentially interfere with or prevent accurate quantification of the intended target sequence.

1. The reaction is prepared as usual, with the addition of fluorescent dsDNA dye.

2. The reaction is run in a Real-time PCR instrument, and after each cycle, the levels of fluorescence are measured with a detector; the dye only fluoresces when bound to the dsDNA (i.c., the PCR product). With reference to a standard dilution, the dsDNA concentration in the PCR can be determined.

21.2.2.4 Fluorescent Reporter Probe Method

Fluorescent reporter probes detect only the DNA containing the probe sequence; therefore, use of the reporter probe significantly increases specificity, and enables quantification even in the presence of non-specific DNA amplification. Fluorescent probes can be used in multiplex assays—for detection of several genes in the same reaction—based on specific probes with different-coloured labels, provided that all targeted genes are amplified with similar efficiency. The specificity of fluorescent reporter probes also prevents interference of measurements caused by primer dimers, which are undesirable potential by-products in PCR. However, fluorescent reporter probes do not prevent the inhibitory effect of the primer dimers, which may depress accumulation of the desired products in the reaction. The method relies on a DNA-based probe with a fluorescent reporter at one end and a quencher of fluorescence at the opposite end of the probe. The close proximity of the reporter to the quencher prevents detection of its fluorescence; breakdown of the probe by the 5' to 3' exonuclease activity of the Taq polymerase breaks the reporter-quencher proximity and thus allows unquenched emission of fluorescence, which can be detected after excitation with a laser. An increase in the product targeted by the reporter probe at each PCR cycle therefore causes a proportional increase in fluorescence due to the breakdown of the probe and release of the reporter.

21.2.2.5 Applications of Real-Time Polymerase Chain Reaction

There are numerous applications for real-time polymerase chain reaction in the laboratory. It is commonly used for both diagnostic and basic research. Diagnostic real-time PCR is applied to rapidly detect nucleic acids that are diagnostic of, for example, infectious diseases, cancer and genetic abnormalities. The introduction of real-time PCR assays to the clinical microbiology laboratory has significantly improved the diagnosis of infectious diseases, and is deployed as a tool to detect newly emerging diseases, such as flu, in diagnostic tests. In research settings, real-time PCR is mainly used to provide quantitative measurements of gene transcription. The technology may be used in determining how the genetic expression of a particular gene changes over time, such as in the response of tissue and cell cultures to an administration of a pharmacological agent, progression of cell differentiation, or in response to changes in climate etc.

21.2.2.6 Real-time Reporters for Multiplex PCR

TaqMan probes, Molecular Beacons and Scorpions allow multiple DNA species to be measured in the same sample (multiplex PCR), since fluorescent dyes with different emission spectra may be attached to the different probes. Multiplex PCR allows internal controls to be co-amplified and permits allele discrimination in single-tube, homogeneous assays. These hybridization probes afford a level of discrimination impossible to obtain with SYBR Green, since they will only hybridize to true targets in a PCR and not to primer-dimers or other spurious products.

Normalization in Real time PCR : Normalization of amount of starting material and differences between tissues or cells in overall transcriptional activity is commonly done by using Internal Control Genes. Internal control genes are most frequently used to normalize the mRNA fraction. This internal control -

often referred to as a housekeeping gene -should not vary in the tissues or cells under investigation, or in response to experimental treatment.

Absolute Quantification: For absolute quantification of a target gene mRNA, a standard curve is first generated by plotting Ct values (Y-axis) against the log of the copy number (on the X-axis). For this known copy numbers (from 10^1 to 10^7 copies) of the same target (in a cloned plasmid) is used as templates for Real time PCR and the respective Ct values are recorded. A line representing the best fit is calculated for a standard curve using the least squares method of linear regression.

$$y = m x + b$$

*where $y = Ct$, $m = slope$, $x = log10$ template amount,*and b = y-intercept

From Fluorescence to Results

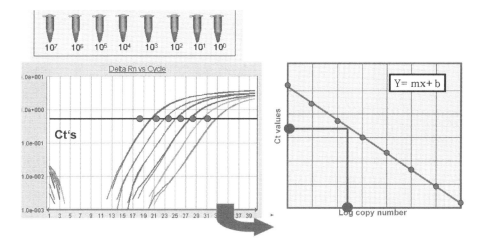

The amount of the template should be expressed in molecules. Convert the mass to molecules using the formula:

$$\frac{\text{Mass (in grams)} \times \text{Avogadro's Number}}{\text{Average mol. wt. of a base} \times \text{template length}} = \text{molecules of DNA}$$

For example, if a synthetic 75-mer oligonucleotide (single-stranded DNA) is used as the template, 0.8 pg would be equal to 2×107 molecules:

$$\frac{0.8 \times 10^{-12} \text{ gm} \times 6.023 \times 10^{23} \text{ molecules/mole}}{330 \text{ gm/mole/base} \times 75 \text{ bases}}$$

$$= 2.0 \times 10^7 \text{ molecules (copies) of DNA}$$

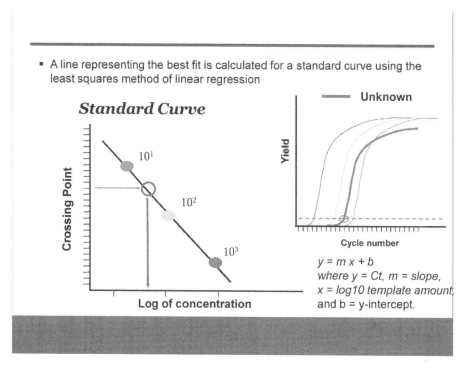

- A line representing the best fit is calculated for a standard curve using the least squares method of linear regression

Standard Curve

$y = m x + b$
where $y = Ct$, $m = slope$,
$x = log10$ template amount,
and $b = $ y-intercept.

For double-stranded templates, use 660 gm/mole/base. To generate a standard curve, prepare seven 10-fold dilutions, starting with 10^7 template copies and ending with 10 copies. For example, if the gene is cloned in a plasmid of 3400bp, instead of 75, write 3400.

21.2.2.7 *Relative Quantification*

Relative quantification relies onthe comparison between expression of a target gene versus a reference gene (to calculate ΔCt); and the expression of same gene in treatment sample versus control (calibrator) samples (to calculate $\Delta\Delta Ct$). Since relative quantification is the goal for most of the realtime PCR experiments, several data analysis procedures have been developed. The method used is referred to as Comparative Ct Method or $\Delta\Delta Ct$ method. This involves comparing the Ct values of the samples of interest (such as treated) with a control or calibrator (such as a non-treated) sample or RNA from normal tissue. The Ct values of both the calibrator and the samples of interest are normalized to an appropriate endogenous housekeeping gene.

The comparative Ct method is also known as the $2^{-\Delta Ct}$ method, where

$\Delta\Delta Ct = \Delta C_{t,sample} - \Delta C_{t,calibrator}$

$\Delta Ct = $ Ct value for the gene of interest $-$Ctvalue of the endogenous control

Here, $\Delta C_{t,sample}$ is the C_t value for any sample normalized to the endogenous housekeeping gene and $\Delta C_{t,calibrator}$ is the C_t value for the calibrator also normalized to the endogenous housekeeping gene (eg. 18 S rRNA gene

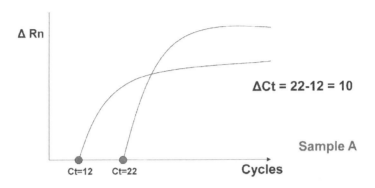

$\Delta Ct = 22\text{-}12 = 10$

Sample A

● **Endogenous Control (eg. 18S rRNA gene)**
● **Target Gene (eg. Hsp70 gene)**

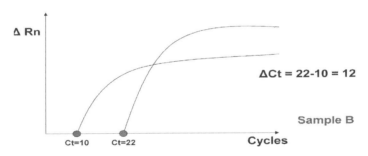

$\Delta Ct = 22\text{-}10 = 12$

Sample B

● **Endogenous Control (eg. 18S rRNA gene)**
● **Target Gene (Hsp70 gene)**

Firstlynormalisation to endogenous control is done and ΔCt value is found out.

$\Delta Ct = Ct$ (Target gene) $-Ct$ (endogenous control)

ΔCt (A) $= 22\text{-} 12 = 10$ & ΔCt (B) $= 22\text{-}10 = 12$

Then normalization to the calibrator is done (suppose in this case A is the calibrator and B is the sample)

$\Delta\Delta Ct = \Delta Ct$ (sample) $-\Delta Ct$ (caibrator)

$= 12\text{-}10 = 2$

Then RQ is found out by the formula, $RQ = 2^{-\Delta\Delta Ct} = 2^{-2}$

Conclusion: Target gene is down-regulated by four folds in the test sample B.

21.3 Transcriptome Analysis Using Next Generation Sequencing (NGS)

NGS encompasses several high-throughput approaches to DNA sequencing; it is also called Multiple Parallel Sequencing (MPS) or second-generation sequencing. These technologies have the capability to produce huge sequence data in a short period as compared to the Sanger method of sequencing. With the development of NGS platforms, sequencing of whole genomes, metagenomes, transcriptomes and metatranscriptomes became faster and cheaper enabling gene expression studies more feasible. Many NGS platforms differ in engineering configurations and sequencing chemistry. They share the technical paradigm of massive parallel sequencing via spatially separated, clonally amplified DNA templates in a flow cell. This design is very different from that of Sanger sequencing—also known as capillary sequencing or first-generation sequencing—that is based on electrophoretic separation of chain-termination products produced in individual sequencing reactions. DNA sequencing with commercially available NGS platforms is generally conducted with the following steps. First, DNA sequencing libraries are generated by clonal amplification by PCR *in vitro*. Second, the DNA is sequenced by synthesis, such that the DNA sequence is determined by the addition of nucleotides to the complementary strand rather through chain-termination chemistry. Third, the spatially segregated, amplified DNA templates are sequenced simultaneously in a massively parallel fashion without the requirement for a physical separation step. While these steps are followed in most NGS platforms, each utilizes a different strategy. NGS parallelization of the sequencing reactions generates hundreds of megabases to gigabases of nucleotide sequence reads in a single instrument run. This has enabled a drastic increase in available sequence data and fundamentally changed genome sequencing approaches in the biomedical sciences.

21.3.1 Template preparation methods

Since most imaging systems cannot detect single fluorescence events, amplification of DNA templates is required. The two most common amplification methods are emulsion PCR (emPCR) and solid-phase amplification.

Emulsion PCR: In emulsionPCR methods, a DNA library is first generated through random fragmentation of genomic DNA. Adapters are ligated on both ends and after size selection and denaturation of the DNA library, single-stranded DNA fragments (templates) are attached to the surface of beads which carry complementary oligonucelotides, and one bead is attached to a single DNA fragment from the DNA library. The beads are then compartmentalized into water-oil emulsion droplets. In the aqueous water-oil emulsion, each of the droplets capturing one bead is a PCR microreactor that produces amplified copies of the single DNA template.

DNA Colony Generation (Bridge amplification): Forward and reverse primers are covalently attached at high-density to the slide in a flow cell. The ratio of the primers to the template on the support defines the surface density of the amplified clusters. The flow cell is exposed to reagents for polymerase-based extension, and priming occurs as the free/distal end of a ligated fragment "bridges" to a complementary oligo on the surface. Repeated denaturation and extension results in localized amplification of DNA fragments in millions of separate locations across the flow cell surface. Solid-phase amplification produces 100–200 million spatially separated template clusters, providing free ends to which a universal sequencing primer is then hybridized to initiate the sequencing reaction.

21.4 NGS Chemistries

21.4.1 Pyrosequencing

Each time, a nucleotide gets incorporated in to the growing chain during complementary strand synthesis, a cascade of enzymatic reactions is triggered which results in a chemiluminiscent signal. The entire process involves five steps. In the first step, ssDNA templates amplified by PCR are hybridized to a sequencing primer and incubated with the enzymes, DNA polymerase, ATP sulfurylase, luciferase and apyrase, and the substrates, adenosine 5' phosphosulfate (APS) and luciferin. This is followed by addition of one of the four dNTPs. If the dNTP is complementary to the template, then the DNA polymerase catalyses the incorporation of dNTP on to the template. This results in the release of PPi (pyrophosphate) equivalent to the amount of dNTP incorporated. In the third step, ATP sulfurylase quantitatively converts PPi released in the earlier step to ATP in the presence of APS. This ATP oxidizes luciferin to oxyluciferin by luciferase leading to generation of visible light in amounts proportional to the amounts of ATP. This light is detected by charge coupled device (CCD) camera, which is subsequently analysed in a program. Each light signal is proportional to the number of nucleotides incorporated. Prior to the next round of incorporation, the nucleotides present in the mixture have to be degraded. This is carried out by apyrase which cleans the solution from all dNTP. The next cycle starts again with new nucleotide addition. The steps are as follows:

$(NA)_n$ + Nucleotide $(NA)_{n+1}$ + PPi

PPi + APS ATP + SO_4

ATP + Luciferin + O_2 AMP + Oxyluciferin + CO_2 + Light

This technique of pyrosequencing has been further developed into one of the fastest methods of sequencing, and is known as Roche 454 sequencing. Using this platform (GS FLX+) one can sequence 700 Mb of raw DNA in a 23 h run. Unlike conventional shotgun cloning, in this method, fragmented, size-fractionated and adapter-ligated DNA molecules are fixed to small DNA capture beads in water in oil emulsion. These DNA fragments in the beads are placed on a picotitre plate of well dimension ~44 μm. A mix of all the enzymes for Pyrosequencing is pre-packed in these wells. As in Pyrosequencing, nucleotides are flowed sequentially in a fixed order across the PicoTiterPlate device during a sequencing run. If there is a flow of nucleotide which is complementary to the template strand, the polymerase extends it which leads to light emission. Therefore, during each nucleotide flow, each of the hundreds of thousands of beads with millions of copies of DNA is sequenced in parallel. This technology is very useful for sequencing short read lengths in a high throughput manner.

21.4.2 Sequencing by reversible terminator chemistry

This approach uses reversible terminator-bound dNTPs in a cyclic method that comprises nucleotide incorporation, fluorescence imaging and cleavage. A fluorescently-labeled terminator is imaged as each dNTP is added and then cleaved to allow incorporation of the next base. These nucleotides are chemically blocked such that each incorporation is a unique event. An imaging step follows each base incorporation step, then the blocked group is chemically removed to prepare each strand for the next incorporation by DNA polymerase. This series of steps continues for a specific number of cycles, as determined by user-defined instrument settings. Sequencing by reversible terminator chemistry can be a four-colour cycle such as used by Illumina/Solexa. The Illumina Hi Seq 2500 platform can generate upto 600 Gb of sequence data in a sighle run of about 11 days.

Semiconductor Sequencing: This approach is based on the H+ ions released during incorporation of a nucleotide during DNA synthesis. The nucleotides are sequentially added in to the system and if incorporated in to the growing DNA strand, the release of H+ causes a drop in the pH of the wells, which is detected by the system and is converted to base calling. This technology has been employed in Ion Torrent and Ion proton NGS platforms of Life Technologies. In their Ion Proton platform with Ion PI chip sequencing data up to 10 Gb can be generated in a single run of 2-4 hours.

21.4.3 Proteomics

Proteomics is the large-scale study of proteins, particularly their structures and functions. The proteome is the entire complement of proteins produced by an organism or a tissue at a given time and it varies with time and distinct requirements, or stresses, that a tissue or organism undergoes. The response of animals to biotic as well as abiotic stress conditions is mediated through deep changes in gene expression which result in changes in composition of transcriptome, proteome and metabolome. Since proteins are directly involved in animal stress response, proteomic studies can significantly contribute to elucidate the possible relationships between variations in protein profile and animal stress response. The two techniques which are commonly used for proteome studies are two dimensional gel electrophoresis (2D GE) and mass spectroscopy.

i) Two-Dimensional Gel Electrophoresis(2D-GE)

2D-GE : is a method for the separation and identification of proteins in a sample by displacement in two dimensions oriented at right angles to one another. First, the proteins are separated according to charge (pI) by isoelectric focusing (IF) and subsequently according to size by SDS-PAGE. 2D GE has the capacity for resolutions of complex mixtures of proteins and is therefore widely used for analysis of thousands of gene products. The 2DGE profiles can be compared between control and test samples for identification of novel and differentially expressed proteins in the test samples, such as samples collected after exposure to a stress condition. The 2D gel electrophoresis in combination with spot picking and peptide sequencing enables identification of proteins including novel ones through bioinformatics analysis.

21.4.4 Metabolomics

The possibility monitoring a complete set of metabolites could largely improve the understanding of many physiological processes. This systematic study, defined as metabolomics, is intended to provide an integrated view of the functional status of an organism. The metabolome represents the downstream result of gene expression and is closer to phenotype than transcript expression or proteins. Extensive knowledge on metabolic flows could allow assessment of genotypic or phenotypic differences between species or among breeds exhibiting differential tolerance to climate change. Techniques such as gas chromatography (GC), liquid chromatography (LC), mass spectrometry (MS) and nuclear magnetic resonance spectroscopy (NMR) are used.

21.4.5 Metagenomics

Metagenomics is the study of metagenomes, genetic material recovered directly from environmental samples. Conventional sequencing begins with a culture of identical cells as a source of DNA. However, early metagenomic studies revealed that there are probably large groups of microorganisms in many environments that cannot be cultured and thus cannot be sequenced. These early studies focused on 16S ribosomalRNA sequences which are relatively short, often conserved within a species, and generally different between species. Many 16S rRNA sequences have been found which do not belong to any known cultured species, indicating that there are numerous non-isolated organisms. These surveys of ribosomal RNA (rRNA) genes taken directly from the environment revealed that cultivation based methods find less than 1% of the bacterial and archaeal species in a sample. Recent studies use massively parallel sequencing of fragmented, adapter ligatedmetagenome DNA to identify the diversity, relative abundance and functions of the microbial communities in different environmental samples. The main sources of metagenomes include soil samples, seawater samples, seabed samples, air samples, medical samples, ancient bones, human microbiome and gut microbiome

21.4.5.1 Rumen Metagenomics

The value of domesticated ruminants comes from their ability to convert forages into high quality, high protein products for human consumption through rumen fermentation. Variation of microorganism communities in the rumen of cattle is of great interest because of possible links to economically or environmentally important traits, such as feed conversion efficiency (Guan *et al.*, 2008; Zhou *et. al.* 2010) or methane emission levels (Zhou *et al.*, 2009; Hegarty, 1999) and more recently to fermentation of biomass for biofuel production (Hess *et al.*, 2011). Rumen microbes, which include representatives from all three domains of life: Eukarya, Bacteriaand Archaea (Deng *et al.*, 2008), provide nutrients, such as volatile fatty acids and bacterial protein to the host animal. A key challenge here is identifying rumen microbial profiles which are associated, and potentially predictive of these traits using methods that are of relatively low cost, so that large numbers of individuals can be profiled for testing associations with the above traits, and repeatable, e.g. similar profiles are generated for the same cow, sheep or goat when two repeat samples are taken at the same time. Methane (CH_4) emissions from ruminants can result in a significant loss of feed efficiency: up to a 12% loss of gross energy intake for forage-fed cattle and 4% for concentrate-fed cattle (Johnson and Johnson, 1995). Because methane is 25 times more potent than carbon dioxide as a greenhouse gas, methane emitted from ruminants amounted to 141 teragrams of CO_2 equivalents (Tg CO_2eq), accounting for 25% of total methane emissions from anthropogenic activities in the United States in 2008. To mitigate the negative impact on climate change and to improve feed efficiency, numerous strategies for reducing methane emission from ruminant livestock have been tested. Plant extracts, vaccines, ionophores, and dietary strategies have been evaluated for their efficacy in reducing ruminal methane emission. The ruminal microbes live in a symbiotic relationship with the host animal and secrete necessary lytic enzymes required for the microbial fermentation of plant polymers that are otherwise not digested by the host. This fundamental relationship allows the ruminant to unlock the nutrient energy in feeds unavailable to mankind. Rumen microbes not only help in digesting the food but also play essential role in shaping of host immune system, detoxification of toxic compounds present in the plant materials as well as limiting the pathogen survival by directly competing with available nutrients. Research on rumen microbes dates back to the 1950's and several microbes have been cultured in the laboratory under strict anaerobic conditions. However, culture methodology accounted for only less than 1 % of the rumen microbiome. With the advent of metagenomics, next generation sequencing and powerful bioinformatics tools, characterizing community microbial populations has become less cumbersome. Metagenomics approach bypasses the requirement of clonal cultures and has become an indispensable and widely affordable tool for studying of yet uncultivable microbes.

Rumen metagenomic studies are gaining importance in the context of climatic change scenario as rumen is a potential source of many novel organisms that are involved in plant fiber utilization, methane emissions and biomass utilization for biofuel production. Recently, identification of the novel cellulase genes have also been identified from termite guts, rabbit cecum, human gut.

21.4.5.2 Shotgun Metagenomics

Advances in bioinformatics, refinements of DNA amplification, and the proliferation of computational power have greatly aided the analysis of DNA sequences recovered from environmental samples, allowing the adaptation of shotgun sequencing to metagenomic samples. This provides information both on which organisms are present and what metabolic processes are possible in the community. This can be helpful in understanding the ecology of a community, particularly if multiple samples are compared to each other. Shotgun metagenomics also is capable of sequencing nearly complete microbial genomes directly from the environment.

21.4.5.3 Comparative Metagenomics

Comparative analyses between metagenomes can provide additional insight into the function of complex microbial communities and their role in host health.[1] Pair wise or multiple comparisons between metagenomes can be made at the level of sequence composition (comparing GC-content or genome size), taxonomic diversity, or functional complement. Comparisons of population structure and phylogenetic diversity can be made on the basis of 16S and other phylogenetic marker genes, or—in the case of low-diversity communities—by genome reconstruction from the metagenomic dataset (Simon *et al,* 2010) Functional comparisons between metagenomes may be made by comparing sequences against reference databases such as COG or KEGG, and tabulating the abundance by category and evaluating any differences for statistical significance (Mitra *et al.,* 2011). This gene-centric approach emphasizes the functional complement of the *community* as a whole rather than taxonomic groups, and shows that the functional complements are analogous under similar environmental conditions(Simon *et al.,* 2010). Consequently, metadata on the environmental context of the metagenomic sample is especially important in comparative analyses, as it provides researchers with the ability to study the effect of habitat upon community structure and function.

21.4.5.4 Metatranscriptomics

Metagenomics allows researchers to access the functional and metabolic diversity of microbial communities, but it cannot show which of these processes are active. The extraction and analysis of metagenomic mRNA (the metatranscriptome) provides information on the regulation and expression profiles of complex communities. Because of the technical difficulties (the short half-life of mRNA, for example) in the collection of environmental RNA there have been relatively few *in situ*metatranscriptomic studies of microbial communities to date. While originally limited to microarray technology, metatranscriptomcs studies have made use of direct high-throughput cDNA sequencing to provide whole-genome expression and quantification of a microbial community as first employed by Leininger *et al.* (2006) in their analysis of ammonia oxidation in soils.

21.5 Application of Genomic Tools and Genomic Selection for Mitigating Climatic Stress

Due to intensive crossbreeding for enhanced production and import of exotic breeds, metabolic heat production has increased in animals which have augmented their susceptibility to heat stress. Because of sporadic and short-term nature of acute heat stress in temperate selection countries, animal survival in hot and humid condition largely depends on managemental practices such as cooling methods, scientific feeding and breeding strategies adopted to cope up with the economic loss due to heat stress. However, under the chronic year round stress conditions, the effect of thermal stress cannot be totally eliminated by management practices alone. Moreover, these corrective managemental practices are expensive, and thus, not economically and technically feasible. Thus, breeding for heat tolerance will be the most appropriate cost-effective tool for mitigating climatic stress (Renaudeau *et al.,* 2012). Different breeds of animals are originated through the process of natural selection based on the adaptability to varying climatic conditions existing in different geographical location. The breeds evolved in tropical climatic conditions are not only thermotolerant, but also able to survive, grow and reproduce in poor nutrition, high parasite and disease pressure. It is true that the performance of indigenous animals is generally lower than that of exotic breeds and crossbreds. Thus, the use of indigenous breeds would be appropriate in very hot and humid conditions where as the identification of well adapted individuals among crossbreds may be an alternative solution in other areas. Thus, genomic tools and selection procedures can be applied either to select animals in stressed conditions or for introgression of heat adaptation genes from available indigenous germplasm pool (Collier et al., 2008). Unlike growth and production traits, tolerance traits are poor in their inheritance and difficult to measure under field conditions. In cattle slick hair coat gene plays an important role in heat tolerance (Oslone et al., 2003). In poultry, traits favoring heat tolerance such as naked–neck gene, QTL for body temperature in the Japanese quail (Minvielle et al., 2005) and in chicken (Nadaf et al., 2010) has been identified. The breeding strategy in general, may favour either selecting animals adapted to each environment of production or choosing robust breeds that are able to maintain a high level of production in most conditions of production prevailing in the region. Since whole genome sequences are now available for almost all domestic species, breed wise availability of complete genomic data will help us to identify genetic markers associated with production traits. Deducing genomic breeding values for production in hot climate with the help of molecular techniques would enable the selection for heat tolerance as genomic breeding value can be estimated for any animal if DNA is available provided reference population and candidates are to be close enough and cost effective genotyping facility are on hand. The functional genomics approach by identification of genes that are up regulated or down regulated during thermal stress shed light about effects of heat stress on production. The highly conserved heat shock responses (HSR) in animals are now identified to be mediated via cascade of gene action involving hundreds of genes differentially regulated through common pathways. This has now emerged as a new technology called expression QTL (eQTL) where transcriptomic data are integrated in to molecular markers which is then used for selection process (Collier et al., 2006). According to the high genetic variability between and within breeds, there is no doubt that it is feasible to select for tolerance to heat stress, although all the practical problems have not yet been solved.

21.6 Transgenic Technology

Cattle from zebu breeds are better able to regulate body temperature, experience less severe reductions in feed intake, growth rate, milk yield and reproductive function in response to heat stress (Hansen, 2004). Once specific genes responsible for thermotolerance in zebu have been identified or mapped, breeding strategies such as marker- assisted selection and transgenics can be applied to further the exploitation of the zebu genotype for cattle production systems (Hansen, 2004). Transgenic approach is experimented in various species in different countries. In pigs, there is an example of the *'Enviro pigs'* (trademarked) developed at Guelph in Canada which has been genetically modified to excrete 60% less phosphorus. Similarly many other genes related to biotic and abiotic stress tolerance could be exploited for transgenesis in the futureto develop improved variety of stress tolerant livestock species.

21.7 Advanced Reproductive Technologies

Biotechnological tools will provide new ways to evaluate reproductive potential in climatic change. Some of the advanced technologies of interest include semen sexing, embryo transfer, methods for early diagnosis of pregnancy and embryonic stem cell technology to revolutionize the livestock production in future.

21.8 Exploiting the Genetic Potential of Native Breeds

During their separate evolution from *Bostauus*, zebu cattle (*Bosindicus*) have acquired genes that confer thermotolerance at physiological and cellular level (Hansen, 2004). Cattle from zebu breeds are better able to regulate body temperature in response to heat stress than are cattle from *B. taurus* breeds. Moreover, exposure to elevated temperature has less deleterious effects on cells from zebu cattle than on cells from European breeds (Muhammed and Aravindakshan, 2012). Superior ability for regulation of body temperature during heat stress is the result of lower metabolic rates as well as increased capacity for heat loss (McManus *et al.*, 2009). Strong investments are needed to identify the unique traits in the indigenous germplasm. Economic evaluation of adaptation technologies, appraisal of farmer's response under different resource conditions and change in regional, national, and international policies relating to farm level adaptation and assessment of the investment requirements for coping with challenges of climate change are important.

21.9 Developing Suitable Breeding Programmes

The zebu genotypes have been utilized in crossbreeding systems to develop cattle for beef and dairy cattle production systems in hot climates but success has been limited by other unfavourable genetic characteristics of these cattle. These include poor meat tenderness, low milk yields and lactation persistency, a long pre pubertal period, short duration of estrus and poor temperament. An alternative scheme is to incorporate specific thermo-tolerance genes from zebu cattle into crossbreds while avoiding undesirable genes. One of the most important interventions needed is for the maintenance of the global animal genetic diversity. The appropriate strategy for breeding programs would be to set suitable selection goals, which match the production system rather than ambitious performance objectives that cannot be reached under the prevailing environment. The Animal Biotechnology Working Group (ABWG) of EU-US Task Force on Biotechnology Research (established in 1990)stresses the urgent need to provide food and resources for the predicted global population of 9 billion by 2050 against a backdrop of environmental challenges, as climate change and water scarcity, in a cost effective and sustainable manner. Towards that aim we need to define critical livestock phenotypes and deploy the latest genomics and biotechnological approaches to control these phenotypes. Recent innovations such as high throughput sequencing provide new opportunities to understand and exploit genetic mechanisms to increase productivity and control diseases. However, the large datasets generated using these technologies provide novel challenges in data storage, processing and analysis.

21.10 Priorities in Genomics Research to Improve Animal Health and Production

Farm animal and poultry exhibit a great deal of intra-species genetic variability owing to the presence of a large number of distinct breeds or genetic groups with a variety of different phenotypes. Analysis and identification of these genetic variations (SNPs and CNVs) using high throughput screening is extremely important in farm animals and poultry. The use of high density SNP chips for screening genetic variations is getting wide acceptance now-a-days. It is also possible to carry out experimental crosses to detect these functional genetic variations. The vast genetic diversity observed in farm animals and poultry are developed through natural adaptations to various adverse conditions such as extreme climate, pathogenic organisms and other stress conditions. Such inherited phenotypes need to be identified and evaluated in order to develop the 'ideal animal of the future' for being 'robust, adapted, andhealthy' and 'producing a safe and nutritious food.' The future of livestock genomics to a great extent depends on establishment of animal resources and infrastructure for research and applications on a global scale.

References

Collier, R.J., Stiening, C.M., Pollard, B.C., VanBaale, M.J., Baumgard, L.H.,Gentry, P.C. and Coussens, P.M. 2006. Use of expression microarrays for evaluating environmental stress tolerance at the cellular level in cattle J. Anim. Sci. 84:1-13.

Deng W, Xi D, Mao H, Wanapat M. 2008. The use of molecular techniques based on ribosomal RNA and DNA for rumen microbial ecosystem studies: a review. MolBiol Rep.35: 265-274.

FAO 2013. Tackling climate change through livestock. A global assessment of emissions and mitigation opportunities.

Guan L.L; Nkrumah, J.D; Basarab, J.A and Moore, S.S. 2008. Linkage of microbial ecology to phenotype: correlation of rumen microbial ecology to cattle's feed efficiency. FEMS Microbiol Lett. 288:85–91.

Hansen, P.J. 2004.Physiological and cellular adaptations of zebu cattle to thermal stress Anim. Reprod. Sci. 82: 349-360.

Hegarty, R. 1999. Reducing rumen methane emissions through elimination of rumen protozoa. Aust J Agric Res. 50:1321–1328.

Johnson, K. A. and Johnson, D. E. 1995. Methane emissions from cattle. J. Anim. Sci.73: 2483–2492.

Leininger, S; Urich, T; Schloter, M; Schwark, L; Qi, J; Nicol, G. W; Prosser, J. I; Schuster, S.C. and Schleper, C. 2006.Archaea predominate among ammonia-oxidizing prokaryotes in soil". Nature 442 (7104): 806–809.

Mc Manus,C., Prescott, E., Paludo, G.R., Bianchini, E., Louvandini, H. and Mariante, A.S. 2009. Heat tolerance in naturalized Brazilian cattle breeds Liv. Sci.120: 256-264.

Minvielle, F., Kayang, B., Miwa, M., Vignal, A., Gourichon, D., and Ito, S. 2005. Microsatellite mapping of QTL affecting growth, feed consumption, production, tonic immobility and body temperature of Japanese quail. BMC genomics 6:87-96.

Muhammed, E.M. and T.V. Aravindakshan. =2012. Physiological and haematiological response to heat stress in Vechur, Kasargode and Crossbred cattle of Kerala. Proceedings of the Indian Veterinary Science Congress- 2012. 224-227.

Nadaf, J., Pitel, F., Gilbert, H., Duclose, M. J., Porter, T. E., Simon, J and Bihan, E. 2009. QTL for several metabolic traits map to loci controlling growth and body composition in an F2 intercross between high and low growth chicken lines Physiological genomics 38:241-149.

Oslon, T.A., Lucena, C., Chase, C.C. and Hammond, A.C. 2003. Evidence of a major gene influencing hair length and heat tolerance in Bostarus cattle J. Anim. Sci. 81: 80-90.

Renaudeau, D., Colllin, A.,Yahav, S., Basllio,V., Gourdine, J. L. and Collier, R.J. 2011. Adaptation to hot climate and strategies to alleviate heat stress in livestock production Anim. 6: 707-728.

Sejian, V., Naqvi,S.M.K., Ezeji, T., Lakrits, J. and Lal, R. 2012. Environmental stress and amelioration in livestock production. Springer, New York. pp568.

Silverstein, J; Lequarre, A.S; Matukumalli, L. and Rohrer, G. 2011.EU-US Animal Biotechnology Working Group. Strategic Priorities to Realize the Promise of Genomics to Animal Health, Well being & Production. Report on a workshop held in Washington DC 8 10 November 2011.

Simon, C. and Daniel, R. 2010. MetagenomicAnalyses: Past and Future Trend. Applied and Environmental Microbiology, 77(4): 1153–1161.

Suparna, M; Rupek, P; Richter, R.D; Urich, T; Gilbert, J.A; Meyer, F; Wilke, A; Huson, D.H. 2011. Functional analysis of metagenomes and metatranscriptomes using SEED and KEGG.BMC Bioinformatics. 12 Suppl 1: S21. doi:10.1186/1471-2105-12-S1-S21. ISSN 1471-2105. PMC 3044276. PMID.

Zhou, M; Hernandez-Sanabria E. and Guan, L.L. 2009. Assessment of the microbial ecology of ruminal methanogens in cattle with different feed efficiencies. Appl Environ Microbiol. 75:6524.

Zhou, M; Hernandez-Sanabria, E. and Guan, L.L. 2010. Characterization of variation in rumen methanogenic communities under different dietary and host feed efficiency conditions, as determined by PCR-denaturing gradient gel electrophoresis analysis. Appl Environ Microbiol. 76:3776–3786.

22

Livestock Insurance – Present and Future

Kolli N.Rao, Senior Advisor

International Reinsurance & Insurance Consultancy & Broking Services Pvt. (IRICBS) Nirmal Bhavan, Nariman Point, Mumbai, India
G. SrinivasaRao, Director eeMAUSAM, Weather Risk Management Services Private Limited, Hyderabad, India

Over the past decades climate change and climate related disaster risks have become more important than ever. The effects of climate change are unprecedented today and thus pose serious risks for human wellbeing, livelihoods and life-supporting systems. They also present serious challenges to society, in addition, increased climate variability significantly undermine the socio-economic development and environment. The consequences of climatic change, such as increase in frequency and extent of droughts, floods, cyclones, mudslides and avalanches have impacts on all sectors of economy including water, agriculture, fisheries, health, forestry, transport, tourism and energy sectors. Amongst these sectors, the agricultural sector is one of the most vulnerable sectors to impacts of climate change, which can have multiplicative effects on other sectors, as well as the economy and overall human being. The projected climate change towards warming may affect not only the crop cultivation but also on agro-climatic conditions for growing pasture vegetation, forming of feed stocks in pastures and conditions for grazing livestock. Favourable conditions for grazing livestock are determined, on the one hand, by sufficient amount of forage in pastures and on the other, by the extreme weather conditions that restrict the use of these fodders.

Rural farmers experience the problems of climate change to a greater extent than that of the urban populace of the country because the consequences of climate change greatly affects the productivity and productive capacity of agricultural producers. Restricted access to resources, and poverty as a result of lower income of farmers have a negative effect on the overall economic development of a country and leads to an increase in the migration of rural population to urban areas.

22.1 History of Cattle Insurance in India

1. Agriculture provides only seasonal employment and so the majority of those engaged in agriculture are unemployed or underemployed except during the crop season. Due to non-availability of irrigation facilities most of the cultivated area is mono-crop area. Most of the farmers still adopt traditional farming practices and so crop yields are poor even during seasons when climate is favourable. The average farmer, who depends exclusively on agriculture for subsistence, continues to be poverty-ridden. Hence, the Government of India has encouraged subsidiary occupations for small, marginal, landless farmers by providing technical training as well as funds for capital and working expenditure and by providing easy access to efficient marketing facilities.

2. Among such subsidiary occupations dairy farming occupies a very important place. In some areas where soil is not suitable for profitable crop production, dairy farming is a whole-time occupation for all people, and especially for landless labourers. This industry does not require much capital and whatever capital is required is provided by way of loans by banks and by way of subsidies to poor farmers by different government agencies. In a few areas this occupation has started to be commercially viable. Nonetheless, milk production in India has been a subsidiary occupation of farmers since time immemorial. Cattle in India have been mostly kept to meet the agriculture draught power requirements of the farmers. Therefore, over the years, the draught power qualities in these animals took over their milk production capacities,milch buffaloes are kept because of their high-fat content in milk, and higher milk yield as compared to indigenous cows.

3. As a result of the pioneering efforts of the General Insurance Corporation (GIC) and its (erstwhile) four subsidiary companies, financial institutions which finance purchase of cattle by dairy farmers, the authorities of the Rural Development Projects, as well as all the well-organized dairy co-operatives in the country realized the valuable support that cattle insurance can provide to their dairy development programs. If the milch animal

financed by them dies or becomes disabled due to accident or disease or natural calamities like floods or cyclone, then all their efforts are set at naught unless the animal is insured. In such a context, the uninsured farmer's indebtedness increases, whereas the insured farmer can purchase another animal promptly with the insurance claim proceeds. It is in this sense that cattle insurance is considered an essential input of great strategic importance to the development of the dairy industry in India.

4. Prior to nationalization of the General Insurance Industry in 1973, some insurance companies made efforts to introduce cattle insurance on a very limited scale, but due to adverse claims experience, these efforts were abandoned. The premium rate was over 6%. Cattle insurance was introduced on a regular countrywide basis by the General Insurance Industry in the year 1974, during which year about 30,000 animals were insured. The main aim was to effectively protect the cattle owners against financial loss due to death of or accident to the cattle. The insurance company's liability to pay compensation to the insured is subject to receipt of satisfactory proof of the cause of loss etc., but it shall not exceed the sum insured.

22.1.1 Market agreement on cattle insurance

5. GIC decided to introduce a Market Agreement on cattle insurance (excluding SFDA / MFAL cattle) for standardizing premium rates and other terms and conditions. The Market Agreement came into force from 01/04/1976. Under the Agreement uniform premium rates, policy terms and conditions and procedures were adopted by all the subsidiaries. Basic premium rate was 4% gross. A Central Cattle Committee consisting of representatives of GIC and companies was set up (a) to ensure smooth implementation of the Market Agreement and (b) to fix special rates and terms for cattle within the purview of major co-operative dairies and those owned by well-managed dairy farms.

6. For the purpose of the Market Agreement on cattle insurance, the word Cattle refers to: (a) Milch cows and buffaloes, (b) Calves/heifers, (c) Stud bulls and (d) Bullocks and male buffaloes (castrated) whether indigenous or exotic or cross-bred. Cattle insurance business is transacted directly by all the four companies, viz., National Insurance Co. Ltd., Kolkata, the New India Assurance Co. Ltd., Mumbai, the Oriental Insurance Co. Ltd., New Delhi, and United India Insurance Co. Ltd., Chennai, GIC's role was limited to overall co-ordination.

7. At present besides the public sector general insurance companies, many insurance companies from private sector are also active in livestock insurance.

22.1.2 Cattle Insurance

The scheme is for underwriting cattle insurance business with in India

1. Scope

The word 'Cattle' refers to:

(i)   Milch Cows and Buffaloes

(ii) Calves/Heifers

(iii) Stud Bulls

(iv) Bullocks (Castrated Bulls) and Castrated Male Buffaloes, whether indigenous, exotic or cross-bred.

Note

- Exotic animal means an animal, whose both parents are of foreign breed. This includes animals born in India as well as those born abroad.

- Cross-bred animal means an animal, one of whose parents is of foreign breed.

2. Age Group

Animals of age in years shown below shall be accepted under the Standard Insurance Scheme.

I	a) Milch cows (Indigenous / Cross-bred / Exotic)	2 years (or age at first calving) to 10 years.
	b) Milch Buffaloes	3 years (or age at first calving) to 12 years.
	c) Stud Bulls	(Cow / Buffalo species) 3 years to 8 years or earlier age at sexual maturity.
	d) Bullock	(castrated bulls and 3 years to 12 years castrated male buffaloes)
II	Indigenous, cross-bred and exotic female calves / heifers.	From 4months up to the date of first calvingor minimum age as in (I) (a) and (b).

Note: At the time of acceptance, the maximum age should not exceed the prescribed age limit (in case of annual policy).

3. Valuation and Sum Insured

(i)   The Market value of cattle varies from breed to breed, from area to area and from time to time. The examining Veterinarians' recommendations shall be considered as the proper guide for acceptance of insurance as well as for settlement of claims. Wherever possible, high valued animals shall be inspected by Insurer's representative.

(ii)   Sum Insured will not exceed 100% of Market Value.

Indemnity

(i)   Sum Insured or market value prior to illness whichever is less. Care should be taken to assess the value of dry animals.

(ii)   In case of scheme animals, the policy is issued as agreed value policy hence claim will be settled for 100% of Sum Insured, subject to terms, conditions and exclusions.

4. Premium Rates (Annual)

For Milch Cows / Milch Buffaloes / female Calves / Heifers / Stud Bulls / Bullocks and Castrated Male Buffaloes:

Species		Covers	
		Death	PTD Extra
1 Scheme Animals (Indigenous / cross–bred)		2.25% (net)	0.85% (net)
2 Non-Scheme Animals (Indigenous /cross–bred)		4.00% (gross)	1.00% (gross)
3 Exotic Animals		4.00% as above plus 2.00 % extra	1.00%

Scheme Animals

For the purpose of concessional insurance rating should cover animals financed/subsidised under various Central/State Government Programmes/Schemes besides IRDP such as:

(i)   Development of Women & Children Schemes in Rural Areas (DWCRA),

(ii)   Schemes for Scheduled Castes / Scheduled Tribes operated by the State Governments,

(iii) Special Livestock Development Programmes

(iv)   Mini-dairy units, and

(v)   Support to training and implementation programme for women, etc.

Malus

The premium rates indicated are the minimum to be charged. In case of adverse claims ratio, following Malus system should be adopted:

% Claim Ratio	Percentage of Malus
100 to 110	20%
111 to 130	33%
131 to 160	60%
161 to 200	100%
Above 200	Premium rate to be adjusted such that Claim Ratio would appear as 90% for the rate

Minimum Premium: Rs. 50 per Policy.

Short period Rates: The short period rates are as follows:

Period (not exceeding)	Proportion of Premium
1 week	1/8 of annual rate
1 month	1/4 of annual rate
2 months	3/8 of annual rate
3 months	1/2 of annual rate
4 months	5/8 of annual rate
5 months	3/4 of annual rate
8 months	7/8 of annual rate
Exceeding 8 months	Full annual premium

5. Discount

(A) Group Discount (For Non-Scheme Only)

A group discount is allowed in case of single /partner ownership and/or single source of premium payment and covered under a single policy (not applicable to bank business):

No. of Animals	Rate of Discount
5 – 10	2. 5%
11 –15	5.0%
16 – 25	7. 5%
26 – 50	10.0%
51 – 100	12.5%
101 - 500	15.0%

(If the Group is over 500 and / or very large, the discount not exceeding 20% may be given with HO approval)

(B) Long Term Discount (For Non- Scheme Animals):

Long Term Discount on	Non – Scheme
3 and 4 years Policies	15%
5 years and above	25%

(i)   Full premium has to be paid in advance

(ii)   No refund of premium will be allowed even if claim arises in the earlier years.

(iii) Age limit should be as per Para 2 above.

(C) Long Term Discount (For Scheme Animals)

For Scheme animals, the basic premium for 3 years policy would be 4.80%. For any additional year over 3 years, the premium would be @ 1.60 % per year. (This rate is applicable only in case of Long Term Policies). If Permanent Total Disability (PTD) cover has to be extended the long term rate applicable for a three year period would be 6.60%. The working would be as follows:

Rate for basic Cover for 3 years   4.80%

Rate for PTD cover for 3 years   1.80%

6.60%

(i)   Full premium has to be paid in advance;

(ii)   Adjustment of premium: In case of premature death of cattle before the expiry of policy period, the premium for the balance period after adjusting the period up to the year in which the animal dies, would be allowed as credit to be adjusted against premium for the new animal acquired by the Insured. *Explanation:* If the animal dies in the first year of the Policy, the premium @ 2.25% should be deducted from the total premium paid by the insured i.e., 4.80% for the purpose of adjustment. Similarly, if the animal dies during the second year of the policy, full premium@ 2.25% for two years should be deducted. Since no premium will be left for adjustment if the animal dies during the third year of the policy, there is no question of any adjustment of premium;

(iii) Age limit should be as per Para 2 above.

6. Insurance Coverage

Underwriting offices shall use standard policy wordings for Cattle Insurance as per the existing specimen clause. The policy shall give indemnity only for death due to:-

(i)   Accident (Inclusive of fire, lightning, flood, inundation, storm, hurricane, earthquake, cyclone, tornado, tempest and famine).

(ii)   Diseases contracted or occurring during the period of this policy.

(iii) Surgical Operations.

(iv)   Riot and Strike.

(v)   The Policy can also be extended to cover PTD on payment of extra premium:

  a) PTD which, in the case of Milch Cattle results in permanent and total incapacity to conceive or yield milk .

  b) PTD which in the case of Stud Bulls results in permanent and total incapacity for breeding purpose.

  c) PTD in case of Bullocks, Calves/Heifers and Castrated male buffaloes results in permanent and total incapacity for the purpose of use mentioned in the proposal form.

22.1.3 Transit cover

(a)   For Scheme animals, No extra premium to be charged for transit of animal from the place of purchase to the place of stabling. For non-scheme animals, the distance is not exceeding 80 kms.

(b)   In case of transfer of animal during currency of policy, transit cover can be extended to the new owner without any additional premium in case the transit is within 80 kms.

(c)   In case the transit is for more than 80 kms, an additional premium of 1% shall be charged and such transit shall only be by road or rail and not by foot.

7. Exclusions

(A) Common Exclusions

i)   Malicious or willful injury or neglect, overloading, unskillful treatment or use of animal for purpose other than stated in the policy without the consent of the Insurer in writing.

ii)   Accidents occurring and /or Disease contracted prior to commencement of risk.

iii) Intentional slaughter of the animal except in cases where destruction is necessary to terminate incurable suffering on humane consideration on the basis of certificate issued by qualified Veterinarian or in cases where destruction is resorted to by the order of lawfully constituted authority.

iv)   Theft and clandestine sale of the insured animal.

v)   War, invasion, act of foreign enemy, hostilities (whether war be declared or not), civil war, rebellion, revolution, insurrection, mutiny, tumult, military or usurped power or any consequences thereof or attempt threat.

vi)   Any accident, loss, destruction, damage or legal liability directly or indirectly caused by or contributed to by or arising from nuclear weapons.

vii)   Consequential loss of whatsoever nature.

viii)   Transport by air and sea.

ix)   Any non-scheme claim arising due to diseases contracted within 15 days from the date of commencement of risk are not covered.

(B) Specific Exclusions

i)   Pleuropneumonia in respect of Cattle in Lakhimpur and Sibasagar Districts and newly carved out districts out of these two districts of Assam.

ii)   All the claims received without ear tag.

8. Additional Policy Conditions

(i)   The provision of 15 days waiting period should be included as policy condition No. 8. The wording may be as under:

"The Insurer is not liable to pay the claim in the event of death of insured animal due to disease occurring within 15 days from the commencement of risk."

(ii)   The provision of No Tag No Claim should be included as policy condition No. 9. The wording may be as under:

"In the event of death of animal/s covered under the policy, claim/s shall not be entertained unless the ear tag/s are surrendered to the insurer. In the event of loss of ear tag/s, it is the responsibility of the Insured to give immediate notice to the Insurer and get the animal retagged."

9. Veterinary Examination

i.   The report of a qualified Veterinarian giving the age, identification marks, health of the cattle must be obtained for each proposal. Such Certificate should be obtained in the prescribed format duly filled in all respects.

ii. Wherever qualified Veterinarians are not available, the underwriting office with HO approval may accept Certificate of Health issued by Livestock Inspectors who are diploma holders. Such certificates will be valid for acceptance of proposals only.

iii. Fresh Veterinary Examination is not necessary for renewal of Cattle Insurance Policy if renewal is made on or before the date of expiry of the policy provided the animal is within insurable age.

iv. Insurers may pay the Veterinarian a fee of Rs.15/- and Rs.5/- towards examination and tagging per large animal of the proposal which has been accepted by the insurer.No Veterinary examination fees and Tagging charges are payable for scheme animals.

v. As regards Veterinary examination fees following the death of animals or for post mortem if required, the same will be payable by the insurer. If qualified Veterinarians are not available in the area concerned for certifying cause of death, the insurer may make suitable alternate arrangements at their discretion. The maximum fees payable may be as under. Ensure that post-mortem is conducted and report obtained before payment of fees.

Schedule of Fees

a	For issue of Death Certificate	Rs. 20/- per animal
b	For conducting post-mortem examination and preparing report	Rs. 75/- per animal

- A panel of Vet. Doctors / Investigators should be made by Regional offices and their services can be utilised at the time of acceptance of risk, settlement of claims and for investigation purpose.

- The performance of Vet. Doctors, investigators should be evaluated through appraisal form on half yearly basis by the Regional offices.

- The scale of fees payable to Veterinary Doctors for investigation of claims/ Professional advice will be:

Professional fees per claim for technical advice	Rs. 100/-
Daily Allowance payable per day	Rs. 100/-
Professional fees for spot investigation per claim	Rs. 150/-
Conveyance – Actual 1st class fare by rail if journey Undertaken by Rail other wise conveyance as applicable for two-wheelers or actual bus fare	
For full investigation of individual cattle claim	Rs. 200/- and Rs. 100/- for additional claim in the same locality, if given together.

10. Identification of Animal

All insured animals should be suitably identified by one or more of the following methods:

(a) Ear tag made of suitable material may be used. The cost of ear-tags and tagging charges will be borne by the Insurer.

(b) Natural Identification marks and color should be clearly noted in the proposal form and Veterinarian's Report.

(c) Photographs of animals may be insisted in case of high value animal.

Tagging Charges

(i) Since Veterinarian fees is inclusive of tagging charges, no further tagging charges are required to be paid.

(ii) Re- tagging charges of Rs10/- per large animal.

10. Methods of Identification

Identification of insured animals is an important aspect of the cattle insurance business. All insured animals should be suitably identified by one or more of the following methods:

i. Metal Ear tags.

ii. Branding with hot iron.

iii. Tattooing

iv. Cold Branding

v. Muzzle Printing or Nose Printing

vi. Radio Frequency Identification Device (RFID)

From the experience of handling the cattle insurance business it has been found by insurers that it is profitable to bear the cost of ear tags and that tagging of animals by ear tags made from durable metal is the easiest and most convenient method of identification. However, Radio Frequency Identification Device (RFID) is also gaining popularity for its unique features and utility in controlling moral hazard. The various methods are briefly described below:

22.1.4 Ear tagging

Ear tags made of durable metal like brass or aluminum are fixed on any one ear of the animal by means of an apparatus known as applicator. This procedure is carried out by trained persons so that the ears of animals are not injured. All the tags are pre-numbered. The tags indicate the name of the insurer branch office code and serial number of the insured animal. This facilitates identification.

22.1.5 Branding with hot iron

This is known as hot branding of animals. In this method a red-hot iron is used for numbering and it's applied to horns or hip portion of the animal for a few seconds. This is a rather painful method, but the numbers remain for a long time. However, care should be taken to avoid using white-hot iron, since it may cause deeper and more extensive damage to the skin of the animal.

22.1.6 Tattooing

Tattooing is done by a tattoo machine, usually in the ears of cows, or on the root of the tail in case of buffaloes. However, this is a labourious method, as two or more persons are required to hold the animal, besides the tattooer. If good quality tattooing ink and machines are used, tattooing marks remain for a long time. However, to read the tattoo number either the ear or tail of the animal is required to be held before the tattoo number can be read.

22.1.7 Cold branding

This is also known as freeze branding or cryogenic branding. Freeze branding is more humane and more easily read than hot branding. The application of extreme cold selectively destroys colour producing cells of the hair, thereby the hair of the branded area would become white or discoloured. Essentials of successful cold branding include properly cooled irons, uniform pressure and correct timing. After cold branding the animal, within 50-70 days white hair begins to appear on the branded site.

22.1.8 Muzzle printing or nose printing

Muzzle prints or nose prints of cattle can be used for identification purposes in the same way as finger prints of human beings. Nose prints of any two cattle can be distinguished from one another by using the finger print technique. In foreign countries lot of research has been carried out and the results appear to be encouraging.

22.1.9 Radio frequency identification device (RFID)

RFID tag used in the cattle insurance product is uniquely coded, making duplication near impossible. Each tag is designed in such a way that if it is removed from the ear of the animal, the chip is destroyed, leading to non-acceptance of the insurance claim. The tag thereby reduces the possibility of fraudulent claims, which, in turn, means that the insurance company can offer a lower premium. When an insured animal dies and a claim is made, concerned person on behalf of the insurer will verify the genuineness of the claim and coordinate the submission of required documents. This is expected to greatly expedite the claims settlement process. The tag can serve other purposes as well. The cattle health and productivity management data, including vaccinations given to a particular animal data can also be updated into the chip.

22.1.10 Transfer of interest

Provided previous notice in writing is given to the Insurer, a Policy may be transferred to an approved new owner or to cover a new (another) animal which is subject to adjustment of premium on a pro-rata basis and requirements such as ear tag and health certificate. Transfer fee of Rs. 15 should be collected.

22.1.11 Claim procedure

(A) Non-Scheme Animals

In the event of death of an animal, immediate intimation should be sent to the Insurers and the following requirements should be furnished:

(a) Duly completed claim form.

(b) Death Certificate obtained from qualified Veterinarian on Insurer's form.

(c) Postmortem examination report.

(d) Ear Tag applied to the animal should be surrendered. The underwriting Offices should follow the principle of 'No Tag- No claim'. Generally claim should not be paid if tag is not submitted. However, in cases of genuine hardship, the higher Competent Authority may consider the claim if the identity of animal is established.

(e) The value of the animal should be established properly keeping in view of age, *etc*.

(B) Scheme Animals

Information of loss/death of animal should be given to the Insurer or Financing Bank immediately, within 7 days. Claimant has to furnish the following requirements within 30 days:

(a) Duly completed and signed claim form along with ear tag.

(b) Certification of death from Veterinary Surgeon or a Certificate jointly by any two of the following:

i) Sarpanch of the village.

ii) President or any other Officer of Co-op. Credit Society

iii) Official of the Milk Collection Centre.

iv) Supervisor/ Inspector/ Officer of any Banking or Credit Institution (other than the financing Bank).

v) DRDA or its authorised nominee.

vi) Secretary and Vice President of Panchayat.

vii) Village Revenue Officer / Village Accountant.

viii) Headmaster of a primary school,subject to their declaration that they have seen the carcass and Ear Tag intact in the ear mentioning number thereof.

(C) Post-mortem Report, if Conducted.

22.2 Claim Procedure for PTD Claim

(i) A certificate from the qualified Veterinarian to be obtained.

(ii) The animal should be inspected by our Veterinary Officer also.

(iii) Complete chart of treatment, medicines used, receipts, etc., should be collected. (The U/W office may engage an independent qualified Veterinarian or another investigator in special circumstances).

(iv) Admissibility of claim to be considered after two months of obtaining Veterinary Doctor / Insurer Doctor's report. The Insurer's Veterinarian should examine the animal and confirm PTD, before settlement of the claim.

(v) The indemnity is limited to 75% of Sum Insured.

22.2.1 Salvage

No salvage will be deducted from claims.

Calf/Heifer Rearing Insurance Scheme for Scheme/Non-Scheme Beneficiaries

1. The Offices may implement this Scheme with caution and separate statistics may be maintained for review of the Scheme.

2. The Sum insured is only indicative and depending on local market conditions, it can be altered. Proportionate premium amount may be charged using the method given below.

3. The minimum period of coverage should not be less than 12 months.

4. In case of Non-Scheme animals, the premium rate would be @ of 4% and accordingly, the premium amount should be computed. The premium rate for Scheme animals would be @ 2.25%.

5. The premium amount is computed from 1 day to 32 months. For example: In respect of Scheme animals, the premium is Rs. 207 for 32 months and for Non-Scheme animals, the premium is Rs. 368 for 32 months as per enclosed *Table*.

6. The formulae adopted for calculation of premium is as under:

			Scheme Animals	Non-Scheme Animals
Add: The aggregate sum insured from 1 day to 32 months	=	Rs	1,10,450.00	1,10,450.00
Average Sum Insured for 32 months	=	Rs	3,451.60 or 3452.00	3,451.60 or 3452.00
Apply premium rate onAverage amount	=	Rs	3,452 x 2.2577.65	3,452 x 4.00138.08
Average for One month	=	Rs	77.66 /12 6.47	138.08 /1211.50
Premium for 32 months	=	Rs	6.47 x 32 207.00	11.50 x 32368.00

If the cover is given from 6 months onwards, the sum insured for earlier months is excluded from aggregate sum insured while calculating the premium.

7. The Scope of cover and exclusions are as per Standard Cattle Insurance Policy.

8. The Claim procedure will be same as under Cattle Insurance Policy. As far as Buffalo Calves are concerned, no changes have been made in the existing scheme.

Calf Rearing Scheme Valuation Chart from 1 to 32 Months:

Age at the Commencement of Insurance	Amount Payable in the event death during corresponding month mentioned in Column 1	Premium Schedule to be collected (Rs.)	
		Scheme Animals @ 2. 25 %	Non-Scheme Animals @ 4.00%
1 day to 1 month	150	207	368
1 to 2 months	200	207	368
2 – 3 months	300	206	367
3 – 4 months	400	206	366
4 – 5 months	600	205	365
5 – 6 months	800	204	363
6 – 7 months	1000	203	360
7 – 8 months	1200	201	357
8 – 9 months	1400	198	353
9 – 10 months	1600	196	348
10 – 11 months	1800	193	343
11– 12 months	2000	189	337
12 – 13 months	2300	186	330
13 – 14 months	2550	181	322
14 – 15 months	2800	177	314
15 – 16 months	3050	171	305
16 – 17 months	3300	166	294
17 – 18 months	3600	159	283
18– 19 months	3900	153	271
19 – 20 months	4200	145	258
20 – 21 months	4500	137	244
21 – 22 months	4800	129	229
22 – 23 months	5100	120	213
23 – 24 months	5400	110	196
24 – 25 months	5700	100	178
25 – 26 months	6000	90	159
26 – 27 months	6300	78	139
27 – 28 months	6600	67	118
28 – 29 months	6900	54	96
29 – 30 months	7100	41	73
30 – 31 months	7400	28	50
31 – 32 months	7500	14	25

Valuation Table for Buffalo Calves

AGE	Valuation For sum insured
4 days to 1 month	Rs 125/-
2nd month	Rs 175/-
3rd month	Rs 200/-
4th month	Rs 250/-
5th month	Rs 325/-
6th month	Rs 400/-
7th month to 12thmonth	Rs 850/-
13thmonth to 24th month	Rs 1500/-
25th month to 36th month	Rs 2000/-
37thmonth to 45th month	Rs 3000/-

Note: Premium @ Rs. 145 per Buffalo calf for the entire period from 4th day to 45 months

22.2.2 Foetus (Unborn Calf) insurance scheme

This cover can be granted as a separate policy in addition to Cattle Insurance Policy.

(i) Scope

The Scheme covers the risk of death of embryo/foetus due to accident or disease contracted through the recipient or directly from external source. The Scheme is applicable to both the embryo transferred from a selected donor to the synchronized recipient or frozen embryo transferred to the recipient and also the embryo/foetus developed by artificial insemination technique. The embryo/foetus in the uterus of the recipient/mother cow can be covered as a separate policy in addition to cattle insurance policy covering the recipient/ mother cow/buffalo.

(ii) Period of Cover

The cover commences from the 60th day of the transfer of live quality embryo/ successful insemination subject to production of confirmed pregnancy certificate from a qualified VeterinarySurgeon particularly the specialist in the field. The cover terminates on 220 (+/-5) days for cow from the date of confirmation of pregnancy or on the date of calving whichever is earlier. For the recipient / mother cow the period of insurance is annual similar to standard cattle insurance policy.

(iii) Perils Covered

a) Still births (If calf has not survived)

b) Abortion of all kinds except otherwise it is proved to be a malafide or induced abortion by the insured. Malafide or induced abortion caused by a third party falls within the scope of cover.

c) Accidental Risks–damages caused due to external violent and visible means including transportation by Road/Rail.

d) Induced abortion carried out under Veterinary advice to save the mother in conditions like downer cow syndrome, prolapse of uterus, torsion of uterus, fracture of limbs.

(iv) Sum Insured/Valuation Table

The valuation table is arrived on the basis of input cost, which includes cost of semen/cost of embryo and cost of medicines used for maintenance of pregnancy.

(A) Embryo/Foetus

Age of the Foetus	Through Embryo Technology	Artificial Insemination
	Rs.	Rs.
60 days	1000/-	500/-
61-90 days	1100/-	600/-
91-120 days	1200/-	700/-
121-150 days	1300/-	800/-
151-180 days	1400/-	900/-
181-210 days	1500/-	1000/-
211-240 days	1650/-	1100/-
241-270 days	1800/-	1200/-
271 and above	2000/-	1500/-

(B) Recipient/Mother Cow

Market value certified by the Veterinary doctor or sum insured whichever is lower.

(v) Premium

(i) Embryo/Foetus: Rs. 75/- (net) irrespective of the stage of pregnancy under artificial insemination method. Rs. 100/- (net) in case of embryo transplants method.

(ii) Recipient/Mother Cow: The rate is as per the Cattle Insurance Scheme in force.

(vi) Exclusions

- Poor quality of embryo

- (The cover shall operate only if the embryo is supplied by the authorized *Embryo Bio–technology (Transfer) Laboratories)*

- Unskillful handling of embryo

- (The embryo transfer carried out by a qualified Veterinarian who has undergone special training for embryo transfer technique only shall be covered by this policy)

- Unskillful and mishandling of foetus at any stage of pregnancy.

- Malicious act leading to death, still birth or abortion of foetus by the insured.

- Abortion /Still birth arising out of negligent /careless act of any person attending on to the health care of pregnant animal.

- Death or ejection of embryo / foetus due to administration of steroids /or due to vaccination which are contra indicated.

(vii) Certification of Pregnancy of Animal

Since the confirmation of pregnancy is a very vital factor, strict adherence of this condition of obtaining pregnancy certificate as documentary evidence, is necessary at the time of proposal. Also all the insemination receipts /embryo transfer receipt should be strictly produced along with proposal to correctly assess the age of the embryo / foetus.

All other terms and conditions of the standard cattle insurance policy holds good. The insured should intimate claim if any within 24 hours to Insurer.

22.3 Central Sector Pilot Scheme on Cattle Insurance

The Livestock Insurance Scheme, a centrally sponsored scheme, which was implemented on a pilot basis during 2005-06 and 2006-07 of the 10th Five Year Plan and 2007-08 of the 11th Five Year Plan in 100 selected districts. The scheme is being implemented on a regular basis from 2008-09 in 100 newly selected districts of the country. Under the scheme, the crossbred and high yielding cattle and buffaloes are being insured at maximum of their current market price. The premium of the insurance is subsidized to the tune of 50%. The entire cost of the subsidy is being borne by the Central Government. The benefit of subsidy is being provided to a maximum of 2 animals per beneficiary for a policy of maximum of three years. The scheme is being implemented in all states except Goa through the State Livestock Development Boards of respective states. The scheme is proposed to be extended to 100 old districts covered during pilot period and more species of livestock including indigenous cattle, yak &mithun (GoI, 2016).

The Livestock Insurance Scheme has been formulated with the twin objective of providing protection mechanizm to the farmers and cattle rearers against any eventual loss of their animals due to death and to demonstrate the benefit of the insurance of livestock to the people and popularize it with the ultimate goal of attaining qualitative improvement in livestock and their products.

22.3.1 Guidelines for implementation of livestock insurance scheme

Livestock Sector is an important sector of national, especially rural economy. The supplemental income derived from rearing of livestock is a great source of support to the farmers facing uncertainties of crop production, apart from providing sustenance to poor and landless farmers.

For promotion of the livestock sector, it has been felt that along with providing more effective steps for disease control and improvement of genetic quality of animals, a mechanism of assured protection to the farmers and cattle rearers needs to be devised against eventual losses of such animals. In this direction, the Government approved a new centrally sponsored scheme on Livestock Insurance which was implemented on pilot basis during the 10th Plan. From 2008-09 onwards, the scheme is being implemented as a regular scheme in the100 newly selected districts till the end of 11th Five Year Plan i.e. 2011-12. The broad guidelines, subject to the plausible discretion of the Chief Executive Officers to be followed by the States for implementing the scheme are detailed below:-

22.3.2 Implementing agency

Department of Animal Husbandry, Dairying & Fisheries is implementing the Centrally Sponsored Scheme of National Project for Cattle and Buffalo Breeding (NPCBB) with the objective of bringing about genetic up-gradation of cattle and buffaloes by artificial insemination as well as acquisition of proven indigenous animals. NPCBB is implemented through State Implementing Agencies (SIAs) like State Livestock Development Boards. In order to bring about synergy between NPCBB and Livestock Insurance, the latter scheme will also be implemented through the SIAs. Almost all the states have opted for NPCBB. In states which are not implementing NPCBB or where there are no SIAs, the livestock insurance scheme will be implemented through the State Animal Husbandry Departments.

22.3.3 Executive authority

The Chief Executive Officer of the State Livestock Development Board will also be the executive authority for this scheme. In those states where no such Boards are in place, the Director, Department of Animal Husbandry will be the Executive Authority of the scheme. The CEO will have to get the scheme implemented in various districts through the senior most officer of the Animal Husbandry Department in the district; the necessary instructions for this purpose will have to be issued by the State Government. The Central funds for premium subsidy, payment of honorarium to the Veterinary Practitioners, awareness creation through Panchayats etc. will be placed with the S.I.A. As Executive Authority of the scheme, the Chief Executive Officers will be responsible for execution, and monitoring of the scheme.

22.4 Innovating to Reduce Risk: Weather Index Insurance

While the basic concept is simple, effective implementation of weather index insurance is not at all simple. The continuing availability of accurate historical weather data is critical. Weather parameters required for conducting the study are temperature both minimum and maximum, rainfall, relative humidity, amount and duration of snowfall etc. It is also necessary to determine whether any of the available weather variables are in fact highly correlated with realized losses and if so, the time periods in which losses are most likely to occur. International experience has also shown that effective implementation requires careful attention to the services currently being provided by local risk aggregators as well as legal and regulatory constraints.

22.4.1 Demand assessment

Before investing in data collection and product development, it is important to assess the potential demand for weather index insurance in a particular area. Personal interviews, focus groups, and surveys can be used to determine answers to the following questions:

- What are the key weather perils of concern?
- How frequently do the perils occur and how significant is the impact?
- Who is affected by these perils?
- What mitigation or informal risk transfer strategies are currently being employed?
- What is the (opportunity) cost of those strategies?
- How much are end users willing and able to pay for an insurance product?

22.4.2 Legal and regulatory framework

To facilitate the offer of weather index insurance, governments must establish an appropriate legal and regulatory framework. The legal framework should address not only the proper regulation of insurance sales but also contract enforcement. In many lower-income countries insurance is so poorly understood that courts often force insurance providers to pay indemnities for losses that were clearly not covered under the contract provisions. Conversely, insurance providers may refuse to pay claims to poor policyholders because they know that the policyholders cannot afford to have an attorney represent them in court. Thus, to protect the interests of small-scale policyholders, some sort of binding arbitration procedure is typically desirable (USAID, 2006). Even in countries where the legal and regulatory system is more highly developed, the existing regulatory standards for traditional insurance products may not be appropriate for index insurance products. Index insurance creates unique regulatory challenges because the indemnities are not based on the actual loss incurred. Also, index insurance is highly exposed to spatially covariate losses; so the minimum capital (or contingent capital) requirements need to be higher than those for traditional insurance (Barnett and Mahul, 2007).

22.4.3 Data collection and management

For weather index insurance to be successful, both the insurer and the policyholder must have confidence that the index is being measured accurately and the data are secure from tampering. To build this confidence, the underlying index should be measured by a trusted government or private source of publicly available weather data. In addition, a sufficient amount of historical (normally daily) data on the underlying weather variable must be available for the insurer to estimate premium rates along with pasture NDVI / biomass production. The amount of historical data (i.e., weather and biomass production) required depends on the frequency of occurrence of the risk. Twenty years of data may be required to set initial premium rates for relatively frequent weather events. Thirty or forty years of data may be required for infrequent but potentially catastrophic weather events. Without having the historical data, it is difficult to calculate base premium rates and hence, the insurer would either refuse to underwrite the risk or add a large premium load to account for uncertainty (Hellmuth et al., 2009). Since weather data have public goods characteristics, they are unlikely to be collected, cleaned, archived, and made publicly available by private-sector companies. Government meteorological bureaus usually provide these services. However, many lower-income countries find it difficult to adequately fund meteorological bureaus or sustain a sufficient network of weather stations. To facilitate the availability of weather index insurance, some donor organizations have provided funding for expanded meteorological services in lower income countries.

21.4.4 Need of designing index based livestock insurance program

The traditional individual livestock insurance (based on individual losses) was ineffective because of high loss adjustment costs due to the spread of animals among vast areas, ex ante moral hazard inducing herders failure to take effective measures to protect their stock, and ex post moral hazard leading herders to falsely report animal deaths are among the key endemic problems that plague the traditional livestock insurance program. Monitoring individual herders is a nearly impossible task. The formal financial insurance products related to livestock mortality are unpopular among both insurance companies and livestock owners and are limited almost entirely to a small number of high value livestock (Mahul and Jerry Skees, 2006). An alternative approach is to develop a collective system for indemnifications: indemnity payments are based on a transparent index designed to reflect the loss incurred by the herders. Such schemes are known as index-based insurance (e.g., area-yield insurance, weather index based insurance). These schemes present some advantages (e.g., reduction of moral hazard and adverse selection, lower administrative costs), but their main impediment is the presence of basis risk, i.e., the index payout may not exactly match the individual livestock loss (Mosleh, 2016). An index-based insurance product to indemnify herders based on the mortality rate of adult animals in a given area was recommended for the first time in Mongolia. The index-based livestock insurance (IBLI) policy pays indemnities whenever the adult mortality rate exceeds a specific threshold for a localized region. This system provides strong incentives to individual herders to continue to manage their herds so as to minimize the impacts of bad weather (i.e., if a better herder has no losses when their neighbors have had large losses, the better herder is rewarded for the extra effort by receiving a payment based on the area losses).

22.4.5 Data description

Normalized Difference Vegetation Index (NDVI) – sometimes referred to as greenness maps, is a satellite-derived indicator of the amount and vigor of vegetation, based on the observed level of photosynthetic activity. The NDVI data that is available reliably at high spatial resolution of 8 km^2 grids and consistent quality from Advanced Very High Resolution Radiometer (AVHRR) on board of the United States National Oceanic and Atmospheric Administration (NOAA) satellite, and have been available in real time every 10 days with the longest temporal profile since late 1981. NDVI data are commonly used to compare the current state of vegetation against the long-term average condition in order to detect anomalies and to anticipate drought and have now been used by many studies that apply remote sensing data to drought management (Hellmuth et al., 2009).

We rely on NDVI data for two reasons as explained by Sommarat *et al* (2011). The first is conceptual. Catastrophic herd loss is a complex, unknown function of rainfall – which affects water and forage availability, as well as disease and predator pressure – and rangeland stocking rates – which affect competition for forage and water as well as disease transmission. Rangeland conditions manifest in vegetative cover reflect the joint state of these key drivers of herd dynamics. When forage is plentiful, disease and predator pressures are typically low and water and nutrients are adequate to prevent significant premature herd mortality. By contrast, when forage is scarce, whether due to overstocking, poor rainfall, excessive competition from wildlife, or other pressures, die-offs become frequent. Thus a vegetation index makes sense conceptually.

The second reason is practical. In countries where, longstanding seasonal or annual livestock census surveys are not available as in the case of Mongolia for computing area average mortality for developing IBLI contract. It is also useful in countries where, consistent weather data series at sufficiently high spatial resolution are not available. Rainfall data from hydromet station may not be representative to settle the claims due to presence of limited number of stations. Hence, rainfall estimates derived from satellite - based remote sensing could be used as an alternative source of data (Tucker, 2005).

22.4.6 Training of insurance suppliers and consumer education

Insurance suppliers in lower-income countries are unlikely to be familiar with weather index insurance. Thus, they require training and capacity building opportunities to build the expertise needed to offer these unique insurance instruments. Similarly, in rural areas of many lower income countries, insurance products are not widely available. Even if potential policyholders are familiar with other types of insurance products, they will almost certainly not be familiar with weather index insurance. To make an informed purchase decision, it is critically important that potential policyholders understand the basis risk inherent with weather index insurance. That is, they need to understand that they may experience a loss but not receive an indemnity. Thus, the successful introduction of weather index insurance will require a significant educational effort. While insurance suppliers will provide some information as part of their sales efforts, potential policyholders also need information from objective sources.

22.4.7 Product development

Once weather index insurance product is developed and offered for sale by an insurance supplier, competitors can easily copy it, since the underlying index is based on publicly available data. This "free rider" problem makes it very unlikely that private-sector insurance suppliers will invest in the research and development required to bring a weather index insurance product to the market. For this reason governments and donors have tended to fund feasibility studies and pilot tests of new weather index insurance products (Barnett and Mahul, 2007).

22.4.7.1 Catastrophic risk-sharing

Local suppliers of weather index insurance policies must be able to transfer their loss exposure outside of the local area. Traditional lines of insurance (e.g., automobile, life, property and casualty) are offered on loss events that are largely uncorrelated, so the law of large numbers reduces the variance in indemnities for local insurance providers. But weather index insurance protects against spatially covariate loss events. When a policyholder collects an indemnity on a weather index insurance product, all other holders of that same policy will be collecting indemnities as well. This implies that, in any given year, indemnities can be very high relative to premiums collected. While in principle it may be possible for insurance suppliers to set aside adequate liquid reserves to cover the potential for large indemnities, in practice this is highly unlikely. There is a high opportunity cost associated with keeping such large amounts of capital in investments that can be readily liquidated. Further, in many countries there are tax disincentives for holding large reserves. Thus, index insurance suppliers generally obtain contingent capital via reinsurance. Catastrophe bonds and contingent loan mechanisms can also be used as sources of contingent capital (Barnett and Mahul, 2007).

22.4.8 Potential benefits of weather index insurance in regions affected by climate change

Multilateral institutions on climate change have highlighted the importance of insurance in the context of adaptation and many are pointing to weather index insurance because of the positive experiences of weather index insurance pilots in lower income countries.

First, weather index insurance pilots have been designed to protect households from catastrophic weather events. Given the likely increase of extreme events associated with climate change impacts in some regions, weather index insurance could play a key role in protecting vulnerable households. In this fashion, weather index insurance increases the resilience, including the adaptive capacity, of the insured (Benjamin et al., 2009). Second, insurance markets are likely to motivate households to adapt through price signals. Insurance has a long history of using price signals to reduce vulnerability. For many risks in lower income countries, insurance provides a first-time estimate of the monetary cost of the risk being insured. Many households may be unaware of the monetary cost of their production risk – many households have likely never been exposed to this way of thinking. The price of weather index insurance may allow households to improve their decision-making process regarding whether they need to adapt, and if they do decide to change their behaviors, how and to what extent they must change. Third, insurance provides cash at opportune times for the insured to adapt. After a catastrophic event occurs, households must decide whether they will continue in their previous livelihood strategies, often requiring households to restock damaged assets, or if they will change livelihood strategies, often requiring capital investments. A major difficulty for lower income populations is that they lack the financial means to adapt. Cash payments from an insurer improve opportunities for farmers to make the capital investments needed to adapt or to maintain their current production strategies.

Fourth, weather index insurance can encourage adaptation by being bundled with new technologies. Bundling weather index insurance with drought-resistant fodder-seed, for example, may increase access to both the seed and insurance for households. Several arguments are advanced about the value of weather index insurance for regions experiencing climate change creates problems for pricing weather index insurance. Relative to traditional insurance products, weather index insurance has several advantages (IPCC, 2012);

- The insurance contract is relatively straightforward, simplifying the sales process.

- Indemnities are paid based solely on the realized value of the underlying index. There is no need to estimate the actual loss experienced by the policyholder.

- Unlike traditional insurance products, there is no need to classify individual policyholders according to their risk exposure.

- There is little reason to believe that the policyholder has better information than the insurer about the underlying index. Thus, there is little potential for adverse selection. Also, there is little potential for *ex ante* moral hazard since the policyholder cannot influence the realization of the underlying weather index.

- Operating costs are low relative to traditional insurance products due to the simplicity of sales and loss adjustment; the fact that policyholders do not have to be classified according to their risk exposure; and the lack of asymmetric information. However, start-up costs can be quite significant. Reliable weather and agricultural production / livestock data and highly skilled agro-meteorological expertise are all critical for the successful design and pricing of weather index insurance products.

Table 22.1: Types of risk and loss - and local capacity to cope (IPCC, 2012)

Type of risk		Type of loss			
Degree of Covariance	Frequency	Life	Assets	Seasonal production/income	Examples
High	Low	Widespread loss of life and injuries from catastrophic weather events such as hurricanes, floods or severe drought Little or no capacity to cope locally; recovery is difficult and slow	Widespread loss of homes and productive assets from catastrophic weather events Little or no capacity to cope locally; recovery is difficult and slow	Impacts of catastrophic weather events on regional production and income can be severe, with limited local coping capacity Recovery can be slow if lives and assets are also lost	Catastrophes such as tsunami, severe drought, flood, hurricane or earthquake
Medium	Medium	Some loss of life and widespread health problems can arise from seasonal malnutrition Moderate capacity to cope with the effects of the shock locally; recovery occurs	Widespread loss of animals from drought or contagious diseases Moderate capacity to cope locally and slow recovery. Some people fall into poverty traps	Loss of income from poor market prices; regional production and income impacts can be widespread owing to shrinkage of the rural non farm economy Moderate capacity to cope locally and quick recovery if assets are not lost as well; some people fall into poverty traps	Less-severe drought or excess rainfall in critical periods, new pest outbreaks and animal diseases

Contd.

Type of risk		Type of loss			
Degree of Covariance	Frequency	Life	Assets	Seasonal production/income	Examples
Low	High to medium	Deaths, accidents and illnesses that affect a predictable share of the population each year Some local capacity to pool these risks, but recovery from losses can be slow for the households involved	Loss, damage or disease of a predictable share of the total stock of homes or productive assets each year Good local capacity to pool these risks, but recovery from losses can be slow for the households involved	Low yields for some farmers due to a variety of localized weather and pest problems Good local capacity to cope with these risks; recovery is usually quick	Localized weather and pest problems (e.g. frost in a particular valley, pest outbreak in certain fields)

- Since no farm-level risk assessment or household loss adjustment is required, the insurance products can be sold and serviced by insurance companies that do not have extensive agricultural expertise.

Table 22.2: The climate-related root causes and the impacts they are likely to have (UNEP, 2012)

Climate-related risks	Impacts
Increase / decrease in temperature	<ul><li>Increase in glacial melting rates, resulting in increased river flow in summer causing floods and soil erosion;</li><li>Increase in outbreaks of agricultural pests;</li><li>Increase in land degradation;</li><li>Decrease in agricultural productivity;</li><li>Increase in the extent of arid and semi-arid areas;</li><li>Expansion of infectious animal diseases and transfer of from sheep and cattle to poultry and pets</li><li>Decrease in biodiversity.</li></ul>
Changing rainfall patterns	<ul><li>Increase in frequency and duration of droughts;</li><li>Increase in number of flood events;</li><li>Changed seasonal river flow patters</li><li>Decrease in water volume in catchments and reservoirs;</li><li>Decrease in agricultural productivity;</li><li>Decrease in livestock productivity;</li><li>Increase in extent of arid and semi-arid areas;</li><li>Increase in infectious human diseases (exacerbated by the increase in temperature), such as enteric infections, tropical fevers, parasitic diseases and malaria[1].</li></ul>
Extreme events a) Heat waves	a) Increase in the frequency of heat waves<ul><li>Decrease in water supply and quality;</li><li>Decrease in grassland vegetation;</li><li>Decreasing livestock productivity, particularly sheep[2];</li><li>Increased incidence of human diseases, such as ischemic heart diseases in elderly people.</li></ul>
b) Intense rainfall	b) Increase in intense rainfall events<ul><li>Increase in floods, avalanches, mudflows, glacial lake bursts;</li><li>Increase in soil erosion & landslides;</li><li>Increase in river bank erosion;</li><li>Damage to infrastructure, *i.e.* hydroelectric reservoirs;</li><li>Pest outbreaks.</li></ul>

22.5 Prerequisites for Program Success and Expansion

However well designed the product, implementation of weather index crop insurance in developing countries requires considerable on-going management and stakeholder inputs. It should be noted that conditions in different countries, or even different regions of the same country, vary widely and it is difficult to imagine one model's being directly applicable to another situation without adaptation. Several prerequisites have been identified as essential elements of promoting a successful livestock insurance program that will form a strong platform. (Shadreck, 2008).Some of these are identified below:

- A competent local project manager must be in place to ensure that all the complexities of the program are effectively handled and stakeholder obligations are met.

- A committed meteorological services authority is absolutely essential to provide timely and reliable data.

- An adequate weather infrastructure must be in place with sufficient operational weather stations.

- The distribution channel must be competent as well as committed to the project. There has to be a sophisticated understanding of technicalities of insurance and animal health, and outside expertise should be brought in supplement distributor knowledge where appropriate. There is a real need to properly research and evaluate any potential project prior to commencement.

- Ever insurance program requires well-capitalized risk carriers who have a clear understanding of the market. An understanding of data provision including the need for proxy data in some cases that increases the margin for error in pricing calculations, and profit limitations and opportunities in the rural agricultural market place.

22.6 Index-Based Mortality Livestock Insurance in Mongolia (reproduced from CDKN, 2013)

Harsh and unpredictable weather, exacerbated by climate change, makes herders in Mongolia vulnerable to mass livestock losses. In 2010, over 50% of Mongolia's herders were affected by extreme weather, with 75,000 herders losing more than half their livestock. About 33% of the country's workforces are herders, leaving many Mongolian households – and the nation's economy – vulnerable to shocks affecting livestock populations. The Government of Mongolia's Index-Based Livestock Insurance (IBLI) Project, supported by the World Bank, developed an innovative index-based mortality livestock insurance now available in every Mongolian province. Index-based insurance programs aim to make payouts based on an index of aggregated criteria, such as livestock losses over a geographic area, rather than households' or businesses' actual, individual losses. IBLI protects Mongolian families from significant livestock loss by providing financial security, while also encouraging herders to adopt practices that build their resilience to extreme weather events. In 2012 alone, herders bought 16,000 insurance policies. This brief demonstrates how an insurance program such as IBLI can be used as part of a strategy to protect populations from climate-based risks.

Breeding and raising livestock for meat, milk and cashmere constitute an integral part of Mongolia's economy. The agriculture sector accounts for about 15% of the country's gross domestic product, and roughly 80% of the value added is livestock. In 2011, Mongolia's National Statistical Office counted about 36 million head of livestock. This large number of animals using common grazing land has led to increasing degradation of grasslands, making livestock more vulnerable to dzuds (harsh weather events that consist of drought, heavy snowfall, extreme cold and windstorms). Degraded grasslands can no longer provide sufficient vegetation, so when dzuds occur there are not enough nutrients to sustain the livestock.

The International Panel on Climate Change (IPCC), 2012 has demonstrated that climate change is exacerbating extreme weather events worldwide and negatively impacting vulnerable populations. The IPCC's Special Report on Managing the Risks of Extreme Events and Disasters to Advance Climate Change Adaptation (SREX) has found that "a changing climate leads to changes in the frequency, intensity, spatial extent, duration and timing of extreme weather and climate events, and can result in unprecedented extreme weather and climate events". The SREX case study on dzuds found that this impact applies to Mongolia. These more severe weather events – droughts, changes in rainfall patterns and floods – have a particularly significant impact on agriculture. An increase in temperature of almost two degrees Celsius in the past 50 years and increasingly unpredictable precipitation patterns has negatively impacted the pastures that Mongolian herders use for their livestock by increasing the intensity of droughts and dzuds. In addition, Mongolian herders themselves have noted how the sudden cold spells and changing rainfall patterns have caused greater livestock deaths. In order to help herders cushion the effects of climate change and desertification, the government of Mongolia is encouraging the transfer of information and technology to herders. The government is also promoting research on sustainable herding methods to protect pasturelands, education of relevant stakeholders on adaptation methods and sustainable practices, and coordination of research and monitoring.

22.7 Establishing Index-Based Livestock Insurance in Mongolia

The dzuds that occurred between 2000 and 2002 prompted the Government of Mongolia and World Bank to explore solutions, which led to the establishment of IBLI in Mongolia. During this period, over 11 million animals died, representing a loss of over US$200 million. It also led to an increase in the poverty rate from approximately 30% in 2000 to over 40% in 2004. This climate-related risk called for a strategy that allows the risk to be shared among communities, insurance companies and the Government. The Government, with help from the World Bank, responded by creating an innovative insurance program – the first to pay out based on livestock mortality rates – that has reduced risk for thousands of Mongolian herders. Drawing on historic data on livestock losses, Mongolia developed a livestock insurance scheme that combines self-insurance, market-based insurance and a social safety net.

Table 22.3: Payments For Losses of Livestock (CDKN, 2013)

Livestock loss	Type of response	Function
Below the trigger point (approximately < 6 %)	None	Herders absorb these livestock losses below the trigger point because they will not affect the viability of their livelihood or reduce their adaptive capacity in the face of climate change.[xv]
Trigger point (approximately 6–30%)	Livestock Risk Insurance (LRI), (replaced the Base Insurance Product (BIP))	Herders purchase LRI for a premium from private insurance companies, which make payments when the livestock mortality rates meet the trigger point, which is the range of loss that affects the viability of herders' livelihood while also being financially feasible for insurance companies. This insurance product is meant to cover severe losses that occur about every 1–5 years. The price that the herder pays for the insurance is based on the risk associated with the species and the location – the higher the risk, the higher the cost of the insurance.
Exhaustion point met (approximately > 30%)	Disaster Response Product (DRP) (replaced by the GCC during the 2009/2010 insurance cycle)	The government provided coverage to herders who purchased the BIP or who just purchased the DRP and covered 100% of livestock losses greater than the exhaustion point. The International Debt Association financed a Contingent Debt Facility (CDF) that made payments under the DRP. The DRP was too complex and fiscally unsustainable.
	Government Catastrophic Coverage (GCC)	GCC provides government-financed insurance that prevents insurance companies from having to absorb the extremely large losses and thus keeps the cost of the insurance down. The government's financial risk is much smaller because it only covers herders who have purchased an LRI policy and only covers the insured livestock, rather than 100% of losses.

22.8 Benefits of Index-Based Livestock Insurance

Weather-related risks deeply affect low-income populations through famine, displacement and devastating financial and/or property losses. As a result, these risks are a disincentive to poor families to invest in their livelihood activities, making it harder to change their economic status. Insurance is an important mechanism that allows poor households to invest in strategies with higher economic growth potential that will help shield them from the impacts of climate change. However, insurance is often not available to low-income populations because insurers do not know or cannot quantify their agricultural risks. This is especially true in locations where climate change is expected to have substantial but uncertain long-term impacts. Even where insurance is available, usually only wealthier population segments can afford it, allowing income disparity to grow.

Index-based insurance can resolve many of the inequities and challenges that are presented by traditional insurance. While traditional insurance pays based on insurance agents' individual assessments of loss, index-based insurance reduces transaction costs and avoids the problem of moral hazard by providing a system of payment that is automatically triggered when a mortality threshold is met. Index-based insurance can also provide a tool to help herders soften the impact of weather events that are made more severe by climate change. Additionally, index-based insurance provides an alternative to a reactive approach to disasters that relies mainly on domestic money diverted from other projects and international donations. This reliance often stems from a lack of understanding about risk and economic incentives and an underdeveloped insurance market. This type of diverted domestic funding can be delayed before disbursement, or can be insufficient and ineffective due to its ad hoc distribution. Even though disbursement of IBLI payouts can also be delayed, IBLI often allows for funds to be readily available as part of the insurance pool and distributed faster than emergency aid (CDKN, 2013).

To be most effective, index-based insurance should be part of a comprehensive risk management strategy that aims to reduce the risk in the livestock sector by establishing sustainable practices to better manage the pastures where livestock graze. For instance, sustainable grazing practices prevent desertification and land degradation, making livestock less vulnerable to harsh weather events. Index-based insurance provides an important method for incentivizing herders to adopt sustainable grazing practices that reduce their risk. Two options for accomplishing this are price signals (e.g. charging lower premiums for diversifying the types of livestock owned) and risk management stipulations (i.e. requiring herders to take certain actions to lower their risk as a condition of the insurance).

References

Barnett, B. J. and Mahul Olivier, 2007. Weather Index Insurance for Agriculture and Rural Areas in Lower-Income Countries. Am J Agr Econ (2007) 89 (5): 1241-1247.

Benjamin Collier, Jerry Skees and Barry Barnett, 2009. Weather Index Insurance and Climate Change: Opportunities and Challenges in Lower Income Countries. The Intern. Associa. For the Study of Insurance and Eco, The Geneva Papers, 34: 401-424.

CDKN, 2013. Report: Index-Based Mortality Livestock Insurance in Mongolia. Climate & Development Knowledge Network.http://cdkn.org/2013/03/inside-story-index-based-mortality-livestock-insurance-in-mongolia/?loclang=en_gb

Government of India, 2016. Livestock Insurance. Ministry of Agriculture. Department of Animal Husbandry, Dairying & Fisheries. http://www.dahd.nic.in/related-links/livestock-insurance-0

Hellmuth M.E., Osgood D.E., Hess U., Moorhead A. and Bhojwani H. (eds) 2009. Index insurance and climate risk: Prospects for development and disaster management. Climate and Society No. 2. International Research Institute for Climate and Society (IRI), Columbia University, New York, USA.

IPCC, 2012. Managing the Risks of Extreme Events and Disasters to Advance Climate Change Adaptation. Special Report of the Intergovernmental Panel on Climate Change.Cambridge University Press. 594 p.

Mahul Olivier and Jerry Skees, 2006. Piloting Index-Based Livestock Insurance in Mangolia.AccessFinance, News Letter Published by the Financial Sector, The World Bank Group, Issues No. 10 March.

Mosleh Ahmed, 2016. Index-Based Livestock Insurance in Mongolia.https://www.linkedin.com/pulse/index-based-livestock-insurance-mongolia-mosleh-ahmed

ShadreckMapfumo, 2008. Weather Index Crop Insurance. White Paper. Implementation, Prodcut Design, Challenges and Successes Lessons Learned in the Field.http://docplayer.net/5909323-Weather-index-crop-insurance-white-paper-implementation-product-design-challenges-and-successes-lessons-learned-in-the-field.html

SommaratChantarat, Andrew Mude, Christopher B. Barrett, Michael R. Carter, 2011. Designing Index Based Livestock Insurance for Managing Asset Risk in Northern Kenya. Journal of Risk and Insurance 80 (1), January 2011.

Tucker, C. J., J. E. Pinzon, M. E. Brown, D. A. Slayback, E. W. Pak, R. Mahoney, E. F. Vermote, and N. E. Saleous, 2005. "An Extended AVHRR 8-km NDVI Data Set, Compatible with MODIS and SPOT Vegetation NDVI Data." International Journal of Remote Sensing 26(14):4485-4498.

UNEP, 2012.Impact of Climate-Related Geoengineering on Biological Diversity.Convention on Biological Diversity. UNEP/CBD/SBSTTA/16/INF/28.

USAID, 2006. Index Insurance for Weather Risk in Lower-Income Countries. USAID. http://pdf.usaid.gov/pdf_docs/Pnadj683.pdf

23

Adaptive Mechanisms and Mitigation Strategies for Bovine Under Tropical Climatic Conditions

Sohan Vir Singh, Simson Soren and Anil Kumar

Climate Resilient Livestock Research Centre (CRLRC), National Innovation of Climate Resilient Agriculture (NICRA), ICAR-National Dairy Research Institute (ICAR-NDRI), Karnal, Haryana-132001, India

Increase in ambient temperature, drought, rainfall, desert, feed and fodder scarcity, water scarcity and other climatic stresses are very common to tropical climatic conditions. The animals living in tropical climate have to undergo some modification for their survival under such extreme conditions. Interestingly, higher percentage of livestock is found in tropical and sub-tropical regions of the world. These areas are rich in animal resources. There are several descriptive and non-descriptive breeds of bovine distributed in different agro-climatic conditions of India. They possess several traits which make them adaptive to such climatic conditions. Identifying those traits and propagation of such traits is beneficial in near future in respect of climate change scenario. Climate change is the major concern for the improvement and sustainability of livestock under tropical climatic conditions. The rising of temperature and humidity is one of the major threats to animal's productive performance that leads to great economic losses. Upadhyay *et al.* (2009) estimated 2% loss of milk production due to thermal stress in India. It has also been expected that 3.2 million tons reduction by 2020 and more than 15 million tons by 2050. The decline in milk production reported to be higher in crossbreds (0.63%) followed by buffaloes (0.5%) and indigenous cattle (0.4%) (Upadhyay *et al.*, 2009). The higher ambient temperature with higher relative humidity in crossbred cattle (Holstein Friesian X Deoni) showed a detrimental effect on milk yield

(Kumar *et al.*, 2014). The higher production losses in crossbred cattle were observed due to the poor adaptability than indigenous cattle under heat stress. The Indian indigenous cattle is popularly known as Zebu cattle. They have several unique traits which make them adaptation to harsh environment under tropical climatic conditions (Singh and Soren, 2017). The high yielder cattle are more vulnerable to heat stress due to their higher metabolic rate. Protection of high yielder animals under stressful conditions is necessary for the development of dairy sector. Several strategies have to adopt to mitigate the heat stress on dairy animals. Another aspect for improvement of dairy sector in to identify several traits in indigenous animals, selection and breeding of such animals might be a suitable approach to achieve the solution to face the climate change scenario under tropical climatic conditions. Some of the adaptive mechanisms and the mitigation strategies are discussed below.

23.1 Adaptation

Adaptability can be defined as the capacity or ability to survive and reproduce to a defined environment (Prayaga and Henshall, 2005). The adaptation to a wide range of environment can be achieved by morphological, anatomical, physiological, genetic, feeding behaviour, metabolism (Silanikove, 2000a; Silanikove and Koluman, 2014). As we know that India is one of the most diverse lands in the world. The livestock farming is closely dependent on natural resources and the environmental conditions. It has been noticed that the livestock breeds and population which are adapted to stressful tropical environments have unique adaptive trails *e.g.* heat tolerance, disease, ability to cope with water scarcity and poor quality feed *etc*. These unique trails enable them to survive and reproduce under these stressful environmental conditions (Mirkena *et al.*, 2010).

23.1.1 Morphological characteristics

The morphological characteristics seen in coat colour, fur depth, hair type, hair density, sweat gland, subcutaneous fat deposition, presence of absence of hump, dewlap, body size, size of the legs etc. Morphological characteristics of cold and heat adapted breeds can be clearly identifies. The higher deposition of subcutaneous fat, absence of hump and higher hair thickness is mostly seen in cold adapted breeds, whereas presence of hump, dewlap, higher and larger sweat glands are seen in heat adapted breeds. The size of the body is compact and small in size in heat adapted breeds than European breeds. Zebu breeds have small size and low body weight with small barrel shaped body and slender legs. They have a hump and a dewlap. The head is held high in most zebu breeds. Since most of these breeds have been developed for draught purpose, long legs with articulate joints provide ample capacity to run and swiftly move even under moist soils. The balanced fore and hind body quarters help them in propelling body and moving forward with loads at moderate speeds. Balanced body is mainly due to small size and low volume of internal organs. Small sized rumen, reticulum, omasum and abomasum, do not distend down belly of these Zebu draught breeds contrary to heavy bodied in Taurine breeds. Some of the Zebu breeds (Tharparkar, Nagori and Sahiwal) are well adapted to hot dry desert conditions are able to reduce their metabolic requirements to minimum and conserve energy for diversion to production (milk and /or work) without extra energy expenditure (Upadhyay *et al.*, 2013). These mechanisms are rarely found in livestock species located in other areas. The evaporative process of heat loss is one of the best way of reducing the internal heat among other way of heat loss process i.e., conduction, convection and radiation. However, the evaporative heat loss depends upon the sweat gland density, size of the sweat gland and function, hair coat density and the regulation of epidermal vascular supply (Collier *et al.*, 2008). The larger salivary glands and higher surface area of absorptive mucosa are also important mechanism for survival in fodder and feed scarcity areas (Silanikove, 2000b).

23.1.2 Coat colour

The absorption and reflection of light depends on the hair and coat colour of the animals. Animals with light coat found to reflect more solar light resulted less heat absorption and the dark colour absorbed more heat. It is commonly stated that the white coloured Tharparkar cattle reflect more solar light and well adapted to extreme heat. Skin is the largest organ of the body that play a significant role by providing a physical barrier against mechanical, chemical and microbial factors (Haake and Holbrook, 1999). It take part in thermo-regulatory mechanism. The epidermal blood circulation and sweating rate support to heat loss, produced in the internal body as a result of metabolism. The hair and skin pigmentation can be an adaptive mechanism in heat tolerant cattle (Maibam *et al.*, 2014). It has been found that the melanin pigmentation in the skin cells act as an antioxidant enzyme and helps in adaption of animals under heat and cold stress. The higher pigmentation of melanin was observed in Tharparkar than Karan Fries cattle (Maibam *et al.* 2014a). The basis of coat colour in mammals including cattle is the presence or the absence of melanin pigment (eumelanin and pheomelanin). Eumelanin is responsible for black and brown colours and pheomelanin for reddish brown (Simon and Peles, 2010). Melanocortin 1 receptor (MC1R) gene is responsible for pigmentation differences in mammals (McRobie *et al.*, 2014). Acquisition of a highly stable MC1R allele promotes black pigmentation which helps in protection from UV damage (Greave, 2014). Another gene, premelanosome (PMEL), encodes a transmembrane protein called pre-melanosomal protein. PMEL is a melanocyte protein necessary for eumelanin deposition (McGlinchey, 2009). Therefore, the above mentioned genes (MC1R and PMEL) divert the pathway of melanin synthesis towards eumelanin (true melanin) rather than pheomelanin. The rate limiting enzyme for melanin synthesis is tyrosinase (Zhang *et al.*, 2010). Eumelanin intensifies skin pigmentation and thus helps in photoprotection because of its efficiency in blocking ultraviolet (UV) rays and scavenging reactive oxygen species (Klungland *et al.*, 1995). The expression of skin colour related genes (MC1R and PMEL) in lymphocytes and plasma tyrosinase activity were found to be significantly higher in Tharparkar than Karan Fries cattle (Maibam *et al.*, 2014a, b). It is a familiar observation that different ecotypes of cattle, whether they are distinguished as species, breeds, or strains, show marked contrasts in coat cover. These differences follow the principle, dignified by Wright (1954) as "Wilson's Rule", of a gradient from thick, woolly coats in cold climates to short coats with bristly hairs lying sleekly against the skin in hot climates. The contrasts between European cattle and Zebus of India clearly represent adaptations to cold and heat. Individual animals also grow shorter coats when transferred from a cold to a hot environment (Berman and Volcani, 1961). The fact that coat genotype seems to have changed fairly rapidly in breeds introduced to the tropics confirms the importance of this trait to adaptation. Verissimo (2002) reported that the tropical breed of cattle had shorter hair length when compared to the crossbred animals. Dowling (1956) showed an association among heat tolerance and performance of different strains of cattle with their coat characters, and Turner and Schleger (1958, 1960) measured the degree of variability of coat type within herds, and assessed the proportion of the variation in growth rate that is accounted for by variation in coat type. Coat characteristics are associated with heat tolerance and performance of animals (Dandage *et al.*, 2010; Collier and Collier, 2012). Skin colour is also associated with the health condition of the individual (Stephen *et al.* 2011). In animals, hair and skin pigmentation is a highly visible trait. Under tropical condition with high levels of solar radiation, animals with a light coloured hair coat and darkly pigmented skin are better adapted (Finch and Western, 1977; Finch, 1984; Uttarani *et al.*, 2014a, b).

23.1.3 Epidermal blood flow

The epidermal circulation provides the nourishment to the skin by supplying the essential nutrients. Under heat stress, the homeorhesis mechanism is activated, which demand the more supply of blood to most important organ of the body. Skin is one of the organ, where more blood supply is required under heat stress. The main purpose of more blood flow in the skin surface is to release the heat from internal to external environment as much as possible. The soft and smooth skin in zebu cattle clearly indicates the superior blood supply in heat tolerant cattle than European cattle. The crossbred (Bos indicus X Bos taurus) showed a poor blood flow rate compared to zebu cattle (Singh and Soren, 2017; Singh *et al.*, 2017). This might be one of the reasons of rough and less soft skin in crossbred cattle than zebu cattle under tropical climatic conditions. It has also been observed that the rate of blood flow is less during winter than summer season. The lower blood flow during winter is to restrict the heat loss from internal body to the environment during winter to maintain their body temperature whereas the higher blood flow during summer is to release the heat load from the internal body to maintain the body temperature in a set points. The blood flow was recorded in different part of the body i.e., dorsal, abdominal and middle ear of zebu (Sahiwal, Tharparkar) and Karan Fries cattle. The results indicated higher blood flow at dorsal, abdominal and middle ear in zebu than crossbred and a positive correlation was showed with skin temperature (Singh *et al.*, 2017). It has been concluded that the higher of blood flow in indigenous cattle is helping the animals to adapt in tropical climatic conditions (Singh *et al.*, 2017; Singh and Soren, 2017). Similarly, the higher blood flow was also observed in buffaloes during summer than winter season (Singh *et al.*, 2014). The blood flow in the skin surface enhance the heat exchange through conduction, convention and evaporation.

23.1.4 Thermoregulation

Bovine relief the overloaded heat by conduction, convection, radiation and evaporation. Evaporation involves in sweating rate and respiratory minute volume (Al-Haidary *et al.*, 2001), this is one of the best way of heat relief when the air temperature reaches to the skin temperature. However, it required energy for its function through evaporation (Kibler and Brody, 1952; Finch, 1986). It has been observed that the evaporative cooling (sweating and panting) is the most important mechanism for body heat dissipation under hot climates (Collier, 2008). Heat loss by panting is also effective for dissipated heat successfully, but the rate of panting also found to differ among the breeds or the genetic makeup of cattle (Robertshaw, 1985). McLean and Calvert (1972) recorded 84% of heat loss by evaporation, of which 65% was lost by sweating and 35% was lost by panting. Cattle utilize evaporative cooling in the form of both sweating and panting in an effort to loose excess body heat when environmental temperatures begin to exceed 35°C and THI of 90 (Collier, 2008). The ability of cattle to maintain body temperature depends on their capacity of thermoregulation based on the balance of heat gain and heat loss through: conduction, convection, radiation, and evaporation (Kadzere *et al.*, 2002). The higher sweating rate was found in zebu cattle due to higher sweat gland density (Schleger and Turner, 1965). Not only the higher density of sweat gland but also baggy shaped i.e. higher in volume (Pan, 1963) and closed to the surface of the skin than those of Bos taurus (Nay and Hayman 1956). Cattle indigenous to tropical regions had a relatively thin hair follicle depth and very often a simple sac-like sweat gland (Jenkinson and Nay, 1975). Zebu cattle have looser and thicker skin, larger ears, and prominent hump that help them to increase surface area per unit of body weight.

23.1.5 Physical adjustment

The physiological responses are the indication and necessity for their survival under extreme climatic conditions. Under thermo-neutral zone (TNZ), bovine maintain the body temperature without much energy utilization and this zone also considered a comfortable zone. The deviation from the set point or little fluctuation of body temperature activates several mechanisms in the body to bring back the body temperature in to a physiological level. The increase of ambient temperature influence physiological response i.e. the respiration rate (RR), pulse rate (PR), rectal temperature (RT) and skin temperature (ST). The RT is considered to be a most effective indicator of heat stress whereas the higher RR indicates the loss of heat through respiratory system. The deviation from TNZ challenged the animals to maintain the body temperature through compromising in metabolism and higher sweating rate with higher epidermal blood circulation. The comfortable range of ambient temperature or TNZ also differ between the species and breeds. The TNZ varies from 15 to 25°C for crossbred cattle and 15-28°C for indigenous cattle (Singh and Upadhyay, 2009). The upper limit of TNZ i.e. upper critical temperature has more significance in tropical and subtropical climate. The information on upper critical temperature for most of the species and breeds of farm animals in India is scanty. The upper critical temperature is lower in exotic breeds and their crosses than indigenous breeds. The upper critical temperature for Haryana bulls is 32.0°C but for its crosses having 50 percent exotic inheritance of Holstein Friesian, Brown Swiss and Jersey is 26.5°C, 27.5°C and 29.0°C, respectively.

The physiological responses are important to cope up the animals to the adverse environments. The RR, PR, RT and ST are recorded higher in crossbred cattle as compared to Zebu cattle (Indu et al., 2016). The relationship between behavioural and physiological indicators can be used to evaluate the adaptive capacity and consequently the "welfare" of animals in relation to different conditions (Broom and Johnson, 1993). Sharma (1974) reported positive correlation between temperature, relative humidity and rainfall with that of pulse rate, respiration rate and body temperature in cattle. This increase in respiration rate may be used as an index of discomfort in large animals (Bhattacharya et al., 1965). The higher rate of RR and RT in crossbred cattle was observed than zebu cattle due to their poor adaptability under tropical climatic conditions (Singh and Upadhyay, 2009). Hansen (2004) also stated that cattle from zebu breeds are better able to regulate body temperature in response to heat stress than those of cattle from a variety of bos taurus breeds of European origin.

23.1.6 Compromise in metabolism and production

Animals gain the heat from internal metabolism and from external sources i.e., from environment. Higher temperature significantly affects the production performance of animal. The ambient temperature around 25°C found to be suitable to achieve maximum productive performance under tropical climatic conditions. Increase of ambient temperature from 25°C showed the effect on animal's performance. The level of heat stress can be classified as mild, medium and severe. Temperature humidity index (THI) is commonly considered as an indicator of thermal climatic conditions in dairy animals. THI can be calculated using the formula $THI = 0.72(W + D) + 40.6$, where "W" is wet bulb temperature and "D" is dry bulb temperature in °C (McDowell, 1972). Mild, medium, and severe stress were classified on the basis of THI range, i.e. 72–80 (mild), 80–90 (medium), and 90–98 (severe). Livestock species are comfortable at THI values between 65 and 72 (Upadhyay et al. 2007). The milk yield of high yielder dairy cattle showed significant decrease when the THI rises from 74. Milk yield was significantly decreased by 0.12 (1.2%) kg for each unit of THI above 74 in Holstein Friesian cows from four herds (Sadek et al. 2015). The threshold level of THI may vary from breeds to breeds, it also depends on the productive level and adaptability of the animals in different agro-climatic conditions. The THI value ranges between70-72 is considered by majority of the researchers to be a warning signal to make the lactating animal cool by modification of micro-climatic conditions. The compromisation of production level in higher THI or under heat stress is the adaptive mechanism to reduce the internal heat production. The more drastic reduction of milk yield was observed in high yielder dairy animals than low yielders. This is due to the high metabolic rate resulted in more heat production. Here the priority is not the production but to maintain the body homeostasis or body temperature for their survival. The small increase of body temperature might be life threathening. The low metabolic rate is one of the adaptive mechanism observed in zebu cattle living in hot humid regions. Zebu cattle can survive, reproduce and can give milk under extreme climatic conditon where the quality and quantity of feed and fodder is low which might be due to their low metabolic. This characteristics of zebu also indicate the better recycling ability of nutrient than temperate breeds (Bayer and Feldmann, 2003). The crossbred animal cannot able to maintain their production performance which reduced drastically during extreme climatic stresses. Under poor quality grass and extreme climatic conditions in tropical climate, Bos taurus cattle cannot able to maintain their health status, the immunity compromised, reproductive failure and production loss. This change is not significantly seen in adapted breeds. The lower expression of thyroid function related genes were observed in thermo tolerant breed (Tharparkar) than crossbred (Karan Fries) cattle (Naidu, 2016). The thyroid hormone, skin temperature and rectal temperature were found to be correlated positively with the expression level of deiodinase type2 (DIO2) gene in peripheral blood mononuclear cells (PBMC) (Naidu, 2016). Similarly, the thyroid hormone receptor protein 11 recorded lower in zebu cattle than crossbred and the positive correlation of this protein with cortisol, rectal temperature, respiration rate and skin temperature (Naidu, 2016). It further gives evidence of their low metabolism and adaptability under tropical climate.

23.1.7 Antioxidant defence system

Under heat stress, free radicals are generated in the body and are neutralized by intracellular antioxidant enzymes produced in the different tissues of the body depending upon their metabolic activity. When reactive oxygen (free radicals) is produced faster than they can be safely neutralized by antioxidants, the oxidative stress results (Bernabucci et al., 2002). An increase in the reactive oxygen species (ROS) production and the promotion of cellular oxidation events, when cells were exposed to heat stress have been observed by Kim et al. (2005). The increase in ROS reflects increased level of thermal stress. Keller et al. (2004) also observed that exposure of animals to elevated temperatures accelerated mitochondrial respiration and increased mitochondrial ROS formation. However, the production of ROS found to be lower in heat tolerant cattle (Tharparkar) than heat sensitive cattle (Karan Fries) under heat stress (Singh et al., 2014). Excess ROS production by intensively respiring mitochondria induces cellular damage. The breed/species differences observed in the levels of the ROS in Tharparkar and Karan-Fries cattle can be attributed to their difference in adaptability to different environmental conditions (Singh et al., 2014; Uttarani et al., 2014a, b). The major defense in detoxification of superoxide anion and hydrogen peroxide resulted from oxidative stress, are Super Oxide Dismutase (SOD), Catalase and Glutathione peroxidase (McCord and Fridovich, 1969; Chance et al., 1979). Erythrocytic SOD and Catalase increased significantly after 3h of exposures in climatic chamber (Lallawmkimi et al., 2012). Kumar (2005) observed a significant positive correlation of THI with the erythrocyte SOD and Catalase activity in Murrah buffalo and KF cattle. The highest increase was registered in KF followed by Murrah.

23.1.8 Molecular chaperones

The molecular chaperones play a key role in assisting the folding or unfolding of proteins and the assembly or disassembly of other macromolecular structures. The climate change scenario compelled us to have a more lucid understanding of the mechanisms of thermal tolerance in livestock species has become imminent (Crozier *et al*. 2008). Different attempts had been made to find out the level of thermotolerance of cattle and buffaloes based on physiological and cardinal reactions. The levels of animal's heat tolerance can be assessed by a group of protein family known as Heat Shock Proteins (HSPs). Cellular tolerance to heat stress is mediated by HSPs. Heat shock proteins are a large protein family consisting of both constitutively expressed and inducible proteins, classified according to their molecular weight (Kregel, 2002). The expression of constitutive HSPs increases in response to stress, whereas inducible HSPs are expressed only after stress. HSPs are representative of stress proteins, and their cellular up-regulation especially that of HSP70, provides resistance to the cell against stressors because the HSPs re-fold or degrade denatured proteins produced by stressors (Morimoto and Santoro, 1998). Stressful conditions in animals elicit HSP synthesis especially the HSP 70. HSP 70 function as molecular chaperones in restoring cellular homeostasis and promoting cell survival (Collier *et al*. 2008).Therefore intracellular levels of HSPs were suggested as indicators of thermotolerance in livestock species. Heat shock proteins involved in these responses are highly conserved and these molecular chaperones encompass several families, play important physiological roles and help in coping with heat stress (Parsell and Lindquist, 1993). HSPs have been considered to play crucial role in environmental stress tolerance and in thermal adaptation (Sorensen *et al*., 2003). Study of Daugaard *et al*. (2007) demonstrated that some of the cytosolic HSP70 family members deal with the cellular stress response and others are involved in tissue- specific and housekeeping biological tasks. Several studies carried out on bovine, mice and human cells gave evidence that constitutive elevation of the inducible HSPs levels in gene and protein expression provides cytoprotection upon thermal stress (Collier *et al*., 2006). Among HSPs family, HSP70 transcription is increased by heat shock as well as other stress stimuli such as oxidative stress, ischemia, inflammation, or aging (Favatier *et al*., 1997). However, continuous temperature rise does not protect cellular damage due to an imbalance between various physiological and cellular functions (Patir and Upadhyay, 2010). Among members of the HSP family, HSP70i (namely, HSPA1A and HSPA2) is the most temperature sensitive and induced by various physiological stressors, pathological stressors and environmental stressors (Beckham *et al*., 2004; Kumar *et al*., 2015). The expression of HSP70i is stress inducible and can only be detected following a significant stress upon the cell or organisms (Satio *et al*., 2004). HSPA1A and HSPA2 play a crucial role in guiding conformational status of the proteins during folding and translocation (Arya *et al*., 2007). In the hot environmental niche, a greater amount of constitutive HSP70.8 (HSPA8) is found during non-stress conditions (Singh *et al*., 2014). The HSPA8 assists in the day to day cell functions of protein folding and unfolding, prevention of polypeptide aggregation, disassembly of large protein complexes, and aid in the translocation of proteins between cellular compartments (Gething, 1997). Kishore *et al*. (2014) reported highest level of expression of HSPs throughout the time period of heat stress in buffaloes, followed by Holstein Friesian (HF) and Sahiwal cows. The higher abundance of HSP70 mRNA at each time point after heat stress showed prolonged effect of heat stress in HF PBMCs. The viability data indicated that HF PBMCs to be the most affected to the heat shock, whereas Sahiwal PBMCs were least affected, indicating its better adaptability during the heat stress condition. This differential expression of HSP70 mRNA in the present study could be attributed to its genetic divergence across the studied species (Sodhi *et al*., 2013). The expression of HSP70 family genes increased significantly in heat stress exposed goat blood mononuclear cells (PBMC) as compared to un-stress cells (Mohanarao *et al*., 2013). Microarray analysis revealed that a total of 460 transcripts were differentially expressed with a fold change of P2 in peripheral blood leukocytes of heat exposed (42 °C, 4 h) Tharparkar (Bos indicus) cattle and the heat stress affects expression of significant number of genes in peripheral blood leukocytes and further analysis is required to understand their functional role in livestock (Kolli *et al*., 2014).

23.2 Mitigation Strategies of Heat Stress Under Tropical Climatic Conditions

23.2.1 Breeding policy

Livestock breeding policy in country needs to be transformed towards the resilience of livestock under climate change scenario. Prioritization should be given to species and breeds most adaptable/suitable for each bioclimatic zone. Livestock production is predominantly owned by small holder farmers of poor resources. Incentives may be given to the farmers for active participation in improvement programme of prioritized breeds in the form of artificial insemination services at farmers' door without charge, health coverage, animal insurance at nominal charges and institutional credit at zero percent interest for procurement of animals of prioritized breeds. Establishment of mother bull farms of prioritized breeds of cattle and buffalo in major climatic zones with public private partnership will be helpful.

23.2.2 Selection, objectives and breeding strategies

Several breeding strategies may be considered to establish new adaptations in breeds of utility: 1) selection within the breeds for the adaptive trait; 2) Identification of a new breed amongst the conserved breeds and select within that breed; 3) indentify the traits that the commercial breed is lacking; detect the genes underlying these traits and introgress these genes from another breed using a combined QTL detection and introgression scheme (Yazdi *et al*. 2010); 4) perform crossbreeding to generate a new, synthetic breed and then select within the synthetic breed; 5) use of genomic selection to speed up the genetic progress (Meuwissen *et al*., 2001). The approach which results in the fastest adaptation to the new production circumstances will be favoured. Genomics will surely play a vital role in all of these areas, as well as in implementation of the results obtained. Increased characterization with high-throughput single nucleotide polymorphism (SNP) assays or genome sequencing will be necessary for unraveling the physiological basis for adaptation. Species-wide HapMap studies (Gibbs *et al*., 2009; Jiang *et al*., 2014) and multi-species studies (Stella, 2014) have represented a valuable first step in understanding the genome and its function in adaptation, but must be expanded to more breeds and geographical areas and augmented with more information on production environments. Metagenomics can provide insight regarding the co-adaptation of Animal Genetic Resources (AnGR) with other organisms in their production environments. Genomic selection has the potential to expedite both pure and crossbreeding programmes for adaptation, assuming phenotypes are available (Hayes *et al*. 2012).

23.2.3 Genetic approach

Although all living organisms are naturally and frequently exposed to many kinds of stressor during their life span, it is the fittest that adapts better and survives to produce and reproduce, a process that occurs over a long period of time. Thermal stress has been shown to affect production in tropical regions (Silanikove, 2000). Nevertheless, well adapted animals have been characterised by the maintenance or minimum loss of production during stress, high reproductive efficiency and disease resistance, as well as longevity and low mortality rates (West, 2003). The genetic approach to mitigate climate change adversity should include measure such as:

- Identifying and strengthening local genotype animals that are resilient to climatic stress/extremes.

- Genetic selection for heat tolerance or bringing in types of animals that already have good heat tolerance and crossbreeding the local genetic population with heat and disease-tolerant breeds.

3. Identifying the genes responsible for unique characteristics like disease tolerance, heat tolerance and ability to survive in low-input conditions, and using these as a basis for the selection of future breeding stock will help to mitigate the adverse effect of climate stress.

23.2.4 Genetic selection for heat tolerance

The advantage of marker-assisted selection, including genomic selection, over traditional methods is greatest where traditional methods are difficult to implement (Meuwissen and Goddard 1996). This usually occurs because the phenotype of interest cannot be observed on selected candidates at the age, when they can first be used for mating. For instance, milk production cannot be observed in bulls, meat tenderness cannot be measured on the live animal and adult wool production cannot be observed in yearling sheep. This limitation of traditional selection is more widespread than sometimes acknowledged. Given the complexity of the traits related to adaptation in tropical environments, the discovery of genes controlling these traits is a very difficult task. One obvious approach to identifying genes associated with acclimation to thermal stress is to utilize gene expression microarrays in models of thermal acclimation to identify changes in gene expression during acute and chronic thermal stress. Another approach would be with single gene deletions exposed to a defined thermal environment. This permits the identification of those genes that are involved in key regulatory pathways for thermal resistance and thermal sensitivity. Finally, gene knockout models in single cells will allow the better delineation of cellular metabolic machinery required for acclimatization to thermal stress. Those genes identified key to the process of thermal acclimation will then need to be mapped their chromosomal location, and the sequences of these genes will need to be determined in order to see if there are single nucleotide polymorphism (SNPs) associated with changes in the coding for gene expression or protein function. Identification of SNPs associated with variation in animal resistance or sensitivity to thermal stress will permit the screening of animals for the presence or absence of desirable or undesirable alleles at an early age.

23.2.5 Single nucleotide polymorphisms (SNPs)

Efforts were made to link the thermal-stress related phenotypes with genotypes to identify single nucleotide polymorphisms (SNPs) in ATP1A1 gene at position (2789) and HSP70.1 at position (895) were associated with the heat tolerance trait in dairy cattle (Liu et al., 2011; Deb et al., 2013). Li et al., (2011) reported SNPs at g.1524G>A, g.3494T>C and g.6601G>A loci of HSP70A1A gene having better thermo-tolerance in Chinese Holstein cattle. In contrast allele T at SNP g.4338>C of HSP90AB1 gene was found to be associated with heat tolerance coefficient in Thai native cattle (Charoensook et al., 2013), Sahiwal and Frieswal cattle in India (Deb et al., 2013; Sajjanar et al., 2015). In Jersey crossbred cows, CC genotype had better thermo-tolerance capacity having SNP at (T17872112C) locus of HSP90AB1 gene (Sailo et al., 2015). Similarly, AA genotype at (C27007790A) locus of ATP1A1 gene in Jersey crossbred cows was desirable for respiration rate and heat tolerance coefficient (Das *et al*. 2015). Sahiwal cows with AA genotype at A1209G locus and CC genotype at A3292C locus in HSP90AA1 gene had better thermo tolerance (Kumar *et al*., 2015). Similarly, in Karan Fries cows, GG genotype at locus A1209G and AC genotype at locus A3292C in HSP90AA1 gene had better thermotoleranace (Kumar *et al*. 2015). Verma *et al*. (2015) reported that GA genotype in ATP1B1 gene at SNP g.507G>A locus is favorable for the heat tolerance in the Sahiwal cattle. Genetic polymorphisms analysis is one of the acceptable tools for selection of breeding bulls for thermo-tolerance in near future.

Two specific genes associated with the effect of heat stress on milk yield, the slick gene (Liu *et al*. 2011) and an allele of ATP1A1 (Olson *et al*., 2003) are associated with lower rectal temperature (RT). Olson *et al*. (2003) reported the presence of a slick hair gene in the bovine genome. This gene is dominant in nature, and cattle carrying the dominant allele of this gene have slick hair and are able to maintain body temperature at lower rates. Slick hair has a positive effect on growth and milk production under dry tropical climatic conditions. The phenomenon of cross-resistance, where exposure to one stressor enhances resistance to other stressor have been reported by Hoffman *et al*. (2003). This suggests that heat tolerant cattle may also be tolerant to other stressors such as disease, poor feed quality, parasites etc. such stress- tolerant cattle may have lower culling rates and thus may stay in herds for longer. Core body temperature during heat stress is a heritable trait varying from 0.15 to 0.31 in dairy cattle (Seath, 1947; Dikmen *et al*., 2012). Reliability of genetic estimates for rectal temperature similar to other genetically controlled traits (VanRanden *et al*., 2009; Hayes *et al*., 2010) should be improved by genome-wide association studies (GWAS) to identify SNPs associated with rectal temperature. Quantitative trait loci (QTL) can be identified for low heritability traits and used to improve the reliability of genetic estimates, despite the gain in reliability being less than for more heritable traits (Cole *et al*., 2011;Wiggans *et al*., 2011).

23.2.6 Residual feed intake and supplementation of vitamins and trace elements

The low residual feed intake (RFI) animal produced less enteric methane than high RFI animals and showed a positive relation with methane emission (Dudi *et al*., 2016). The selection of low RFI animals having good productivity can be one of the strategies against heat stress. Supplementation of Vitamins, minerals and amino acids play significant role to reduce the adverse effects of heat stress. Supplementation of Zn (80 and 120 ppm) found to reduce the postpartum estrus interval, days to first insemination, service period, services per conception and increase the conception rate in Karan Fries cows (Patel *et al*., 2016). In vitro studies revealed that the zinc supplementation (0.01mM) to heat stressed

PBMCs down regulate the HSPs and reduce the IL-10 concentration (Sheikh *et al.* 2016). The down regulation of HSP in zinc treated PBMCs might be due to decrease in free radical production at the cellular levels, which might protect the cellular damage. The Zn supplementation also reduces the concentration of superoxide dismutase (SOD), catalase (CAT) in heat stress (42°C) PBMC cells of Karan Fries and Sahiwal cows suggesting an ameliorative measures of heat stress and immune-modulator in periparturient cows (Sheikh *et al.* 2016). The combination of vitamin-E and Zn supplementation showed an improvement in immunity during peripartum period of Sahiwal cows. The higher concentration of total immunoglobin and interleukin-2 were recorded in the plasma of Sahiwal cows after calving as compared to non-supplemented cows (Chandra *et al.* 2014). Vitamin-E supplementation to murrah buffalo calves also improved the growth rate (higher average daily gain compared to control), metabolic and endocrine profile (Singh *et al.*, 2012). The studies carried out on heifer and lactating buffaloes during thermal stress (heat and cold) demonstrated the ameliorative effect of vitamin-E against thermal stress and improved the performance of lactating buffalo (Lallawmkimi *et al.*, 2013). The combination of Zn and vitamin-E supplemented combat the lipid peroxidation, non-esterified fatty acid and improved the milk yield in Sahiwal cows (Chandra *et al.*, 2013). Vitamin C supplementation has been found to ameliorate the adverse effect of heat stress and worked as immune-modulator (Ganaie *et al.* 2013). Apart from being one of the most potent sources of vitamin C, amla is rich in amino acids, tannins and flavinoids, that are known to protect the body against free radicals. Therefore, use of amla powder as an antioxidant can be of practical importance to ameliorate the adverse effect of heat stress in buffaloes (Lakhani *et al.* 2016). Administration of betaine (a trimethyl form of glycine) shown to reduce the adverse effect of heat stress, the lower expression of HSPs was observed in treated goat (Dangi *et al.*, 2015). Betaine can be utilized as methyl donor by mammals and it participate in protein and lipid metabolism, it can also use as an organic cellular osmo-protectant.

23.2.7 Bicarbonate and direct fed microbials

Elevated environmental temperature negatively affects the physiological mechanism of rumen that might increase the metabolic disorder in ruminants (Soriani *et al.*, 2013). Feeding of more fermentable carbohydrate during heat stress may lead to acidosis and laminitis (Lettat *et al.*, 2012). The less intake of roughages and more production of propionate and butyrate may alter the rumen pH resulting decrease in rumen motility. Lower production of saliva affects the buffering capacity in the rumen. Therefore, supplementation of bicarbonate stimulates saliva production and feeding of direct fed microbes or yeast also helpful for maintaining rumen pH in heat stress cattle and buffalos.

Dietary Fat: Feeding of dietary rumen bypass fat is an effective source of energy during summer to combat the negative energy balance. The heat increment of fat is minimum i.e. approximately 50% less than typical forages. Supplementation of fat shown to increase the milk yield (Wang *et al.*, 2010). Supplemental fat at 5% found to enhance lactation performance under thermo-neutral and heat stress conditions (Knapp and Grummer, 1991).

Glucose: Higher production of glucose precursors i.e. propionate in the rumen could be effective for maintaining production. But it alters the rumen p H and motility. Therefore, safe and effective way is advisable for maintaining the milk production. Supplementation of monensin found to stabilize the rumen pH during stress (Schelling 1984). Propylene glycol is typically fed in early lactation, but may also be an effective method of increasing propionate production during heat stress. With the increasing demand for biofuels and subsequent supply of glycerol, it will be of interest to evaluate glycerols efficacy and safety in ruminant diets during the summer months.

Protein: Heat stress ruminant showed to undergo negative nitrogen balance (Kundu *et al.*, 2013). Bypass protein (formaldehyde treated mustard cake) showed to increase 15% milk production (Walli *et al.*, 2005). Supplementation of bypass protein is beneficial for maintaining energy requirement during heat stress (Kundu *et al.*, 2013).

23.2.8 Effect of cooling during summer

The significant impact of evaporative cooling was observed during late gestation in murrah buffaloes during summer season under tropical climatic conditions. The physiological response (RR, RT, and PR) and skin temperature at thorax were lower in cooled Murrah buffalo than non-cooled buffaloes (Aarif and Aggarwal, 2016). Blood pCO_2, pO_2, PCV, Hb were found to be higher in cool buffalo, similarly, the dry matter intake (DMI) also increased. The milk yield, FCM, fat yield, lactose yield and total solid yield were recorded higher in cooled buffaloes, indicates the importance of cooling during summer season. The mRNA expression of prolactin receptor gene (PRL-R) was higher, whereas cytokine signalling gene 1(SOCS-1) and interleukin-6 was lower in cooled parturition buffaloes (Aarif and Aggarwal, 2015). The level of interferon gama was significantly higher at -20 and +20 days of parturition in cooled animals than non-cooled animals. Verma *et al.* (2016) also found the improvement in conception rate in cooled buffaloes during summer season. The cooling effect using fans with mist showed significant improvement in milk yield in lactating Holstein Friesian cows as compared to control groups. The milk yield of cooled cows for 3 h AM and 3 h PM increased the milk yield by 2 kg per day as compared control (providing only one hour cooling) (Reyes *et al.*, 2010). The output of milk energy also shown to increase in treatment group and the milk yield was recorded 21.12 kg/day in cooled group (3h AM+ 3h PM), 19.1 Kg/ day was recorded in control group (1 h cooled) (Reyes *et al.* 2010). There are many method of cooling system during hot and dry conditions, but water spray and fans are often used. Evaporative cooling is the effective method, which includes mist, fog and sprinkling (Armstrong, 1994). Cows housed in pens and cooled by water spray and fans showed a great improvement in milk production, milk fat and postpartum reproductive performance (conception rate and days open), calf birth weight compared to non-cooled Holstein Friesian cows in hot and dry conditions (Reyes *et al.*, 2006). Reyes *et al.* (2006) reported increase in milk yield by 2.61kg/day in cooled cows compared to non-cooled cows (Reyes *et al.* 2006). Similarly, Wolfenson *et al.* (1995) reported improvement in milk yield by 3.5 kg/day in cows cooled with water spray and fans during dry period.

23.2.9 Shade

Shade provides protection from direct sunlight and allows cooling effect of wind. Availability of shade affects the production of animals. Providing shade to Holstein Frisian during summer showed 3% higher milk production as compared to cows providing no shade (Fischer *et al.*, 2010). When lactating cows were providing adequate shade, their milk production and reproductive performance increased. Milk yield of shaded cows was recorded 16.6 kg/day than non-shade cows who yielded 15.0 kg/day (Roman-Ponce, 1977). Adequate shade also improved the conception rate. Conception rate was significantly higher for shaded cows (44.4%) than no shade cows (25.3%) (Roman–Ponce, 1977). Provision of shade helps in maintaining the productive performance and reduces the radiant heat load upto 30% (Blackshaw *et al.*, 1994).

23.2.9.1 Supplementation as Package

Supplementation of niacin, yeast, edible oil and provided with curtains, additional ceiling fans, and mist showed lower physiological response (RR, RT) and higher average milk production, total fat and SNF in treatment group of lactating buffaloes (Das *et al.*, 2014). This study indicated that the nutrient supplementations, microclimate modifications, and management alterations together in the form of one package helped in reducing heat stress.

23.3 Animal Health

The evaporative cooling showed a beneficial effect to the health of animals and reduced the incidence of reproductive disorders. The incidences of retention of placenta, metritis and endometritis was 37.25 %, 25% and 12.25%, respectively in non-cooled animals whereas only retention of placenta was observed in cooled (12.5%) Murrah buffaloes (Aarif and Aggarwal, 2015). Strengthening the diagnostic laboratories, veterinary dispensaris, hospital, veterinary institute including the awareness camps against climate change, vaccination, preventive measures and extension camps among the farmer should be carried out time to time. Strengthening the animal health infrastructure and ensuring surveillance for diseases, most likely to be influenced by climate. This difficulty could be overcome by establishing a coordinated circumpolar infectious disease surveillance system. If coordinated with appropriate climate data, such a network could also be used to monitor the emergence of climate-sensitive infectious diseases providing both early detection opportunities and precautions for livestock health intervention. Early warning systems should be developed to generate preventive health messages. Timely vaccination of livestock is advisable as a precautive measure. Spread of disease vectors can be reduced by using flies or insect repellants. There should be provision of clean water supply to livestock as water is a good medium for the spread of infectious diseases.

23.4 Strengthening of Existing Capacities

Capacity building and capacity development are among the most urgent requirements for addressing climate risk, particularly at local level. Community capacity to understand climate risk issues and effectively use the available information is important. Development of necessary institutions and networks, plan and build appropriate climate change actions is an essential prerequisite for effective adaptation.

23.5 Development of Early Warning System

Meaningful warning information can be disseminated timely to enable individuals, communities and organizations threatened by a hazard to prepare and to act appropriately and in sufficient time to reduce the possibility of harm or loss due to climatic disaster.

23.6 Conclusion

Responding to the challenges of global warming necessitates a paradigm shift in the practice of agriculture and in the role of livestock within farming systems. Efficient and suitable management practices should be adopted as per the breed and agro-climatic conditions. Integrating grain crops with pasture plants and livestock could result in a more diversified system that will be more resilient to higher temperatures and elevated carbon dioxide levels expected in future. The lack of scientifically back up information, indigenous knowledge about cattle genetic diversity has been revealed to offer an important knowledge base for selective breeding. Indigenous knowledge on animal breeding is a valuable resource about the existence of cattle breeds and their adaptive traits. Increase in temperature causes severe damage to the physiology, the metabolism and to the healthiness of animals. There is considerable research evidence showing a substantial decline in animal performance depicting heavy economic losses when subjected to heat stress. With the progress of molecular biotechnologies, effort in selecting animals that up to now has been primarily oriented toward productive traits, from now on, must be oriented toward fitness, and above all adaptability to heat stress. In this way molecular biology allow to directly achieve genotypes with the necessary phenotypic characteristics. These tools will enable an improved accuracy and efficiency of selection for heat tolerant animals. As livestock is an important foundation of livelihood, it is essential to find appropriate solutions not only to maintain this industry as an economically feasible enterprise, but also to boost up the profitability and decrease environmental pollutants by reducing the ill-effects of climate change. Animal agriculture will witness more tough challenges in many fields in the 21st century. Decision makers, extension services and research institutions have to support livestock activities to handle as best with less loss of production, deteriorating of animal products, expansion of land desertification and the worsening of animal health under the effects of the climate change we expect in the coming decades.

References

Aarif, O and Aggarwal A. 2015. Evaporative cooling in late-gestation Murrah buffaloes potentiates immunity around transition period and overcomes reproductive disorders. Theriogenology 84 (7): 1197-1205.

Al-Haidary A, Spiers DE, Rottinghaus GE, Garner GB and Ellersieck MR. 2001. Thermoregulatory ability of beef heifers following intake of endophyte-infected tall fescue during controlled heat challenge. Journal of Animal Science **79**: 1780–1788.

Armstrong DV. 1994. Heat stress interaction with shade and cooling. J Dairy Sci., 77:2044–2050

Arya R, Mallik M and Lakhotia SC. 2007. Heat shock genes—integrating cell survival and death. Bioscience Journal 32: 595–610.

Barker JSF. 2009. Defining fitness in natural and domesticated populations. In: Adaptation and Fitness in Animal Populations (Ed. by J. van der Werf, H. Grazer, R. Frankham & C. Gondro), pp. 3 – 14. Springer, New York, NY.

Bayer, W. and Feldmann, A. 2003. Diversity of animals adapted to smallholder system. Conservation and Sustainable Use of Agricultural Biodiversity. http://www.eseap.cipotato.org/UPWARD/Agrobio-sourcebook.htm.

Beckham JT, Mackanos MA, Crooke C, Takahashi T, O'Connell-Rodwell C, Contag CH and Jansen ED. 2004. Assessment of cellular response to thermal laser injury through bioluminescence imaging of heat shock protein 70. Photochem Photobiol 79: 76–85.

Berman A and Volcani R. 1961. Seasonal and regional variations in coat characteristics of dairy cattle. Aust. J. Agric. Res., 12: 528–538.

Bernabucci U, Ronchi B, Lacetera N and Nardone A. 2002. Markers of oxidative status in plasma and erythrocytes of transition dairy cows during hot season. J Dairy Sci., 85:2173-2179.

Bhattacharya, S., Acharya, A. K., Choudhary, T. M. and Deb, N. C. 1965. Seasonal variations in body temperature, pulse rate, respiration rate and Hb concentration of blood in different breeds of Indian heifers and growing bulls. Indian J. Vet. Sci. A.H. 35 (1):47.

Blackshaw, Judith K., and A. W. Blackshaw. 1994. Heat stress in cattle and the effect of shade on production and behaviour: a review." Australian Journal of Experimental Agriculture 34 (2): 285-295.

Broom, D.M. and K.G. Johnson. 1993. Stress and animal welfare. Chapman and hall, London, 211.

Chance B, Sies H and Boveris A. 1979. Hydroperoxide metabolism in mammalian organs. Physiol Rev., 59(3):527-605.

Chandra G, Aggarwal A, Kumar M, Singh AK, Sharma VK and Upadhyay RC. 2014. Effect of additional vitamin E and zinc supplementation on immunological changes in peripartum Sahiwal cows. Journal of animal physiology and animal nutrition 98 (6): 1166-1175.

Chandra G, Aggarwal A, Singh AK, Kumar M and Upadhyay RC. 2013. Effect of Vitamin E and Zinc Supplementation on Energy Metabolites, Lipid Peroxidation, and Milk Production in Peripartum Sahiwal Cows. Asian Australas. J. Anim. Sci., 26 (11):1569-1576.

Charoensook R, Knorr C, Brenig B and Gatphayak K. 2013. Thai pigs and cattle production, genetic diversity of livestock and strategies for preserving animal genetic resources. Maejo Int. J. Sci. Technol., 7(01): 113-132.

Cole M, Lindeque P, Halsband C and Galloway TS. 2011. Microplastics as contaminants in the marine environment: A review. Mar. Pollut. Bull., doi:10.1016/j.marpolbul.2011.09.025.

Collier RJ and Collier JL. 2012. Environmental Physiology of Livestock, First Edition. Edited by RJ Collier and JL Collier. John Wiley & Sons, Inc. New York.

Collier RJ, Collier JL, Rhoads RP and Baumgard LH. 2008. Genes involved in the bovine heat stress response. Journal of Dairy Science 91: 445.

Collier RJ, Dahl GE and VanBaale M J. 2006. Major advances associated with environmental effects on dairy cattle. Journal of Dairy Science (Centennial Issue) 89: 1244–1253.

Crozier LG, Hendry AP, Lawson PW, Quinn TP, Mantua NJ, Battin J, Shaw RG and Huey RB. 2008. Potential responses to climate change in organisms with complex life histories: evolution and plasticity in Pacific salmon. Evolutionary Journal compilation @ 2008 Blackwell Publishing Ltd 1:252–27.

Dandage SD. 2009. Estimates of thermal load and heat exchange in cattle and buffaloes. M.V.Sc. Thesis submitted to NDRI Deemed University, Karnal (Haryana), India.

Dangi SS, Dangi SK, Chouhan VS, Verma MR, Kumar P, Singh G and Sarkar M. 2015. Modulatory effect of betaine on expression dynamics of HSPs during heat stress acclimation in goat (Capra hircus). Gene 575 (2): 543-550.

Das KS, Singh JK, Singh G, Upadhyay RC, Malik R and Oberoi PS. 2014. Heat stress alleviation in lactating buffaloes: Effect on physiological response, metabolic hormone, milk production and composition. Indian J. Anim. Sci., 84 (3): 275-280.

Das U, Vinayachandran PN and Behara A. 2015. Formation of the southern Bay of Bengal cold pool. Clim. Dyn., 1-15.

Daugaard M, Rohde M and Jaattela M. 2007. The heat shock protein 70 family: Highly homologous proteins with overlapping and distinct functions. FEBS Lett., 581(19): 3702-10.

Deb R, Sajjanar B, Devi K, Reddy KM, Prasad R, Kumar S and Sharma A. 2013. Feeding animals with GM crops: Boon or bane? Indian Journal of Biotechnology 12: 311-322.

Dikmen S, Cole JB, Null DJ, and Hansen PJ. 2012. Heritability of rectal temperature and genetic correlations with production and reproduction traits in dairy cattle. J Dairy Sci., 95(6):3401-5. doi: 10.3168/jds.2011-4306.

Dowling DF. 1956. An experimental study of heat tolerance of cattle. Australian Journal of Agricultural Research 7: 469.

Dudi K, Datt C, Mohini M, Devi I, Sharma, VK, Kundu SS and Singh SV. 2016. Nutrient utilisation and enteric methane emissions in female Sahiwal calves of varying residual feed intake. Indian J. Dairy Sci., 69: 555-558.

Favatier F, Bornman L, Hightower LE, Gunther E, Polla BS. 1997. Variation in hsp gene expression and hsp polymorphism: do they contribute to differential disease susceptibility and stress tolerance? Cell Stress Chaperones 2:141–155.

Finch J. 1984. It's Great to Have Someone to Talk to": The Ethics and Politics of Interviewing Women', in C. Bell and H. Roberts (eds.) Social Researching: Politics, Problems, Practice. London: Routledge and Kegan Paul.

Finch VA and Western D. 1977. Cattle colors in pastoral herds: natural selection or social preference? Ecology 58: 1384–1392.

Finch VA. 1985. Comparison of non-evaporative heat transfer in different cattle breeds. Crop and Pasture Science 36(3): 497-508.

Fischer J, Koch L, Emmerling C, Vierkotten J, Peters T, Brüning JC and Rüther U. 2010. Fischer et al. reply. Nature 464(7289): E2-E2.

Ganaie AH, Shanker G, Nazir AB, Ghasura RS, Mir NA, Wani SA and Dudhatra GB. 2013. Biochemica and Physiological Changes during Thermal Stress in Bovines. J Veterinar Sci Technol 4:126. doi:10.4172/2157-7579.1000126.

Gething MJ. 1997. Guidebook to molecular chaperones and protein folding catalysts. Oxford University Press, Oxford.

Gibbs R A, Taylor J F, Van Tassell C P, Barendse W, Eversole K A, Gill C A, Green R D, Hamernik D L, Kappes S M, Lien S, Matukumalli L K, McEwan J C, Nazareth L V, Schnabel R D, Weinstock G M, Wheeler D A, Ajmone-Marsan P, Boettcher P J, Caetano A R, Garcia J F, Hanotte O, Mariani P, Skow L C, Sonstegard T S, Williams J L, Diallo B, Hailemariam L, Martinez M L, Morris C A, Silva L O, Spelman R J, Mulatu W, Zhao K, Abbey C A, Agaba M, Araujo F R, Bunch R J, Burton J, Gorni C, Olivier H, Harrison B E, Luff B, Machado M A, Mwakaya J, Plastow G, Sim W, Smith T, Thomas M B, Valentini A, Williams P, Womack J, Woolliams J A, Liu Y, Qin X, Worley K C, Gao C, Jiang H, Moore S S, Ren Y, Song X Z, Bustamante C D, Hernandez R D, Muzny D M, Patil S, San Lucas A, Fu Q, Kent M P, Vega R, Matukumalli A, McWilliam S, Sclep G, Bryc K, Choi J, Gao H, Grefenstette J J, Murdoch B, Stella A, Villa-Angulo R, Wright M, Aerts J, Jann O, Negrini R, Goddard M E, Hayes B J, Bradley D G, Barbosa da Silva M, Lau L P, Liu G E, Lynn D J, Panzitta F and Dodds KG. 2009. Genome-wide survey of SNP variation uncovers the genetic structure of cattle breeds. Science 324(5926):528-532.doi: 10.1126/science.1167936.

Greaves M. 2014. Was skin cancer a selective force for black pigmentation in early hominin evolution? Proceedings of the Royal Society of London B: Biological Sciences 281(1781): 20132955.

Haake and Holbrook. 1999. A.R. Haake, K. Holbrook, et al. The structure and development of skin I.M. Freedberg, A.Z. Eisen, K. Wolff (Eds.), Dermatology in General Medicine, McGraw-Hill, New York (1999), pp. 70–114.

Hansen PJ. 2004. Supplemental Antioxidants to Enhance Fertility in Dairy Cattle. dairy.ifas.ufl.edu/rns/2010/14-Hansen.pdf.

Hayes BJ, Lewin HA and Goddard ME. 2012. The future of livestock breeding: genomic selection for efficiency, reduced emissions intensity, and adaptation. Trends Genet 29, 206–214. 10.1016/j.tig.2012.11.009.

Hayes SM, Murray S, Dupuis M, Dawes M, Hawes IA and Barkun AN. 2010. Barriers to the Implementation of Practice Guidelines in Managing Patients with Nonvariceal Upper Gastrointestinal Bleeding: A Qualitative Approach. Canadian Journal of Gastroenterology 24 (5): 289-296. http://dx.doi.org/10.1155/2010/878135.

Hoffman M, Gora M and Rytka J. 2003. Identification of rate-limiting steps in yeast heme biosynthesis. Biochem. Biophys. Res. Comm. 310(4):1247-1253.

Indu B, Hooda OK and Upadhyay RC. 2016. Effect of thermal stress on physiological, hormonal and haematological parameter in Tharparkar and Karan Fries calves. Indian Journal of Dairy Science 69(4): 467-472.

Jenkinson D McE and Nay T. 1975. The sweat glands and hair follicles of different species of Bovidae. Aust. J. Biol. Sci., 28: 55-68.

Jiang W, Chen SY, Wang H, Li DZ and Wiens JJ. 2014. Should genes with missing data be excluded from phylogenetic analyses? Mol. Phylogenet Evol., 80:308-318.

Kadzere CT, Murphy MR, Silanikove N and Maltz E. 2002. Heat stress in lactating dairy cows: a review. Livestock Production Science 77:59.

Keller H, Yovsi R, Borke J, Ka J, Jensen H, and Papaligoura Z. 2004. Developmental Consequences of Early Parenting Experiences: Self-Recognition and Self-Regulation in Three Cultural Communities. Child Development 75 (6): 1745 – 1760.

Kibler HH and Brody S. 1952. Relative efficiency of surface evaporative, respiratory evaporative and non-evaporative cooling in relation to heat production in Jersey, Holstein, Brown Swiss and Brahman cattle, 50 to 1050 F. Res. Bull. Mo. agric. Exp. Sta. no. 497.

Kim W. Spear ED and Ng DT. 2005. Yos9p detects and targets misfolded glycoproteins for ER-associated degradation. Mol. Cell 19(6):753-764.

Kishore A, Vail A, Majid A, Dawson J, Lees KR, Tyrrell PJ and Smith CJ. 2014. Detection of atrial fibrillation after ischemic stroke or transient ischemic attack: a systematic review and meta-analysis. Stroke 45(2):520-526. doi: 10.1161/STROKEAHA.113.003433.

Klungland H, Vage DI, Gomez-Raya L, Adalsteinsson S. and Lien S.1995. The role of melanocyte-stimulating hormone (MSH) receptor in bovine coat color determination. Mammalian genome 6(9): 636-639.

Knapp DM and Ric R. Grummer. 1991. Response of lactating dairy cows to fat supplementation during heat stress. Journal of Dairy Science 74 (8): 2573-2579.

Kolli SK, Nakhi A, Archana S, Saridena M, Deora GS, Yellanki S, Medisetti R, Kulkarni P, Ramesh Raju R and Pal M. 2014. Ligand-free Pd-catalyzed C-N cross-coupling/cyclization strategy: An unprecedented access to 1-thienyl pyrroloquinoxalines for the new approach towards apoptosis. European Journal of Medicinal Chemistry 86C: 270-278.

Kregel KC. 1985. Heat shock proteins: modifying factors in physiological stress responses and acquired thermotolerance. J. Appl. Physiol., 92(5):2177-86.

Kumar A, Ashraf S, Goud T S, Grewal A, Singh SV, Yadav BR and Upadhyay RC. 2015. Expression profiling of major heat shock protein genes during different seasons in cattle (Bos indicus) and buffalo (Bubalus bubalis) under tropical climatic condition. Journal of Thermal Biology 51:55–64.

Kumar A. 2005. Status of oxidative stress markers in erythrocytes of heat exposed cattle and buffaloes. M VSc Thesis submitted to National Dairy Research Institute-ICAR, Karnal, Haryana.

Kumar S, Dagar SS, Ebrahimi SH, Malik RK, Upadhyay RC and Puniya AK. 2014. Prospective use of bacteriocinogenic pediococcus pentosaceus as direct-fed microbial having methane reducing potential. J. Integrative Agri. Advance., 14 (3): 561-566.

Kundu P, Brenowitz ND, Voon V, Worbe Y, Vertes PE, Inati SJ, Saad ZS, Bandettini PA and Bullmore ET. 2013. Integrated strategy for improving functional connectivity mapping using multiecho fMRI. Proc. Natl. Acad. Sci. U.S.A., 110: 16187-16192.

Lakhani, Preeti, Rajesh Jindal, and Shashi Nayyar. 2016. Effect of heat stress on humoral immunity and its amelioration by amla powder ('Emblica officinalis') supplementation in buffaloes." Journal of Animal Research 6 (3): 401.

Lallawmkimi MC, Singh SV, Hooda OK and Upadhyay RC. 2012. HSP 72 expression and antioxidant enzymes in Murrah buffaloes. Indian Journal of Animal Science 82:268–273.

Lallawmkimi MC, Singh SV, Upadhyay RC and De S. 2013. Impact of vitamin E supplementation on heat shock protein 72 and antioxidant enzymes in different stages of Murrah buffaloes during seasonal stress. Indian Journal of Animal Sciences 83 (9): 909–915.

Lettat A, Nozière P, Silberberg M, Morgavi DP, Berger C and Martin C .2012. Rumen microbial and fermentation characteristics are affected differently by bacterial probiotic supplementation during induced lactic and subacute acidosis in sheep. BMC Microbiology 12: 1–12.

Liu L, Xiao J, Peng Z H and Chen Y. 2011. In vitro metabolism of glycyrrhetic acid by human cytochrome P450. Acta Pharmaceutica Sinica 46(1):81-87.

Maibam U, Singh SV, Singh AK, Kumar S and Upadhyay RC. 2014a. Expression of skin colour genes in lymphocytes of Karan Fries cattle and seasonal relationship with tyrosinase and cortisol. Tropical Ani. Health and Prod. 46 (7):1155-1160.

Maibam U, Singh SV, Upadhyay RC, Kumar Suresh, Beenam and Singh AK. 2014b. Expression of skin colour gene in Tharparkar cattle during summer and winter season. Journal of Environmental Research and Development 9(1):113-119.

McCord JM and Fridovich I. 1969. Superoxide Dismutase. An enzymic function for erythrocuprein (hemocuprein). J. Biol. Chem., 25; 244 (22): 6049-55.

McDowell RE.1972. Improvement of Livestock Production in Warm Climates. San Francisco, CA, USA: W.H. Freeman and Company Publishers. pp. 51-53.

McGlinchey RP, Shewmaker F, McPhie P, Monterroso B, Thurber K and Wickner RB. 2009. The repeat domain of the melanosome fibril protein Pmel17 forms the amyloid core promoting melanin synthesis. Proceedings of the National Academy of Sciences, 106(33), 13731-13736.

McLean JA and Calvert DT. 1972. Influence of air humidity on the partition of heat exchange of cattle. Journal of Agricultural Science 78:303.

McRobie, HR, King LM, Fanutti C, Coussons PJ, Moncrief ND and Thomas APM. 2014. Melanocortin 1 Receptor (MC1R) Gene Sequence Variation and Melanism in the Gray (Sciurus carolinensis), Fox (Sciurus niger), and Red (Sciurus vulgaris) Squirrel. Journal of Heredity. doi:10.1093/jhered/esu006.

Meuwissen THE and Goddard ME. 1996. The use of marker haplotypes in animal breeding schemes. Gen.. Sel. Evol., 28:161-176.

Meuwissen THE, Hayes BJ and Goddard ME. 2001. Prediction of Total Genetic Value Using Genome-Wide Dense Marker Maps. Genetics 157:1819–1829.

Mirkena T, Duguma G, Haile A, Tibbo M, Okeyo AM, Wurzinger M and Sölkner J. 2010. Genetics of adaptation in domestic farm animals: A review. Livestock Science 132(1): 1-12.

Mohanaraoa GJ, Mukherjee A, Banerjee D, Gohain M, Dass G, Brahma B, Datta TK, Upadhyaya RC and De S. 2013. HSP70 family genes and HSP27 expression in response to heat and cold stress in vitro in peripheral blood mononuclear cells of goat (Capra hircus). Small Ruminant Research. doi: 10.1016/j.smallrumres.2013.10.014.

Morimoto RI and Santoro MG. 1998. Stress-inducible responses and heat shock proteins: new pharmacologic targets for cytoprotection. Nat. Biotechnol., 16(9):833-8.

Naidu CK. 2016. Metabolic profile and expression pattern of some genes in Tharparkar and Karan Fries (Tharparkar x Holstein Friesian) heifers during different seasons. M.V.Sc, thesis submitted to ICAR-NDRI, Karnal, Haryana.

Nay T and Hayman R H. 1956. Sweat glands in Zebu (Bos indicus L.) and European (23. taurus L.) cattle. I. Size of individual glands the denseness of their population and their depth below the skin surface. Australian Journal of Agricultural Research 7: 482.

Pan, Y.S. 1963. Quantitative and morphological variation of sweat glands, skin thickness, and skin shrinkage over various body regions of Sahiwal Zebu and Jersey cattle. Aust. J. Agric. Res. 14: 424–437.

Parsell, DA and Lindquist S. 1993. The function of heat-shock proteins in stress tolerance: degradation and reactivation of damaged proteins. Annual Review of Genetics 27:437–496.

Patel MV, Zhu JY, Jiang Z, Richman A, Van Berkum MF, Han Z. 2016. Gia/Mthl5 is an aorta specific GPCR required for Drosophila heart tube morphology and normal pericardial cell positioning. Dev. Biol., 414(1): 100—107.

Patir H and Upadhyay R C. 2010. Purification, characterization and expression kinetics of heat shock protein 70 from Bubalus bubalis. Research in Veterinary Science 88 (2): 258–262.

Prayaga KC, Henshall JM. 2005. Adaptability in tropical beef cattle: Genetic parameters of growth, adaptive and temperament traits in a crossbred population. Aust. J. Exp. Agric., 45: 971- 983.doi:10.1071/EA05045.

Reyes AL, Alvarez-Valenzuela FD, Correa- Calderon A, Saucedo-Quintero JS, Robinson PH and Fadel JG. 2006. Effect of cooling Holstein cows during the dry period on postpartum performance under heat stress conditions. Livestock Sci., 105: 198-206.

Reyes AL, Álvarez-Valenzuela FD, Correa-Calderón A, Algándar-Sandoval A, Rodríguez-González E, Pérez-Velázquez R and Fadel JG. 2010. Comparison of three cooling management systems to reduce heat stress in lactating Holstein cows during hot and dry ambient conditions. Livestock Science, 132 (1): 48-52.

Robertshaw D.1985. Heat loss of cattle In Volume 1 Stress physiology in Livestock Basic Principles, PP SS (M K Yousef) Florida CRC Press.

Roman-Ponce H, Thatcher WW, Buffington DE, Wilcox CJ, Van Horn HH. 1977. Physiological and production responses of dairy cattle to a shade structure in a subtropical environment. J. Dairy Sci., 60:424 – 430.

Sadek B , Khanian SS , Ashoor A , Prytkova T , Ghattas MA , Atatreh N , Nurulain SM , Yang KH , Howarth FC , Oz M. 2015. Effects of antihistamines on the function of human á7-nicotinic acetylcholine receptors. Eur. J. Pharmacol., 746: 308-16.

Sailor DJ, Georgescu M, Milne J and Hart M. 2015. Development of a national anthropogenic heating database with an extrapolation for international cities. Atmospheric Environment, 118: 7-18.

Saito K, Fujimura-Kamada K, Furuta N, Kato U, Umeda M and Tanaka K. 2004. Cdc50p, a protein required for polarized growth, associates with the Drs2p P-type ATPase implicated in phospholipid translocation in Saccharomyces cerevisiae. Mol. Biol. Cell., 15(7): 3418-32.

Sajjanar B, Deb R, Singh U, Kumar S, Brahmane M, Nirmale A, Bal S K and Minhas P S. 2015. Identification of SNP in HSP90AB1 and Its Association With the Relative Thermotolerance and Milk Production Traits in Indian Dairy Cattle. Anim. Biotechnol., 26 (1): 45-50.

Schelling, Gerald T. 1984. Monensin mode of action in the rumen. Journal of Animal Science 58 (6): 1518-1527.

Schleger AV and Turner HG. 1965. Sweating rates of cattle in the field and their reaction to diurnal and seasonal changes. Australian Journal of Agricultural Research 16: 92-106.

Seath DM. 1947. Heritability of heat tolerance in dairy cattle. J. Dairy Sci., 30:137–144.

Sharma SC. 1974. M.Sc. Thesis submitted to Punjab University, Chandigarh.

Sheikh, Aasif Ahmad, Anjali Aggarwal, and Ovais Aarif. 2016. Effect of in vitro zinc supplementation on HSPs expression and Interleukin 10 production in heat treated peripheral blood mononuclear cells of transition Sahiwal and Karan Fries cows. Journal of thermal biology 56: 68-76.

Silanikove N and Koluman (Darcan) N. 2014. Impact of climate change on the dairy industry in temperate zones: Predications on the overall negative impact and on the positive role of dairy goats in adaptation to earth warming. Small Rumin Res http://dx.doi.org/10.1016/j.smallrumres.2014.11.005.

Silanikove N. 2000a. The physiological basis of adaptation in goats to harsh environments. Small Ruminant Research 35 (6): 181-193.

Silanikove, N. 2000. Effects of heat stress on the welfare of extensively managed domestic ruminants. Livestock production science 67(1): 1-18.

Simon JD and Peles DN. 2010. The red and the black. Accounts of chemical research, 43(11): 1452-1460.

Singh AK, Upadhyay RC, Malakar D, Kumar S and Singh SV. 2014. Effect of thermal stress on HSP70 expression in dermal fibroblast of zebu (Tharparkar) and crossbred (Karan-Fries) cattle. Journal of Thermal Biology 43: 46-53.

Singh S, Kushwaha BP, Nag SK, Mishra AK, Singh A and Anele UY. 2012. In vitro ruminal fermentation, protein and carbohydrate fractionation, methane production and prediction of twelve commonly used Indian green forages. Anim. Feed Sci. Technol., 178 (1/2): 2-11.

Singh SV and Soren S. 2017. Unique traits of zebu cattle under tropical climate conditions. In Book: Agriculture under climate change, threats, strategies and policies edited by Belavadi, V.V., Nataraja Karaba, N. and Gangadharappa, N.R. pp. 231-236.

Singh SV and Upadhyay R C. 2009. Thermal stress on physiological functions, thermal balance and milk production in Karan Fries and Sahiwal cows. Indian Veterinary Journal 86 (2)141-144.

Singh SV, Devi R, Kumar Y, Renuka and Upadhyay RC. 2017. Seasonal variation in skin temperature, blood flow and physiological function in zebu (Bos indicus) and Karan fries (Tharparkar X Holstein Friesian) cattle.70 (1) : 96-103.

Singh SV, Hooda OK, Narwade B, Beenam and Updhyay RC .2014. Effect of cooling system on feed and water intake, body weight gain and physiological responses of Murrah buffaloes during summer conditions. Indian J. Dairy Sci. 67(5): 426-431.

Sodhi M, Mukesh M, Kishore A, Mishra BP, Kataria RS, Joshi BK. Novel polymorphisms in UTR and coding region of inducible heat shock protein 70.1 gene in tropically adapted Indian zebu cattle (Bos indicus) and riverine buffalo (Bubalus bubalis) Gene, 527:606–615. doi: 10.1016/j.gene.2013.05.078.

Sørensen JG, Kristensen TN and Loeschcke V. 2003. The evolutionary and ecological role of heat shock proteins. Ecology Letters, 6:1025–1037. doi: 10.1046/j.1461-0248.2003.00528.

Soriani N, Panella G and Calamari L. 2013. Rumination time during the summer season and its relationships with metabolic conditions and milk production. J Dairy Sci., 96(8):5082-94. doi: 10.3168/jds.2013-6620. Epub 2013 Jun 19.

Stella A. 2014. Insights into the interaction of goat breeds and their environment, in Proceedings of the 10th World Congress on Genetics Applied to Livestock Production (Vancouver, BC:).

Stephen ID, Coetzee V and Perrett DI. 2011. Skin carotenoid and melanin coloration affect perceived health of human faces. Evolution and Human Behavior, 32: 216 –227.

Turner HG and Schleger AV. 1958. Field observations on associations between coat type and performance in cattle. Proceedings of the Australian Society of Animal Production 2: 112.

Turner HG and Schleger AV. 1960. The significance of coat type in cattle. Australian Journal of Agricultural Research 11: 645.

Upadhyay R C, Singh S V, Gupta A K and Ashutosh S K. 2007. Impact of climate change on milk production of Murrah buffaloes. Italian Journal of Animal Science 6:1329-1332.

Upadhyay RC, Hooda OK, Aggarwal A and Singh SV. 2013. Indian livestock production has resilience for climate change. In: "Climate Resilient Livestock and Production System" edited by Singh S V, Upadhyay R C, Sirohi S and Singh, A K. National Dairy Research Institute, Karnal, Haryana-132001 (India), 286pp.

Upadhyay RC, Sirohi S, Ashutosh, Singh SV, Kumar A and Gupta S K. 2009. Impact of climate change on milk production in India. In: Global Climate Change and Indian Agriculture (Edited by P.K Aggarwal). Published by ICAR, New Delhi. Pp. 104-106.

Van Randen A, Lameris W, Nio C Y et al. 2009. Inter-observal agreement for abdominal CT in unselected patients with acute abdominal pain. Eur. Radiol., 19: 1394-1407.

Veríssimo CJ. 1993. Prejuízos causados pelo carrapato Boophilus microplus. Zootecnia 31: 97-106.

Verma KK, Prasa S, Mohanty TK, Kumaresan A, Layek SS, Patbandha TK, Datta TK and Chand S. 2016. Effect of short-term cooling on core body temperature, plasma cortisol and conception rate in Murrah buffalo heifers during hot-humid season. Journal of Applied Animal Research 44 (1): 281-286. doi.10.1080/09712119.2015.1031782.

Verma P K, Raina R, Sultana M, Prawez S and Singh M. 2015. Polyphenolic constituents and antioxidant/antiradical activity in different extracts of Alstonia scholaris (Linn.). Afr. J. Biotechnol., 14(47): 3190-3197.DOI: 10.5897/AJB2015.14708.

Wang J P., et al. "Effect of saturated fatty acid supplementation on production and metabolism indices in heat-stressed mid-lactation dairy cows." Journal of dairy science 93.9 (2010): 4121-4127.

West JW. 2003. Effects of heat stress on production in dairy cattle. Journal of Dairy Science 86: 2131-2144.

Wiggins T, Stubbs BM, Boone D and Engledow A. 2011. Disseminated leiomyomatosis peritonei. BMJ Case Reports. doi:10.1136/bcr.04.2011.4145.

Wolfenson D, Thatcher WW, Badinga L, Savio JD, Meidan R, Lew BJ, Braw-Tal R, and Berman A. 1995. Effect of heat stress on follicular development during the estrous cycle in lactating dairy cattle. Biol. Reprod., 52:1106-1113.

Wright NC. 1954. The ecology of domesticated animals. Pages 191–251 in Progress in the physiology of farm animals. Vol. 1. J. Hammond, ed. Butterworths Pub., Ltd., London, UK.

Yazdi AS, Guarda G, Riteau N, Drexler SK, Tardivel A, Couillin I and Tschopp J. 2010. Nanoparticles activate the NLR pyrin domain containing 3 (Nlrp3) inflammasome and cause pulmonary inflammation through release of IL-1á and IL-1â. Proc. Natl. Acad. Sci. U S A. 107(45):19449-54. doi: 10.1073/pnas.1008155107.

Zhang JQ, Chen H, Sun ZJ, Liu XL, Qiang-Ba YZ and Gu YL. 2010. Flesh color association with polymorphism of the tyrosinase gene in different Chinese chicken breeds. Molecular biology reports 37(1): 165-169.

Printed and bound by CPI Group (UK) Ltd, Croydon, CR0 4YY

17/10/2024

01775682-0007